Springer Series in Statistics

Advisors:
P. Bickel, P. Diggle, S. Fienberg, K. Krickeberg,
I. Olkin, N. Wermuth, S. Zeger

Springer Science+Business Media, LLC

Springer Series in Statistics

Andersen/Borgan/Gill/Keiding: Statistical Models Based on Counting Processes.
Andrews/Herzberg: Data: A Collection of Problems from Many Fields for the Student and Research Worker.
Anscombe: Computing in Statistical Science through APL.
Berger: Statistical Decision Theory and Bayesian Analysis, 2nd edition.
Bolfarine/Zacks: Prediction Theory for Finite Populations.
Borg/Groenen: Modern Multidimensional Scaling: Theory and Applications
Brémaud: Point Processes and Queues: Martingale Dynamics.
Brockwell/Davis: Time Series: Theory and Methods, 2nd edition.
Daley/Vere-Jones: An Introduction to the Theory of Point Processes.
Dzhaparidze: Parameter Estimation and Hypothesis Testing in Spectral Analysis of Stationary Time Series.
Fahrmeir/Tutz: Multivariate Statistical Modelling Based on Generalized Linear Models.
Farrell: Multivariate Calculation.
Federer: Statistical Design and Analysis for Intercropping Experiments.
Fienberg/Hoaglin/Kruskal/Tanur (Eds.): A Statistical Model: Frederick Mosteller's Contributions to Statistics, Science and Public Policy.
Fisher/Sen: The Collected Works of Wassily Hoeffding.
Good: Permutation Tests: A Practical Guide to Resampling Methods for Testing Hypotheses.
Goodman/Kruskal: Measures of Association for Cross Classifications.
Gouriéroux: ARCH Models and Financial Applications.
Grandell: Aspects of Risk Theory.
Haberman: Advanced Statistics, Volume I: Description of Populations.
Hall: The Bootstrap and Edgeworth Expansion.
Härdle: Smoothing Techniques: With Implementation in S.
Hart: Nonparametric Smoothing and Lack-of-Fit Tests.
Hartigan: Bayes Theory.
Heyer: Theory of Statistical Experiments.
Huet/Bouvier/Gruet/Jolivet: Statistical Tools for Nonlinear Regression: A Practical Guide with S-PLUS Examples.
Jolliffe: Principal Component Analysis.
Kolen/Brennan: Test Equating: Methods and Practices.
Kotz/Johnson (Eds.): Breakthroughs in Statistics Volume I.
Kotz/Johnson (Eds.): Breakthroughs in Statistics Volume II.
Kres: Statistical Tables for Multivariate Analysis.
Küchler/Sørensen: Exponential Families of Stochastic Processes.
Le Cam: Asymptotic Methods in Statistical Decision Theory.
Le Cam/Yang: Asymptotics in Statistics: Some Basic Concepts.
Longford: Models for Uncertainty in Educational Testing.
Manoukian: Modern Concepts and Theorems of Mathematical Statistics.
Miller, Jr.: Simultaneous Statistical Inference, 2nd edition.
Mosteller/Wallace: Applied Bayesian and Classical Inference: The Case of *The Federalist Papers.*

(continued after index)

Uwe Küchler Michael Sørensen

Exponential Families of Stochastic Processes

Springer

Uwe Küchler
Humboldt-Universität zu Berlin
Institute of Mathematics
Unter den Linden 6
D-10099 Berlin, Germany

Michael Sørensen
Aarhus University
Institute of Mathematics
Ny Munkegade
DK-8000 Aarhus C, Denmark

Library of Congress Cataloging in Publication Data
Küchler, Uwe.
Exponential families of stochastic processes / Uwe Küchler, Michael Sørensen.
p. cm. — (Springer series in statistics)
Includes bibliographical references and index.
ISBN 978-1-4757-7100-8 ISBN 978-0-387-22765-8 (eBook)
DOI 10.1007/978-0-387-22765-8
1. Stochastic processes. 2. Exponential families (Statistics)
I. Sørensen, Michael. II. Title. III. Series.
QA274.K83 1997
519.2—dc21 97-9300
CIP

Printed on acid-free paper.

Production managed by Terry Kornak; manufacturing supervised by Jacqui Ashri.
Camera-ready copy prepared from the authors' LaTeX files.

9 8 7 6 5 4 3 2 1

ISBN 978-1-4757-7100-8 SPIN 10568319

Preface

Exponential families of stochastic processes are parametric stochastic process models for which the likelihood function exists at all finite times and has an exponential representation where the dimension of the canonical statistic is finite and independent of time. This definition not only covers many practically important stochastic process models, it also gives rise to a rather rich theory. This book aims at showing both aspects of exponential families of stochastic processes.

Exponential families of stochastic processes are tractable from an analytical as well as a probabilistic point of view. Therefore, and because the theory covers many important models, they form a good starting point for an investigation of the statistics of stochastic processes and cast interesting light on basic inference problems for stochastic processes.

Exponential models play a central role in classical statistical theory for independent observations, where it has often turned out to be informative and advantageous to view statistical problems from the general perspective of exponential families rather than studying individually specific exponential families of probability distributions. The same is true of stochastic process models. Thus several published results on the statistics of particular process models can be presented in a unified way within the framework of exponential families of stochastic processes.

The exponential form of the likelihood function implies several probabilistic as well as statistical properties. A considerable portion of the book is focused on clarifying such structure of exponential models. Other main themes are asymptotic likelihood theory and sequential maximum likelihood estimation. These areas of statistical inference for stochastic processes are basically different from the similar problems for independent observations. In particular, in the asymptotic likelihood theory, the dependence structure of a process class can imply a wealth of interesting new situations.

In recent years, stochastic calculus has been used increasingly to study inference problems for stochastic processes, and there is scope for using this powerful tool to a much larger extent. A major obstacle to this development is that stochastic calculus is not widely known among statisticians. It is hoped that this book will assist graduate students as well as researchers in statistics not only in getting into the problems of inference for stochastic processes by studying the most tractable type of models, but also if necessary in learning to solve the problems by the tools of stochastic calculus. To attain this goal, the first chapters use only classical stochastic process methods, while tools from stochastic calculus are used at the end. The necessary tools from stochastic calculus are reviewed in an appendix. Most chapters include exercises to support the learning process. We also hope that students and researchers in probability can use the book to get acquainted with the problems of statistical inference.

The authors acknowledge with gratitude the financial support to their collaboration on this book from the EU programme Human Capital and Mobility and from Sonderforschungsbereich 373 at the Humboldt University in Berlin, which among other things made it possible for Michael Sørensen to stay for a longer period in Berlin. We are particularly grateful for the support from the Volkswagen-Stiftung through the programme "Research in Pairs" at the Mathematical Research Institute Oberwolfach, without which it would have taken a lot longer to finish this book. In fact, a large portion of this book was written at Oberwolfach. At an early stage in the writing of the book, an invitation to Michael Sørensen to give a series of lectures on the subject of the book at IMPA in Rio de Janeiro resulted in a pleasant stay that gave much inspiraton to go on with the project.

We would also like to thank the many colleagues who have contributed to our book project by writing papers or otherwise collaborating with us, by sending us preprints, by discussing the topic with us, or by giving helpful comments on our papers and on the book manuscript in its various stages. Whatever errors there remain are, of course, our responsibility. We hope the reader will contact us with any questions, comments, or criticisms she or he might have.

Four secretaries, Ms. O. Wethelund and Ms. H. Damgaard at the University of Aarhus and Ms. S. Bergmann and Ms. A. Fiebig at the Humboldt University in Berlin, typed the first version of most of the book into the computer and assisted us in organizing the activities related to the book project. We are grateful for their invaluable work of outstanding quality.

Finally, we dedicate this book to Ingeborg and Ulla for their care, support, and tolerance.

Uwe Küchler and Michael Sørensen

Contents

1

Introduction

Exponential models play a central role in classical statistical theory for independent observations, where it has often turned out to be informative and advantageous to view statistical problems from the general perspective of exponential families rather than studying individually specific exponential families of probability distributions. The same is true of stochastic process models. One of the authors experienced this in his own research when it turned out that results on sequential maximum likelihood estimation for certain diffusion models (Sørensen, 1983) could be generalized to several other types of stochastic processes (Sørensen, 1986) by viewing the problem from the exponential family perspective.

There are at least three good reasons for studying the classical exponential families of probability distributions. They are analytically tractable models where, e.g., Laplace transform methods can readily be applied; they have several nice statistical properties; and a large part of the models used in statistical practice are exponential families. In this book we study what role exponential families play in statistical inference for stochastic processes.

A large number of stochastic process models with discrete as well as continuous time are exponential families in the sense studied in this book. In Chapter 3 we give a long series of examples including time series models, models for counting processes, Lévy processes, diffusion processes (Markovian as well as non-Markovian), diffusions with jumps, and random fields.

Several definitions of exponential families of stochastic processes, each taking a particular property of classical exponential families for independent observations as the defining property, have been proposed. In this book we concentrate on models where the likelihood function $L_t(\theta)$ at all finite times t exists and has an exponential representation

$$\log L_t(\theta) = \sum_{i=1}^{k} \gamma_t^{(i)}(\theta) B_t^{(i)} - \phi_t(\theta), \tag{1.1.1}$$

where $B^{(1)}, \cdots, B^{(k)}$ are real stochastic processes depending only on the observed stochastic process up to time t, and where $\gamma^{(1)}, \cdots, \gamma^{(k)}$ and ϕ are non-random functions of θ and t. We call $B_t = (B_t^{(1)}, \ldots, B_t^{(k)})$ the canonical process. It is essential that the dimension k of the canonical statistic B_t is finite and does not depend on the time t. This definition not only covers many practically important stochastic process models; it also gives rise to a rather rich theory. The book aims at showing both aspects of exponential families of stochastic processes.

It is perhaps at this point worthwhile mentioning that an exponential family is really a family of probability measures. What is special here is that the data are a realization of a stochastic process. It is therefore natural to talk about an exponential family of stochastic processes. We shall discuss this point further in Chapter 3.

Exponential families of stochastic processes are relatively easy to handle from an analytical as well as a probabilistic point of view. Therefore, and because the theory covers many important models, they form a good starting point for an investigation of the statistics of stochastic processes and cast interesting light on basic inference problems for stochastic processes. Several published results on the statistics of particular process models can be presented in a unified way within the framework of exponential families of stochastic processes.

The exponential structure of the likelihood function implies several probabilistic properties of the processes in an exponential family and statistical properties of the model. A large part of this book is concerned with a thorough study of such structure of exponential models rather than with problems of statistical inference. Stochastic processes differ from sequences of independent random variables in at least two respects. First, there is stochastic dependence between the individual observations, which usually has a certain structure given, for instance, by the Markov property, by the martingale property, or for Gaussian processes by a covariance function. Second, for continuous time processes the sample paths are functions of time that have properties like continuity, right-continuity, or being piecewise constant. On one hand, properties of an exponential likelihood function can imply properties of the sample paths or the dependence structure of the process. On the other hand, properties of the processes can imply restrictions on which types of exponential models are possible.

We also give a thorough discussion of asymptotic likelihood theory and of sequential maximum likelihood estimation. These areas of statistical inference for stochastic processes are basically different from the similar problems for independent observations. In particular, in the asymptotic theory the dependence structure of a process class can imply a wealth of interesting new situations.

In recent years stochastic calculus has been used increasingly to study inference problems for continuous time stochastic processes, and there is scope for using this powerful tool to a much larger extent. A major obstacle to this development is that stochastic calculus is not widely known among statisticians. It is hoped that this book will assist researchers in statistics not only in getting into the problems of inference for stochastic processes by studying the simplest type of models, but also, if necessary, in learning to solve the problems by the tools of stochastic calculus. To attain this goal, the first chapters use only classical stochastic process methods, while tools from stochastic calculus are used later in the book. The necessary tools from stochastic calculus are presented without proofs in an appendix. In all cases we give references to books and articles where the reader can find the proofs.

Classical results on exponential families of probability distributions and on inference from i.i.d. observations from such families will not be reviewed in this book. Instead we refer to the books by Barndorff-Nielsen (1978), Johansen (1979), Brown (1986), and Hoffmann-Jørgensen (1994).

The historical starting point for the theory of exponential families of stochastic processes is the class of processes with independent stationary increments. We call a process from this class a Lévy process if its sample paths are right continuous with limits from the left. Exponential families of Lévy processes for which the canonical statistic at time t is the last observation X_t are studied in Chapter 2. From a statistical point of view these models are completely analogous to exponential models for i.i.d. observations. Therefore, they are a good starting point for a study of the general exponential families of stochastic processes. A main result on the structure of such models gives the dependence of the Lévy characteristics on the statistical parameters. The Lévy characteristics describes the local behaviour of the process. Results about existence, consistency, and asymptotic normality of the maximum likelihood estimator follow immediately from similar results for i.i.d. observations from an exponential family.

In Chapter 3 the definition of a general exponential family of stochastic processes is given and discussed in detail. Further, a long list of examples is given covering a broad range of stochastic process models ranging from Markov chains with a finite state space to random field models.

Central notions from the classical theory of exponential families of probability distributions are generalized to the setting of stochastic processes in Chapter 4, where also some basic results on general exponential families of stochastic processes are given. It turns out that an exponential family of processes where the canonical process B (cf. (1.1.1)) is not a process with independent increments is necessarily a curved exponential family. This result, which is a first indication that the exponential families of stochastic processes form a world rather different from the classical exponential families, obviously has consequences for the statistical properties of the families and is the background for much of the theory developed in the subsequent chapters.

In some important particular cases, which are mainly curved $(k, k-1)$ exponential families, a random time transformation can turn a curved exponential family of stochastic processes into a non-curved exponential family of Lévy processes. This observation is utilized in Chapter 5 to generalize statistical results for classical exponential families, mainly via the results in Chapter 2.

Chapter 6 treats exponential families of Markov processes. A main result, which follows from the general theory of Markov processes, is that the observed process is Markovian under all measures in an exponential family of processes if and only if the canonical process B in (1.1.1) is an additive functional.

For a general exponential family of processes the family of probability measures corresponding to observing the process up to a fixed time point t is, apart from the case of a canonical process with independent increments, a curved exponential family. It can therefore always be extended to a strictly larger full exponential family of probability measures with a parameter space that may depend on the time t. This full family is called the envelope family at time t. The envelope families at different time points are not consistent families of probability measures, so a stochastic process interpretation is not straightforward. The structure of the envelope families and their relation to the observed process are discussed in Chapter 7.

In the subsequent three chapters the focus is on statistical properties. In the classical case of i.i.d. observations from a full exponential family of distributions, simple conditions for the existence of the maximum likelihood estimator can be given. Moreover, the maximum likelihood estimator is sufficient, consistent, and asymptotically normal. For exponential families of processes the likelihood theory is more complex. This is investigated in Chapter 8. For instance, the maximum likelihood estimator is usually not sufficient, and it is often not asymptotically normal. In many cases, defined precisely in Chapter 8, the asymptotic distribution is a variance mixture of normal distributions. In such situations normalization of the maximum likelihood estimator with a random information matrix circumvents the problem and yields an asymptotic normal distribution. A general discussion of information processes is given.

Chapter 9 presents a very illuminating example of an exponential family of continuous processes given as solutions of a two-parameter linear stochastic differential delay equation. The maximum likelihood estimator behaves asymptotically very differently depending on the true parameter values. It is shown that the distribution of the maximum likelihood estimator may asymptotically be normal, a variance mixture of normal distributions, a distribution determined by integrals of a Wiener process, or even more exotic.

In Chapter 10 we consider again curved exponential families of the type studied in Chapter 5. Here we investigate random stopping rules, i.e., observation up to a stopping time, for which a non-curved exponential family is obtained, exactly or approximately. Thus classical exponential family methods can be applied to derive exact or approximate results concerning the existence of the maximum likelihood estimator and about the distribution of the canonical statistic and the maximum likelihood estimator. Also some results about the stopping times are given.

Many stochastic processes are semimartingales. In fact, most processes used in applications are. For semimartingales the methods of stochastic calculus gives an efficient tool to study properties of exponential families. In particular, the properties of the sample paths can be specified by the so-called local characteristics, which describe the dynamics of the changes of the process, whether these are continuous or by jumps. In Chapter 11 we study how the local characteristics (and thus the sample path properties) vary within an exponential family of semimartingales. For instance, we show how the sample path properties of an exponential family obtained by a random time transformation are related to those of the original family. This gives a method for constructing exponential families with particular path properties. Another result is that within an exponential family of semimartingales with a time-continuous likelihood function only the drift changes. Finally, methods from stochastic calculus are applied to study likelihood theory for exponential families of semimartingales in more depth than was possible without the assumption that the processes are semimartingales.

For the processes with independent stationary increments studied in Chapter 2 the defining exponential representation (1.1.1) is equivalent to the property that the last observation X_t is a sufficient statistic, and also equivalent to the property that the one-dimensional marginal distributions form exponential families. In general, these equivalences do not hold. Thus, any of these three properties could be taken as definition of the notion of exponential families of stochastic processes. Indeed, this has been done by different authors. Chapter 12 gives a review of these alternative attempts to define exponential families of stochastic processes.

2

Natural Exponential Families of Lévy Processes

The exponential families of Lévy processes are, from a statistical point of view, only a slight generalization of repeated sampling from a classical exponential family of distributions. We shall see that classical results for exponential families of distributions also hold for exponential families of Lévy processes. There are, moreover, interesting probabilistic properties to study. In particular, we study how the path structure varies within an exponential family. The investigation of this question for Lévy processes is a prelude to the study in full generality for exponential families of semimartingales in Chapter 11.

2.1 Definition and probabilistic properties

A real-valued stochastic process X is said to have independent stationary increments if $X_{t_i} - X_{t_{i-1}}$, $i = 1, \ldots, n$, are independent random variables for all values of n and $0 \leq t_1 < t_2 < \cdots < t_{n-1} < t_n$ and if, for fixed $s > 0$, the distribution of $X_{t+s} - X_t$ is the same for all $t > 0$. A continuous time process with independent stationary increments is called a *Lévy process* if its sample paths are right continuous with limits from the left and $X_0 = 0$. Under weak conditions, a process with independent stationary increments has a modification which is a Lévy process. This is, for instance, the case if the process is continuous in probability. Another sufficient condition is that the mapping $t \rightarrow E(\exp(isX_t))$ is Borel measurable for every fixed $s \in \mathbb{R}$.

First we consider the natural exponential family generated by a one-dimensional Lévy process X defined on a probability space $(\Omega, \mathcal{F}, P)$. We assume that $X_0 = 0$. Denote by F_t the distribution function of X_t and by $\kappa(s)$ the cumulant transform of F_1, i.e.,

$$\kappa(s) = \log\ E(e^{sX_1}). \tag{2.1.1}$$

Moreover, let Θ be the domain of κ, i.e., the set of values of s for which the expectation in (2.1.1) is finite. This is an interval containing 0. In the sequel we assume that $\Theta \neq \{0\}$ to avoid cases where a trivial exponential family consisting of only one distribution is generated. An example where this happens is the Cauchy process. Under this assumption the cumulant transform κ determines F_1. The cumulant transform of F_t is $\kappa(s)t$. This follows from the fact that

$$\log\ E(e^{sX_{t+u}}) = \log\ E(e^{sX_t}) + \log\ E(e^{sX_u}), \tag{2.1.2}$$

provided that the mapping $t \to E(\exp(sX_t))$ is Borel measurable for fixed $s \in \Theta$. We see that $t\kappa(s)$ is a cumulant transform for all $t > 0$. A cumulant transform κ (and the corresponding distribution F_1) with this property is called *infinitely divisible.*

From the fact that the cumulant transform of F_t is $\kappa(s)t$ it follows that the process $L_t(\theta) = \exp(\theta X_t - \kappa(\theta)t)$ is a martingale for every $\theta \in \Theta$. Consider the right-continuous filtration $\{\mathcal{F}_t\}$ generated by X. The σ-algebra $\mathcal{F}_t$ is defined as

$$\mathcal{F}_t = \bigcap_{s>0} \mathcal{G}_{t+s},$$

where for all $t > 0$

$$\mathcal{G}_t = \sigma(X_s : s \leq t)$$

is the smallest σ-algebra with respect to which all X_s with $s \leq t$ are measurable. For each $t > 0$ we can define a class $\{P_\theta^t : \theta \in \Theta\}$ of probability measures on $(\Omega, \mathcal{F}_t)$ by

$$\frac{dP_\theta^t}{dP^t} = L_t(\theta) = \exp[\theta X_t - t\kappa(\theta)]. \tag{2.1.3}$$

Since $L_t(\theta)$ is a P-martingale, the family of probability measures $\{P_\theta^t : t \geq 0\}$, where θ is fixed, is consistent. We assume that $\mathcal{F} = \sigma(\mathcal{F}_t : t \geq 0)$ and that $\{P_\theta^t : t \geq 0\}$ can be extended to a probability measure P_θ on $(\Omega, \mathcal{F})$ such that P_θ^t is the restriction of P_θ to $\mathcal{F}_t$. This can, for example, be done if $(\Omega, \mathcal{F})$ is standard measurable, see Ikeda and Watanabe (1981, p. 176). This is not a restriction in the applications we have in mind.

We call the class $\mathcal{P} = \{P_\theta : \theta \in \Theta\}$ the *natural exponential family* generated by X. We can also think of $\mathcal{P}$ as being generated by the infinitely divisible cumulant transform κ. Under P_θ, the process X has independent stationary increments. The cumulant transform of X_t under P_θ is $t\kappa_\theta(s)$, where $\kappa_\theta(s) = \kappa(\theta+s) - \kappa(\theta)$, and the distribution of the increment $X_{t+u} - X_u$ is given by

$$F_t^\theta(dx) = \frac{e^{\theta x} F_t(dx)}{\exp(\kappa(\theta)t)}. \tag{2.1.4}$$

Thus the class $\{F_t^\theta : \theta \in \Theta\}$ is the exponential family of distributions generated by F_t. In particular, $E_\theta X_t = t\dot{\kappa}(\theta)$ and $V_\theta X_t = t\ddot{\kappa}(\theta)$ for $\theta \in \operatorname{int}\Theta$, where E_θ and V_θ denote expectation and variance under P_θ. Note that the last observation X_t is minimal sufficient for $\mathcal{P}$ with respect to $\mathcal{F}_t$.

Example 2.1.1 Suppose X under P is a *Poisson process* with intensity one. Then $\kappa(s) = \exp(s) - 1$ and $\Theta = \mathbb{R}$. Under P_θ we find that X is a Poisson process with intensity $\lambda = e^\theta$ and that

$$L_t(\theta) = \exp[\theta X_t - (e^\theta - 1)t]. \tag{2.1.5}$$

□

Example 2.1.2 Let X be a standard *Wiener process* under P so that $\kappa(s) = \frac{1}{2}s^2$ and $\Theta = \mathbb{R}$. Then we obtain the class of Wiener processes with drift θ and find that

$$L_t(\theta) = \exp[\theta X_t - \tfrac{1}{2}\theta^2 t]. \tag{2.1.6}$$

□

Let us first consider the relatively simple case where X is a *compound Poisson process* under P. This is a Lévy process of the form

$$X_t = \sum_{i=1}^{N_t} Z_i,$$

where N is a Poisson process with intensity λ, independent of the random variables Z_i, $i = 1, 2, \cdots$, (the sizes of the jumps) that are mutually independent and identically distributed. The number of jumps in the time interval $[0, t]$, N_t, is Poisson distributed with mean λt. We denote by φ the Laplace transform of the common probability distribution of the jump sizes. Then it easily follows that $\kappa(s) = \lambda(\varphi(s) - 1)$, so the cumulant transform $\kappa_\theta(s)$ of X_1 under P_θ is given by

$$\kappa_\theta(s) = \kappa(\theta + s) - \kappa(\theta) = \lambda\varphi(\theta)[\varphi(\theta + s)/\varphi(\theta) - 1]. \tag{2.1.7}$$

This is again the cumulant transform of a compound Poisson process, but now the Poisson process N has intensity $\lambda\varphi(\theta)t$, and the jump size distribution has Laplace transform $\varphi(\theta + s)/\varphi(\theta)$. Thus the family of jump size distributions (for $\theta \in \Theta$) is also an exponential family, the natural exponential family generated by the original jump size distribution under P. The distributions of the waiting times between jumps are exponential distributions with parameter $\lambda\varphi(\theta)t$, i.e., they also form an exponential family.

Thus one recognizes that exponential families of stochastic processes may have an inner structure, where the exponential property is repeated. This phenomenon was studied in a more general setting in Küchler (1993).

In order to study the behaviour of X under P_θ for more general Lévy processes, we consider the so-called *Lévy characteristics* of X. Every infinitely divisible cumulant transform κ can be written as

$$\kappa(s) = \delta s + \tfrac{1}{2}\sigma^2 s^2 + \int_E [e^{sx} - 1 - sx/(1+x^2)]\nu(dx), \tag{2.1.8}$$

where $\delta \in \mathbb{R}$ and $\sigma^2 \geq 0$, and where ν is a measure on $E = \mathbb{R}\backslash\{0\}$ satisfying

$$\int_E \frac{x^2}{1+x^2}\nu(dx) < \infty. \tag{2.1.9}$$

A measure ν on E satisfying (2.1.9) is called a *Lévy measure.* The domain of κ is the set of real numbers s for which the integral in (2.1.8) converges. Since the integrand behaves as x^2 near zero, and since $1 + sx/(1+x^2)$ is bounded for $|x| \geq 1$, it follows that

$$\Theta = \operatorname{dom}\kappa = \left\{\theta : \int_E \frac{x^2}{1+x^2}e^{\theta x}\nu(dx) < \infty\right\}. \tag{2.1.10}$$

This is exactly the set of θ's for which $e^{\theta x}\nu(dx)$ defines a Lévy measure.

When $\Theta \neq \{0\}$, as we have assumed above, there is a one-to-one correspondence between κ and the triple (δ, σ^2, ν). If κ is the cumulant transform of X_1 under P, we call the triple (δ, σ^2, ν) the Lévy characteristics of the process X under P. Since (δ, σ^2, ν) determine κ uniquely, we can think of $\mathcal{P}$ as being generated by the Lévy characteristics (δ, σ^2, ν).

The first characteristic δ is connected to the drift of the process X, whereas σ^2 is the infinitesimal variance of the Brownian motion part of X, and ν determines the probabilistic character of the jumps of X. To be more specific, let $N_t(A)$ denote the number of jumps of X in the time interval $[0, t]$ with jump sizes in the Borel set A that has a strictly positive distance to zero. Under P the random variable $N_t(A)$ is Poisson distributed with mean value $t\nu(A)$. In fact, $N_t(\cdot)$ is a so-called Poisson random measure: For fixed t, $N_t(A)$ is a measure in A, and if $A_1, \cdots, A_n$ are disjoint Borel subsets of E all with a strictly positive distance to zero, then $N_t(A_1), \cdots, N_t(A_n)$ are independent. For fixed A the process $N_t(A)$ is a Poisson process. Under P the Lévy process X has the decomposition

$$X_t = \delta t + \sigma W_t + X_t^d, \tag{2.1.11}$$

where W and X^d are independent stochastic processes, W is a standard Wiener process, and

$$X_t^d = \lim_{n\to\infty}\left(\int_{|u|>n^{-1}} u\, N_t(du) - t\int_{|u|>n^{-1}} u/(1+u^2)\nu(du)\right). \tag{2.1.12}$$

The convergence is almost sure and uniform for t in a bounded interval.

The process X is a compound Poisson process if and only if $\delta = \sigma^2 = 0$ and ν is a finite measure. In this case, $\nu/\nu(E)$ is the distribution of the jump size, and $\nu(E)$ is the intensity of the Poisson process. In particular, X is a Poisson process if $\nu = \lambda\epsilon_1$, where ϵ_1 denotes the Dirac measure at 1. When ν is not a finite measure, X is not a compound Poisson process. In this situation X has infinitely many jumps in every finite time interval of strictly positive length.

Proposition 2.1.3 *The Lévy characteristics of* X *under* P_θ *are*

$$\begin{aligned} \nu_\theta(dx) &= e^{\theta x}\nu(dx), \\ \sigma_\theta^2 &= \sigma^2, \\ \delta_\theta = \delta + \theta\sigma^2 &- \int_E \frac{x}{1+x^2}(1-e^{\theta x})\nu(dx). \end{aligned} \tag{2.1.13}$$

Proof. By direct calculation it is seen that the cumulant transform of X_1 under P_θ has the form

$$\kappa(\theta+s) - \kappa(\theta) = \delta_\theta s + \tfrac{1}{2}\sigma^2 s + \int_E [e^{sx} - 1 - sx/(1+x^2)]\nu_\theta(dx).$$

□

Note that for every $\epsilon > 0$ and $t > 0$ the conditional probabilities given $\{|\Delta X_t| > \epsilon\}$,

$$P_\theta(\Delta X_t \in A \mid |\Delta X_t| > \epsilon) = \nu_\theta(A\backslash[-\epsilon,\epsilon])/\nu_\theta(\mathbb{R}\backslash[-\epsilon,\epsilon]), \; \theta \in \Theta, \tag{2.1.14}$$

form an exponential family of distributions on $\mathbb{R}\backslash[-\epsilon,\epsilon]$. Here ΔX_t is the jump of X at time t, i.e., $\Delta X_t = X_t - X_{t-}$, where $X_{t-} = \lim_{s\uparrow t} X_s$. The process Y^ϵ defined by observing only the jumps of X that are numerically larger than ϵ,

$$Y_t^\epsilon = \sum_{s\leq t} \Delta X_t 1_{\{|\Delta X_t|>\epsilon\}}, \tag{2.1.15}$$

is a compound Poisson process under P_θ with Lévy measure $\nu_\theta^\epsilon(A) = \nu_\theta(A\backslash[-\epsilon,\epsilon])$ and with jump size distribution given by (2.1.14).

From Proposition 2.1.3 it follows that for $\sigma^2 = \nu = 0$ we have $\Theta = \mathbb{R}$, but $P_\theta = P$ for all $\theta \in \mathbb{R}$, so this exponential family is trivial. If, on the other hand,

$$\sigma^2 + \int_E \frac{x^2}{1+x^2}\nu(dx) > 0, \tag{2.1.16}$$

then $\theta_1 \neq \theta_2$ implies $P_{\theta_1} \neq P_{\theta_2}$.

The method used in this chapter to study the behaviour of X under P_θ is a special case of the method that we will use in Chapter 11 for more general exponential families of stochastic processes. There we will study a generalization of the Lévy characteristics, the so-called local characteristics of semimartingales.

Example 2.1.4 Consider the Lévy measure of the gamma type

$$\nu_{\rho,\beta}(dx) = \rho x^{\beta-1} e^{-x} dx, \; x > 0, \tag{2.1.17}$$

where $\rho > 0$ and $\beta > -2$. Define

$$\delta_\beta = \mu_\beta + \int_0^\infty \frac{x^{\beta+2}}{1+x^2} e^{-x} dx$$

with

$$\mu_\beta = \begin{cases} -(\beta+1)^{-1} & \text{for } \beta \neq -1 \\ 1 & \text{for } \beta = -1. \end{cases}$$

The cumulant transform corresponding to the Lévy characteristics $(\rho\delta_\beta, 0, \nu_{\rho,\beta})$ is by (2.1.8) given by

$$\kappa_{\rho,\beta}(s) = \rho(\beta-1)[(1-s)^{-\beta} - 1], \; s < 1, \tag{2.1.18}$$

for $\beta \neq -1$ and $\beta \neq 0$. For $\beta = 0$,

$$\kappa_{\rho,0}(s) = -\rho \; \log(1-s), \; s < 1, \tag{2.1.19}$$

and for $\beta = -1$,

$$\kappa_{\rho,-1}(s) = \rho(1-s)\log(1-s), \; s < 1. \tag{2.1.20}$$

Now consider the natural exponential family of Lévy processes generated by $(\rho\delta_\beta, 0, \nu_{\rho,\beta})$ or by $\kappa_{\rho,\beta}$. The Lévy measures are

$$\nu_{\theta,\rho,\beta}(dx) = \rho x^{\beta-1} e^{(\theta-1)x}, \; x > 0. \tag{2.1.21}$$

For $\beta \geq 0$ the possible values of θ are $\theta < 1$, and for $\beta < 0$ it is $\theta \leq 1$. The cumulant transform of X_t under $P_{\theta,\rho,\beta}$ is

$$t[\kappa_{\rho,\beta}(\theta + s) - \kappa_{\rho,\beta}(\theta)], \; s < -\theta. \tag{2.1.22}$$

For $\beta > 0$ the Lévy measure (2.1.21) is finite, so in this case the process X is under $P_{\theta,\rho,\beta}$ a compound Poisson process with jump intensity $\rho\Gamma(\beta)(1-\theta)^{-\beta}$ and with the gamma distribution with shape parameter β and scale parameter $(1-\theta)$ as the jump size distribution. For $\beta \leq 0$ the measure (2.1.21) is not finite, so X is not a compound Poisson process under $P_{\theta,\rho,\beta}$ in this case. For $\beta = 0$, X is a gamma process, and X_t is gamma distributed with shape parameter $t\rho$ and scale parameter $1-\theta$. For $\beta < 0$ the natural exponential family corresponding to (2.1.21) equals the natural exponential family generated by the extreme stable process of order $-\beta$. An extreme stable distribution of order $\alpha \in (0, 2]$ has Lévy measure of the form $ax^{-(\alpha+1)}$, $x > 0$, and Laplace transform of the form

$$\varphi_\alpha(s) = \begin{cases} e^{-b(1-\alpha)s^\alpha} & ,\alpha \neq 1, b > 0, s \geq 0 \\ s^{cs} & ,\alpha = 1, c > 0, s \geq 0, \end{cases}$$

see, e.g., Eaton, Morris, and Rubin (1971). For $0 < \alpha < 1$ the support of the distribution is $(0, \infty)$; for $1 \leq \alpha \leq 2$ the support is the whole real line. For $\alpha = \frac{1}{2}$ the distribution of X_t belongs to the family of inverse Gaussian distributions, and X is called an inverse Gaussian process.

The exponential family of distributions of X_t for ρ and β fixed all have a variance function that is a power function, except for $\beta = -1$, where the variance function is exponential. The variance function of an exponential family is the function that gives the variance as a function of the mean value. Apart from the exponential families of distributions considered in this example, the only exponential families with power variance function are the family of Poisson distributions and the family of normal distributions with fixed variance. These two families can be obtained as weak limits of the families considered here ($\beta \to \infty$ and $\beta \to -2$); see Jørgensen (1992). □

It is easy to generalize the results about natural exponential families generated by one-dimensional Lévy processes to higher dimensions. In this book, T denotes transposition. Suppose $\kappa(s_1, \cdots, s_k)$ is a k-dimensional infinitely divisible cumulant transform, i.e., $t\kappa$ is a cumulant transform for all $t > 0$. Then there exists a probability space $(\Omega, \mathcal{F}, P)$ and a process $X = (X^{(1)}, \cdots, X^{(k)})^T$ with independent stationary increments such that the cumulant transform of $X_t - X_u$ is $(t-u)\kappa$, $t > u$. We can assume that X is right-continuous with limits from the left. The natural exponential family of processes generated by κ (or X) is given by

$$\frac{dP_\theta^t}{dP^t} = \exp\{\theta^T X_t - t\kappa(\theta)\}, \ \theta \in \operatorname{dom} \kappa. \tag{2.1.23}$$

To avoid trivial complications we assume that $\operatorname{int} \operatorname{dom} \kappa \neq \emptyset$. The Lévy characteristics of X under P_θ are given by

$$\begin{aligned} \nu_\theta(dx) &= e^{\theta^T x}\nu(dx), \\ \Sigma_\theta &= \Sigma, \\ \delta_\theta &= \delta + \Sigma\theta + \int_{E^k} \tau(x)(e^{\theta^T x} - 1)\nu(dx). \end{aligned} \tag{2.1.24}$$

Here

$$\tau_i(x) = x_i/(1 + x_i^2), \ i = 1, \cdots, k,$$

and (δ, Σ, ν) are the Lévy characteristics of X under P, i.e.,

$$\kappa(s) = \delta^T s + \tfrac{1}{2} s^T \Sigma s + \int_{E^k} [e^{s^T x} - 1 - s^T \tau(x)]\nu(dx), \tag{2.1.25}$$

where $s = (s_1, \ldots, s_k)$. In particular, δ is a vector; Σ is a symmetric, positive definite $k \times k$-matrix; and ν is a measure on E^k satisfying

$$\int_{E^k} \frac{x^T x}{1 + x^T x} \nu(dx) < \infty.$$

2.2 Maximum likelihood estimation

Maximum likelihood estimation for the model (2.1.23) follows the pattern of classical exponential families, see, e.g., Barndorff-Nielsen (1978) or Brown (1986). We need some regularity conditions on the cumulant transform κ. First of all, we assume that the affine span of the distribution corresponding to κ is $\mathbb{R}^k$, or in other words, that κ is strictly convex. This is no restriction of the generality. The set $\operatorname{dom}\kappa$ is convex. We will, as earlier, assume that $\operatorname{int}\operatorname{dom}\kappa \neq \emptyset$.

The concept of a steep convex function is useful. A cumulant transform κ is called *steep* if for all $\theta_1 \in \operatorname{dom}\kappa\backslash\operatorname{int}\operatorname{dom}\kappa$ and all $\theta_0 \in \operatorname{int}\operatorname{dom}\kappa$,

$$\frac{d}{d\rho}\kappa(\theta_\rho) \to \infty \text{ as } \quad \rho \uparrow 1,$$

where $\theta_\rho = (1-\rho)\theta_0 + \rho\theta_1$, $0 < \rho < 1$. A necessary and sufficient condition that κ is steep is that $E_\theta(\|X_1\|) = \infty$ for all $\theta \in \operatorname{dom}\kappa\backslash\operatorname{int}\operatorname{dom}\kappa$. Obviously, a sufficient condition that κ is steep is that $\operatorname{dom}\kappa$ is open. Let C denote the closure of the convex hull of the support of the random variable X_1. Then C is also equal to the closed convex support of the random variable $\bar{X}_t = X_t/t$ for every $t > 0$. This is obvious when κ is steep. A proof that also covers the non-steep case can be found in Casalis and Letac (1994). If κ is steep, $\dot{\kappa}$ is a homeomorphism of $\operatorname{int}\operatorname{dom}\kappa$ and $\operatorname{int} C$.

Theorem 2.2.1 *Suppose κ is steep and that the parameter space Θ equals $\operatorname{dom}\kappa$. Then the maximum likelihood estimator $\hat{\theta}_t$ based on observation in the time interval $[0,t]$ exists and is uniquely given by $\hat{\theta}_t = \dot{\kappa}^{-1}(\bar{X}_t)$ if and only if $\bar{X}_t \in \operatorname{int} C$.*

Suppose $\theta \in \operatorname{int}\Theta$. Then under P_θ the maximum likelihood estimator exists and is unique for t sufficiently large,

$$\hat{\theta}_t \to \theta \textit{ almost surely}$$

and

$$\sqrt{t}(\hat{\theta}_t - \theta) \to N(0, \ddot{\kappa}(\theta)^{-1}) \textit{ weakly}$$

as $t \to \infty$.

Let $Q_t = \sup_{\beta \in B} L_t(h(\beta))/\sup_{\theta\in\Theta} L_t(\theta)$ be the likelihood ratio test statistic for the hypothesis that the true parameter value θ belongs to $h(B)$, where $B \subseteq \mathbb{R}^l$ $(l < k)$ and $h : B \mapsto \operatorname{int}\Theta$ is a differentiable function for which the matrix $\{\partial h/\partial\beta\}$ has full rank for all $\beta \in B$. If $\theta \in h(B)$, then

$$-2\log Q_t \to \chi^2(k-l) \textit{ weakly}$$

under P_θ as $t \to \infty$.

Proof. The existence and uniqueness of $\hat{\theta}_t$ follow from classical exponential family results. The asymptotic results can be proved by classical arguments because X has stationary independent increments. Note in particular that the cumulant transform of X_t under P_θ is $s \mapsto t(\kappa(\theta+s)-\kappa(\theta))$, which for $t=n$ is the cumulant transform of the canonical statistic corresponding to n independent identically distributed observations of X_1. □

2.3 Exercises

2.1 Show (2.1.2) and deduce that the cumulant transform of F_t is given by $t\kappa(s)$ (provided that the mapping $t \to E(\exp(sX_t))$ is Borel measurable for all $s \in \Theta$). Then prove that $L_t(\theta)$ given by (2.1.3) is a P-martingale.

2.2 Show that X has independent stationary increments under P_θ given by (2.1.3). Hint: Consider the simultaneous Laplace transform of the increments.

2.3 Prove that the cumulant transform of X_t for a compound Poisson process with intensity λ and jump size distribution with Laplace transform φ is $\lambda t(\varphi(s)-1)$.

2.4 Verify the formulae (2.1.18), (2.1.19), and (2.1.20). Show that the distribution of X_t under $P_{\theta,\rho,\beta}$ (given by the cumulant transform (2.1.22)) for $\beta > 0$ has an atom at zero given by

$$P_{\theta,\rho,\beta}(X_t=0)=\exp(-t\rho\Gamma(\beta)(1-\theta)^{-\beta}),$$

and that the rest of the distribution has a density with respect to the Lebesgue measure given by

$$a_t(x;\rho,\beta)e^{\theta x - t\rho\Gamma(\beta)(1-\theta)^{-\beta}}, x>0,$$

where

$$a_t(x;\rho,\beta)=\frac{e^{-x}}{x}\sum_{n=1}^{\infty}\frac{(t\rho\Gamma(\beta)x^{\beta})^n}{\Gamma(n\beta)n!}.$$

For $\beta=1$ (exponentially distributed jumps) we can express the function a in terms of the modified Bessel function of the first kind I_1 :

$$a_t(x;\rho,1)=e^{-x}\sqrt{t\rho/x}I_1(2\sqrt{xt\rho}).$$

2.5 Let the price of a stock S_t at time t be modeled as

$$S_t = s_0\exp(X_t),$$

where s_0 is a constant and X is a Lévy process starting at zero for which the cumulant transform of X_1 is $\kappa(u)$. Further, let r be the interest rate of a risk free asset. Gerber and Shiu (1994) proposed to value a European call option with strike price K and exercise time t by

$$\pi = \exp(-rt)E_{\theta^*}[(S_t - K)_+],$$

where E_{θ^*} is the expectation under the probability measure P_{θ^*} in the natural exponential family generated by X under which the discounted stock price $e^{-rt}S_t$ is a martingale. As usual, $a_+ = \max\{a, 0\}$.

Find an equation that determines θ^* (when it exists), and show that

$$\pi = s_0[1 - F_t^{\theta^*+1}(\log(K/s_0))] - e^{-rt}K[1 - F_t^{\theta^*}(\log(K/s_0))],$$

where F_t^θ is the distribution function given by (2.1.4). Find π under the assumption that $\kappa(u) = \frac{1}{2}\sigma^2 u$. In this case, where the stock price is modeled by a geometric Brownian motion, the formula for the price π is the celebrated Black-Scholes formula.

Finally, suppose that κ is given by (2.1.19), and prove that in this case $\theta^* = (1 - \exp(r/\rho))^{-1}$. Show that for $\rho = 0.02$ and $t = 100$ the Gerber-Shiu price is

$$\begin{aligned}\pi &= s_0[1 - (1 - e^{r/\rho})^{-1}\log(K/s_0)](K/s_0)^{(1-e^{r/\rho})^{-1}} \\ &\quad - e^{-rt}K[1 + (1 - e^{-r/\rho})^{-1}\log(K/s_0)](K/s_0)^{-(1-e^{-r/\rho})^{-1}},\end{aligned}$$

provided that $K \geq s_0$.

2.4 Bibliographic notes

For a detailed account of the theory of Lévy processes and their Lévy characteristics, see, e.g., Ito (1969), Gihman and Skorohod (1975), or Bertoin (1996).

Exponential families of Lévy processes were introduced independently by Michalevič (1961), Magiera (1974), and Franz and Winkler (1976). They were studied further by Franz (1977), Winkler and Franz (1979), Küchler and Küchler (1981), Winkler, Franz, and Küchler (1982), Stefanov (1985), Jacod and Shiryaev (1987, p.555), Küchler and Lauritzen (1989), and Küchler and Sørensen (1989, 1994a). Küchler and Küchler (1981) studied the Lévy characteristics of natural exponential families of Lévy processes. In Küchler and Lauritzen (1989) it was shown that the extreme point model generated by a natural exponential family of Lévy processes is essentially equal to the exponential family; see also Chapter 12. The class of marginal distributions $\{F_t^\theta : \theta \in \Theta, t > 0\}$, with F_t^θ defined by (2.1.4), is an exponential dispersion model; see Jørgensen (1992) and Jørgensen (1997).

Exponential families of distributions with a power variance function, the classical analogue of the processes considered in Example 2.1.4, were introduced by Tweedie (1984) and have been studied by, e.g., Morris (1981),

Hougaard (1986), Bar-Lev and Enis (1986), and Jørgensen (1997). The stochastic processes studied in Example 2.1.4 were also considered in Lee and Whitmore (1993), and the compound Poisson processes were used to model the claim process in an analysis of Brazilian car insurance data in Jørgensen and Souza (1994).

3

Definitions and Examples

3.1 Basic definitions

Let $(\Omega, \mathcal{F}, \{\mathcal{F}_t\})$ be a filtered space where the filtration $\{\mathcal{F}_t : t \geq 0\}$ is right-continuous. The filtration is indexed by $t \in [0, \infty)$, and we will think of t as time. By the definition $\mathcal{F}_t = \mathcal{F}_{[t]}$, where $[t]$ denotes the integer part of t, also discrete time is covered. Consider a class

$$\mathcal{P} = \{P_\theta : \theta \in \Theta\}, \quad \Theta \subseteq \mathbb{R}^k,$$

of probability measures on $(\Omega, \mathcal{F})$. We will denote by P_θ^t the restriction of P_θ to the σ-algebra $\mathcal{F}_t$. We call a filtered space equipped with a class of probability measures a filtered statistical space.

The class $\mathcal{P}$ is called an *exponential family* on the filtered space $(\Omega, \mathcal{F}, \{\mathcal{F}_t\})$ if there exists a σ-finite measure μ on $(\Omega, \mathcal{F})$ such that for all $\theta \in \Theta$

$$P_\theta^t \ll \mu^t, \quad t \geq 0,$$

and

$$\frac{dP_\theta^t}{d\mu^t} = a_t(\theta) q_t \exp(\gamma_t(\theta)^T B_t), \quad t \geq 0, \quad \theta \in \Theta. \tag{3.1.1}$$

Considered as a function of θ, this Radon-Nikodym derivative is the likelihood function corresponding to observation of events in $\mathcal{F}_t$. In (3.1.1), $a > 0$ and $\gamma^{(i)}$, $i = 1, \ldots, m$, are non-random real-valued functions of θ and t that are right-continuous functions of t with limits from the left, while $q_t \geq 0$ and $B_t^{(i)}$, $i = 1, \ldots, m$, are real-valued right-continuous stochastic processes with limits from the left and adapted to $\{\mathcal{F}_t\}$. Moreover,

$\gamma_t(\theta) = (\gamma_t^{(1)}(\theta), \ldots, \gamma_t^{(m)}(\theta))^T$ and $B_t = (B_t^{(1)}, \ldots, B_t^{(m)})^T$, where T denotes transposition. We denote (3.1.1) and other versions of the likelihood function by $L_t(\theta)$. By a version of the likelihood function is meant a function of θ proportional to (3.1.1). The constant of proportionality will often be random.

We call a representation of the form (3.1.1) of the class $\{dP_\theta^t/dP^t : \theta \in \Theta, t \geq 0\}$ of Radon-Nikodym derivatives an *exponential representation* of the family $\mathcal{P}$. An exponential representation is obviously not unique: Other locally dominating measures could be used, and, what is more important, other functions $\gamma_t(\theta)$ or other processes B_t could be chosen. If an exponential representation exists with γ independent of t, we call the exponential family *time-homogeneous.* The vast majority of exponential families met in statistical practice are time-homogeneous. We call a process B appearing in an exponential representation of the form (3.1.1) a *canonical process.*

The exponential form (3.1.1) of the Radon-Nikodym derivatives implies that for all $\theta \in \Theta$, all $t \geq 0$, and all $A \in \mathcal{F}_t$,

$$P_\theta^t(A) = 0 \Leftrightarrow \mu^t(A \cap \{q_t > 0\}) = 0.$$

Hence, for fixed t, all P_θ^t are equivalent, and for a fixed $\theta_0 \in \Theta$ we have the following version of the likelihood function:

$$\frac{dP_\theta^t}{dP_{\theta_0}^t} = \frac{a_t(\theta)}{a_t(\theta_0)} \exp((\gamma_t(\theta) - \gamma_t(\theta_0))^T B_t).$$

Without loss of generality we can assume $\theta_0 = 0$ and $\gamma_t(0) = 0$, so that

$$\frac{dP_\theta^t}{dP_0^t} = \exp(\gamma_t(\theta)^T B_t - \phi_t(\theta)), \tag{3.1.2}$$

where

$$\phi_t(\theta) = \log(a_t(\theta)/a_t(\theta_0)). \tag{3.1.3}$$

From (3.1.2) we see, by the factorization criterion (see, e.g., Lehmann, 1983, p. 39), that for fixed $t > 0$ the random vector B_t is sufficient for $\{P_\theta^t : \theta \in \Theta\}$.

A natural exponential family of Lévy processes, as defined in Chapter 2, is an example of a time-homogeneous exponential family, where $B_t = X_t$, $\gamma_t(\theta) = \theta$, and $\phi_t(\theta) = t\kappa(\theta)$ with the notation of Chapter 2.

We will only consider statistical models for which an exponential representation exists, where the canonical process B is right-continuous with limits from the left, and we will only use such representations. This is no essential restriction, because the stochastic process $\{dP_\theta^t/dP_0^t : t \geq 0\}$ is a P_0-martingale; see Chapter 8. Concerning construction of a modification of $\{dP_\theta^t/dP_0^t\}$ that is right-continuous with limits from the left, see Rogers and Williams (1994, II.65-68) and Jacod and Shiryaev (1987, p. 10).

The existence of the exponential representation (3.1.1) and its particular form depends, of course, on the particular filtration $\{\mathcal{F}_t\}$. For a different filtration $\{\mathcal{G}_t\}$ on $(\Omega, \mathcal{F})$, the measures $P_\theta|\mathcal{G}_t$ will typically not have Radon-Nikodym derivatives of exponential form. If they do, the exponential form will, in general, be different from any of the exponential representations of $P_\theta|\mathcal{F}_t$. The measures $P_\theta|\mathcal{G}_t$ might even be singular. The importance of the filtration is illustrated by two examples in Section 8 of this chapter. When we need to emphasize the filtration, we shall speak of $\mathcal{P}$ as an *exponential family with respect to* $\{\mathcal{F}_t\}$.

Very often, the filtration $\{\mathcal{F}_t\}$ is generated by observation of a stochastic process X. In order to obtain a right-continuous filtration, we have to define $\mathcal{F}_t$ as

$$\mathcal{F}_t = \cap_{s>t}\sigma(X_u : u \leq s).$$

We call $\{\mathcal{F}_t\}$ the *right-continuous filtration generated by* X. In this case, we speak about an exponential family of stochastic processes. For fixed t and considered a function of θ, (3.1.1) is then the likelihood function corresponding to observing X continuously in $[0, t]$. Every exponential family on a filtered space can be thought of as an exponential family of stochastic processes because with respect to the right-continuous filtration generated by a canonical process B, the family $\mathcal{P}$ is obviously an exponential family with representation (3.1.2).

When studying exponential families of stochastic processes, it is often useful to consider the filtered space, where Ω is the space of functions ω from $[0, \infty)$ into $\mathbb{R}^d$ that are right-continuous with limits from the left, and where the filtration is given by

$$\mathcal{F}_t = \cap_{s>t}\sigma(\omega(u) : u \leq s).$$

We call this particular filtered space the *canonical space*. Consider a class of right-continuous d-dimensional processes $\{X^\theta : \theta \in \Theta\}$ with limits from the left defined on some probability space $(\tilde{\Omega}, \tilde{\mathcal{F}}, P)$. Every X^θ induces a probability measure P_θ on the canonical space by the mapping $\tilde{\omega} \in \tilde{\Omega} \rightarrow X^\theta(\tilde{\omega}) \in \Omega$. Under P_θ the coordinate process $X_t(\omega) = \omega(t)$, $\omega \in \Omega$, is a process with the same distribution as X^θ. If the class of probability measures $\{P_\theta : \theta \in \Theta\}$ induced on the canonical space forms an exponential family, we shall also refer to the class $\{X^\theta : \theta \in \Theta\}$ as an exponential family of stochastic processes.

In the rest of this chapter we give several examples of exponential families of stochastic processes. In all of these examples $L_t(\theta)$ denotes the likelihood function corresponding to observing the process under discussion continuously in the time interval $[0, t]$. The exponential representation will in all examples be of the form (3.1.2) with $\theta_0 = 0$, $\gamma(0) = 0$, and $\phi_t(0) = 0$. We also give an example of an exponential family of random fields and examples illustrating the importance of the filtration.

3.2 Markov processes in discrete time

Let X be a d-dimensional Markov chain with one-step transition densities (with respect to some dominating measures on $\mathbb{R}^d$) of the form

$$f_{X_t|X_{t-1}}(y;\theta|x) = b_t(x,y)\exp[\theta^T m_t(y,x) - \gamma(\theta)^T h_t(x)], \tag{3.2.1}$$

where $\theta \in \Theta$ and $m_t(y,x)$ are k-dimensional vectors, while $\gamma(\theta)$ and $h_t(x)$ are $(m-k)$-dimensional, $m > k > 0$. The set of possible parameters θ depends, in general, on x and t. We suppose that a non-empty common parameter set Θ can be found for all t and x. Assume, moreover, that the distribution of X_0 belongs to an exponential family or is known. Then the model is an exponential family of processes. For simplicity of exposition, we assume that

$$f_{X_0}(y;\theta) = b_0(y)\exp[\theta^T m_0(y) - \gamma(\theta)^T h_0], \quad \theta \in \Theta, \tag{3.2.2}$$

with h_0 a constant vector. (This also covers a known distribution of $X_0 : m_0(y) = 0$, $h_0 = 0$, $b_0(y) = 1$, and a probability measure as dominating measure). The likelihood function has the following exponential representation:

$$L_t(\theta) = \exp\left[\theta^T \sum_{i=0}^{t} m_i(X_i, X_{i-1}) - \gamma(\theta)^T \sum_{i=0}^{t} h_i(X_{i-1})\right], \tag{3.2.3}$$

with $m_0(y,x) = m_0(y)$ and $h_0(x) = h_0$.

A trivial example is repeated sampling of independent random variables the distribution of which belongs to an exponential family. Another simple, but quite different, example is the class of discrete time *Markov chains* with finite state space $I = \{1,\dots,m\}$ or any sub-class thereof. The family of all Markov chains with state space I is parametrized by the set of $m \times m$ matrices $\pi = \{p_{ij}\}$ satisfying $p_{i1} + \cdots + p_{im} = 1$, $i \in I$. The one-step transition distribution is given by

$$P_\pi(X_t = j|X_{t-1} = i) = p_{ij}, \quad i,j \in I, \tag{3.2.4}$$

and the likelihood function is

$$L_t(\theta) = \exp\left(\sum_{i,j=1}^{m} \theta_{ij} N_t^{(i,j)}\right), \tag{3.2.5}$$

where $\theta_{ij} = \ln(p_{ij}\}$, and $N_t^{(i,j)}$ denotes the number of one-step transitions from state i to state j in the time interval $[0,t]$. Here we have for simplicity assumed that $p_{ij} > 0$, $i,j \in I$, and that the initial state X_0 is given. An exponential family is also obtained if we allow X_0 to be random with a distribution belonging to any class of distributions on $\{1,\ldots,m\}$. For a sub-model with $p_{ij}=0$, the process $N^{(i,j)}$ is identically equal to zero.

A third example is the *Gaussian autoregressive process* of order one defined by

$$X_i = \theta\, X_{i-1} + Z_i, \quad i = 1,2,\ldots. \tag{3.2.6}$$

Here $\theta \in \mathbb{R}$, $X_0 = x_0$, and the Z_i are independent random variables with a standard normal distribution. In this example, $m_i(y,x) = yx$, $h_i(x) = x^2$ $(i = 1,2,\ldots)$, and $\gamma(\theta) = \frac{1}{2}\theta^2$. The likelihood function is given by

$$L_t(\theta) = \exp\left[\theta \sum_{i=1}^{t} X_i X_{i-1} - \frac{1}{2}\theta^2 \sum_{i=1}^{t} X_{i-1}^2\right]. \tag{3.2.7}$$

A final example is a *branching process* Z_t with $Z_0 = 1$ and offspring distribution belonging to an exponential family on $\{0,1,2,\ldots\}$, i.e.,

$$P_\alpha(Z_1 = j) = a_j\alpha^j/A(\alpha), \quad j = 0,1,2,\ldots, \tag{3.2.8}$$

where $a_j \geq 0$; $j = 0,1,2,\ldots$; $0 < \alpha < r$; $A(\alpha) = \sum a_j\alpha^j > 0$; and r is the radius of convergence for this power series. Under this assumption, the one-step transition densities satisfy

$$\frac{P_\alpha(Z_{k+1} = y|Z_k = x)}{P_1(Z_{k+1} = y|Z_k = x)} = \alpha^y[A(\alpha)/A(1)]^{-x}. \tag{3.2.9}$$

Note that we can always choose the a_j's such that $\alpha = 1$ is a possible parameter value. By comparing (3.2.9) to (3.2.1) we see that $\theta = \log\alpha$, $m_i(y,x) = y$, $h_i(x) = x$, and $\gamma(\theta) = \log[A(e^\theta)/A(1)]$. The likelihood function is

$$L_t(\theta) = \exp\left\{\theta \sum_{i=1}^{t} Z_i - \log[A(e^\theta)/A(1)] \sum_{i=1}^{t} Z_{i-1}\right\}. \tag{3.2.10}$$

3.3 More general discrete time models

The Markov property is of no significance in the derivation of the likelihood function (3.2.3) in the previous subsection. In fact, consider a d-dimensional stochastic process X for which the distribution of X_t conditional on $\mathcal{F}_{t-1} = \sigma(X_0, X_1, \ldots, X_{t-1})$ has a density of the form

$$f_t(y;\theta) = b_t(y)\exp[\theta^T T_t(y) - \gamma(\theta)^T U_t] \tag{3.3.1}$$

with respect to some dominating measure for $t = 1,2,\ldots$ and $\theta \in \Theta$. For given y and t, the random variables $b_t(y)$, $T_t(y)$, and U_t are $\mathcal{F}_{t-1}$-measurable. The dimension of θ and T_t is k, while $\gamma(\theta)$ and U_t are $(m-k)$-dimensional, $m > k > 0$. The distribution of X_0 should also belong to an exponential family or be fixed. For simplicity of exposition we assume that the distribution of X_0 has the form of (3.3.1) with $b_0(x)$, $T_0(x)$, and U_0 non-random. A model of this type is an exponential family of processes with likelihood function

$$L_t(\theta) = \exp\left[\theta^T \sum_{i=0}^{t} T_i(X_i) - \gamma(\theta)^T \sum_{i=0}^{t} U_i\right]. \tag{3.3.2}$$

The model type in the previous section is a particular case.

An example is the *Gaussian autoregression* of order k defined by

$$X_t = \theta_1 X_{t-1} + \cdots + \theta_k X_{t-k} + Z_t, \quad t = 1, 2, \ldots, \tag{3.3.3}$$

where $\theta_i \in \mathbb{R}$, where $X_0 = x_0, X_{-1} = x_{-1}, \ldots, X_{-k+1} = x_{-k+1}$ are given, and where the Z_t are independent standard normal distributed random variables. For this model $\theta = (\theta_1, \ldots, \theta_k)$, $T_t(y) = (yX_{t-1}, \ldots, yX_{t-k})$, $m = \frac{1}{2}k(k+3)$, $\gamma(\theta) = (\frac{1}{2}\theta_i^2 : i = 1, \ldots, k;\ \theta_i\theta_j : i,j = 1, \ldots, k, i < j)$, and $U_t = \{X_{t-i}X_{t-j} : i,j = 1, \ldots, k, i \leq j)$. The likelihood function is

$$L_t(\theta) = \exp\left[\sum_{i=1}^{k} \theta_i \sum_{s=1}^{t} X_s X_{s-i} - \sum_{i=1}^{k}\sum_{j=1}^{k} \frac{1}{2}\theta_i\theta_j \sum_{s=1}^{t} X_{s-i}X_{s-j}\right]. \tag{3.3.4}$$

Another example will be studied in Section 5.5.

3.4 Counting processes and marked counting processes

We shall first consider classes of counting processes with the following structure. The filtration is generated by the counting process N itself, and the model is defined by assuming that the predictable intensity is of the form

$$\lambda_t = \rho H_t, \quad \rho > 0, \quad t > 0, \tag{3.4.1}$$

where ρ is a parameter and H is a given positive predictable stochastic process. We assume that H_t has finite expectation for all $t < \infty$, so that N_t is finite for $t < \infty$. For simplicity we assume that N_0 is fixed or has a known distribution. This model has an exponential likelihood function of the form

$$L_t(\alpha) = \exp\left(\alpha(N_t - N_0) - (e^\alpha - 1)\int_0^t H_s ds\right), \tag{3.4.2}$$

where $\alpha = \log\ \rho$.

A very simple example is the class of *Poisson processes*, where $H = 1$. Another simple example is the class of *pure birth processes*, where $H_t = N_{t-}$, and the likelihood function is

$$L_t(\alpha) = \exp\left[\alpha(N_t - N_0) - (e^\alpha - 1)\int_0^t N_s ds\right]. \qquad (3.4.3)$$

A similar example is the *logistic birth process*, where $H_t = N_{t-}(M - N_{t-})^+$. Here M is a constant that is interpreted as the carrying capacity of the environment.

Let us next consider *marked counting processes*. Let N be a counting process with $N_0 = 0$ and with intensity given by (3.4.1) where H satisfies the conditions above. However, now we allow that the filtration is generated by the observed marked counting process X and not only by N. Let τ_i denote the time of the ith jump of N. Then we define a marked counting process by

$$X_t = \sum_{i=1}^{N_t} Y_i, \qquad (3.4.4)$$

where the distribution of Y_i given $\{\tau_1, \ldots, \tau_i, Y_1, \ldots, Y_{i-1}\}$ has a density with respect to some dominating measure of the form

$$f_i(y;\theta) = b_i(y)\exp[\theta^T T_i(y) - \gamma(\theta)^T U_i], \qquad (3.4.5)$$

$i = 1, 2, \ldots$, $\theta \in \Theta \subseteq \mathbb{R}^k$. Here $b_i(y)$, $T_i(y)$ (k-dimensional), and U_i (($m - k$)-dimensional) depend on $\tau_1, \ldots, \tau_i$ and $Y_1, \ldots, Y_{i-1}$, or a subset of these. The model thus defined is an exponential family of processes with

$$\begin{aligned} L_t(\alpha, \theta) &= \exp\left[\alpha N_t - (e^\alpha - 1)\int_0^t H_s ds \right. \qquad (3.4.6) \\ &\quad \left. + \theta^T \sum_{i=1}^{N_t} T_{\tau_i}(\Delta X_{\tau_i}) - \gamma(\theta)^T \sum_{i=1}^{N_t} U_{\tau_i}\right], \end{aligned}$$

where $\alpha = \log\,\rho$ and $\Delta X_{\tau_i} = X_{\tau_i} - X_{\tau_{i-1}} (= Y_i)$. A particular case is a class of compound Poisson processes with intensity $\lambda \in (0, \infty)$ and with jump distribution from an exponential family, which was studied in Chapter 2.

A different example is the class of *linear birth-and-death processes* with birth-rate $\lambda > 0$ and death-rate $\mu > 0$. Let X denote the observed process. The counting process N is in this example the number of births and deaths in the time interval $[0, t]$ and $\lambda_t = (\lambda + \mu)X_{t-}$. The distribution of ΔX_{τ_i} belongs to the family of all probability distributions concentrated on the set $\{-1, 1\}$, which is an exponential family. We find the likelihood function

$$L_t(\theta) = \exp\left\{\theta_1 B_t + \theta_2 D_t - (e^{\theta_1} + e^{\theta_2} - 2)S_t\right\}. \qquad (3.4.7)$$

Here $\theta = (\theta_1, \theta_2) = (\log\ \lambda,\ \log\ \mu)$, B_t is the number of births in $[0, t]$, D_t is the number of deaths in $[0, t]$, and

$$S_t = \int_0^t X_s ds$$

is the total time lived in the population before time t.

The linear birth-and-death process can be modified by adding immigration if the population dies out. Assume that if the population size becomes zero a single immigrant will arrive after an exponentially distributed waiting time with mean γ^{-1}. When the population size is strictly positive, no immigration is allowed. This modified model is also an exponential family. Its likelihood function is the product of (3.4.7) and

$$\exp(\theta_3 M_t - (e^{\theta_3} - 1)T_t),$$

where $\theta_3 = \log(\gamma)$, M_t is the number of immigrants in $[0, t]$, and T_t is the total time before time t during which the population size was zero.

An exponential family of stochastic processes related to those discussed above is the class of *continuous time Markov processes with finite state space* $\{1, \ldots, m\}$. Denote the intensity matrix by $\{\alpha_{ij}\}$, and set

$$\beta_i = \sum_{j \neq i} \alpha_{ij}.$$

For simplicity we assume that $\alpha_{ij} > 0$ for $i \neq j$. In this example the counting process N counts the number of transitions. Its intensity is

$$\lambda_t = \beta_{X_{t-}} = \sum_{i=1}^{m} 1_{\{X_{t-}=i\}} (\sum_{j \neq i} \alpha_{ij}), \tag{3.4.8}$$

which is of a more general type than the intensities of the form (3.4.1). The distribution of X_{τ_k} depends on $X_{\tau_{k-1}}$ and belongs to the family of all distributions with support $\{1, \ldots, m\}$, which is an exponential family. If π_{ij} denotes the probability of $\{X_{\tau_k} = j\}$ given $\{X_{\tau_{k-1}} = i\}$, then $\alpha_{ij} = \pi_{ij}\beta_i$, and after some calculation we find the likelihood function

$$L_t(\theta) = \exp\left\{ \sum_{i=1}^{m} \left[\sum_{j \neq i} \theta_{ij} K_t^{(i,j)} - \left(\sum_{j \neq i} e^{\theta_{ij}} + 1 - m \right) S_t^{(i)} \right] \right\} \tag{3.4.9}$$

with $\theta_{ij} = \log\ \alpha_{ij}$. The process $K_t^{(i,j)}$ is the number of transitions from state i to state j in the time interval $[0, t]$, and $S_t^{(i)}$ is the time spent in state i before time t. For detatils, see, e.g., Jacobsen (1983). Submodels where $\alpha_{ij} = 0$ for certain values of (i, j) are exponential families of processes too. In this situation the corresponding processes $K^{(i,j)}$ are identically equal to zero.

3.5 Diffusion-type processes

Consider the class of stochastic differential equations

$$dX_t^\theta = (a_t(X^\theta) + b_t(X^\theta)\theta)dt + c_t(X^\theta)dW_t, \quad t > 0, \tag{3.5.1}$$

with initial condition $X_0^\theta = x_0$, and where $\theta \in \Theta \subseteq \mathbb{R}^k$. The process W is a p-dimensional standard Wiener process, while a_t, b_t, and c_t are functionals depending on $\{X_s^\theta : s \leq t\}$ only. The process X^θ and the functional a_t are d-dimensional, b_t is a $d \times k$-matrix, and c_t is a $d \times p$-matrix. We assume that a unique (strong) solution X^θ exists for every $\theta \in \Theta$. It is not a restriction to assume that $0 \in \Theta$. A solution of an equation of the form (3.5.1) is called a *diffusion-type process.* If for all $t > 0$ the functionals a_t, b_t, and c_t depend on X^θ only through X_t^θ, the process X^θ is a Markov process and is called a *diffusion process.*

The class of processes $\{X^\theta : \theta \in \Theta\}$ induces a family of probability measures $\{P_\theta : \theta \in \Theta\}$ on the canonical space of continuous functions from $[0, \infty)$ into $\mathbb{R}^d$. We denote by X the canonical process. For every fixed $t \geq 0$ the measures $\{P_\theta^t : \theta \in \Theta\}$ are equivalent, provided that the matrix $c_t(X)c_t(X)^T$ is almost surely regular under P_θ for all $\theta \in \Theta$ and $t > 0$, and that the matrix of processes

$$B_t = \int_0^t b_s(X)^T[c_s(X)c_s(X)^T]^{-1}b_s(X)ds \tag{3.5.2}$$

satisfies

$$P_\theta(B_t < \infty) = 1 \tag{3.5.3}$$

for all $\theta \in \Theta$ and all $t > 0$. By (3.5.3) is meant that all k^2 entries of B_t are finite. Under these conditions, $\{P_\theta : \theta \in \Theta\}$ is an exponential family with

$$L_t(\theta) = \frac{dP_\theta^t}{dP_0^t} = \exp[\theta^T A_t - \tfrac{1}{2}\theta^T B_t \theta], \tag{3.5.4}$$

where

$$A_t = \int_0^t b_s(X)^T[c_s(X)c_s(X)^T]^{-1}d\tilde{X}_s \tag{3.5.5}$$

with

$$\tilde{X}_t = X_t - x_0 - \int_0^t a_s(X)ds. \tag{3.5.6}$$

Note that $\gamma(\theta) = \{\frac{1}{2}\theta_i^2 : i = 1, \ldots, k;\ \theta_i\theta_j : i, j = 1, \ldots, k,\ i < j\}$.

A simple example is the class of one-dimensional Wiener processes with drift, which corresponds to $a_t(X) = 0$ and $b_t(X) = c_t(X) = 1$. Here $\Theta = \mathbb{R}$, and the likelihood function is given by (2.1.6).

An important example is the class of one-dimensional *Ornstein-Uhlenbeck* processes, which is obtained for $a_t(X) = 0$, $b_t(X) = X_t$, and $c_t(X) = 1$. The likelihood function is

$$L_t(\theta) = \exp\left\{\theta \int_0^t X_s dX_s - \tfrac{1}{2}\theta^2 \int_0^t X_s^2 ds\right\}, \quad \theta \in \mathbb{R}. \tag{3.5.7}$$

A one-dimensional *non-Markovian* diffusion-type process that generalizes the Ornstein-Uhlenbeck process is defined by

$$dX_t^\theta = (\theta_1 X_t^\theta + \theta_2 X_{t-r}^\theta)dt + dW_t, \quad X_t^\theta = \bar{X}_t, \; -r \le t \le 0, \tag{3.5.8}$$

where $\theta = (\theta_1, \theta_2) \in \mathbb{R}^2$ and where $\{\bar{X}_t : -r \le t \le 0\}$ is a given initial process. Finally, r is a known positive constant. Conditionally on $\{X_s : 0 \ge s \ge -r\}$ the class of solutions is an exponential family with

$$\begin{aligned} L_t(\theta) \;=\; & \exp\left\{\theta_1 \int_0^t X_s dX_s + \theta_2 \int_0^t X_{s-r} dX_s \right. \qquad (3.5.9) \\ & \left. - \tfrac{1}{2}\theta_1^2 \int_0^t X_s^2 ds - \tfrac{1}{2}\theta_2^2 \int_0^t X_{s-r}^2 ds - \theta_1\theta_2 \int_0^t X_s X_{s-r} ds \right\}. \end{aligned}$$

An example of an exponential family of diffusion processes that is not of the type (3.5.1) is given by

$$dX_t^\theta = h(\theta, t) X_t^\theta dt + dW_t, \quad t > 0, \quad X_0^\theta = x_0, \tag{3.5.10}$$

where $\theta = (\theta_1, \theta_2)$, $\theta_2 \ge 0$, $-\theta_2 \le \theta_1 < \theta_2 \coth(\theta_2)$, and

$$h(\theta, t) = \theta_1 + \frac{\theta_1^2 - \theta_2^2}{\theta_2 \, \coth((1-t)\theta_2) - \theta_1}. \tag{3.5.11}$$

The likelihood function for this model is

$$L_t(\theta) = \exp\left(\tfrac{1}{2}\left[h(\theta, t) X_t^2 - \theta_2^2 \int_0^t X_s^2 ds - h(\theta, 1) x_0^2 + m(\theta, t)\right]\right), \tag{3.5.12}$$

where

$$m(\theta, t) = \log\left(\frac{\cosh(\theta_2) - \theta_1 \sinh(\theta_2)/\theta_2}{\cosh((1-t)\theta_2) - \theta_1 \sinh((1-t)\theta_2)/\theta_2}\right). \tag{3.5.13}$$

We see that this exponential family is not time-homogeneous. For $\theta_2 = |\theta_1|$ a homogeneous Ornstein-Uhlenbeck process is obtained. We shall consider this model in more detail in Chapter 7.

3.6 Diffusion processes with jumps

An important generalization of the diffusion processes is the class of diffusion processes with jumps. In this section we consider the special case

$$dX_t^\theta = (a_t(X_t^\theta) + b_t(X_t^\theta)\theta_1)dt + c_t(X_{t-}^\theta)dW_t + d_t(X_{t-}^\theta)dZ_t, \qquad (3.6.1)$$

$t > 0$, $X_0^\theta = x_0$. Here W is a d-dimensional standard Wiener process, and $\theta^T = (\theta_1^T, \theta_2, \theta_3^T)$, $\theta_1 \in \Theta_1 \subseteq \mathbb{R}^{k_1}$, $\theta_2 \in \mathbb{R}$, $\theta_3 \in \Theta_3 \subseteq \mathbb{R}^{k_3}$. The dimensions of the process X^θ and of the functions a_t, b_t, and c_t are as in the previous section (with k replaced by k_1), and d_t is an invertible $d \times d$-matrix. The process Z is independent of W and is supposed to belong to an exponential family of compound Poisson processes. Specifically, we define Z by

$$Z_t = \sum_{i=1}^{N_t} Y_i, \qquad (3.6.2)$$

where N is a Poisson process with intensity $\exp(\theta_2)$ and the Y_i's are mutually independent identically distributed d-dimensional random vectors independent of N. The distribution of Y_1 is assumed to belong to an exponential family with density

$$f_Y(y) = \exp[g(y)^T\theta_3 - \kappa(\theta_3)] \qquad (3.6.3)$$

with respect to some dominating measure. Here g is a k_3-dimensional vector. We assume that a unique solution X^θ exists for every $\theta \in \Theta = \Theta_1 \times \mathbb{R} \times \Theta_3$ and that $0 \in \Theta$.

The class of processes $\{X^\theta : \theta \in \Theta\}$ induces a family of probability measures $\{P_\theta : \theta \in \Theta\}$ on the canonical space of functions from $[0, \infty)$ into $\mathbb{R}^d$ that are continuous from the right with limits from the left. The canonical process is denoted by X. For every $t \geq 0$ the measures $\{P_\theta^t : \theta \in \Theta\}$ are equivalent, provided that the conditions imposed on b and c in Section 3.5 are satisfied. Let B_t and A_t be defined by (3.5.2) and (3.5.5) with the process $\tilde{X}$ given by

$$\tilde{X}_t = X_t - x_0 - \sum_{s \leq t} \Delta X_s - \int_0^t a_s(X)ds \qquad (3.6.4)$$

rather than by (3.5.6). Then the likelihood function is given by

$$\begin{aligned} L_t(\theta) &= \exp\Bigg[\theta_1^T A_t - \tfrac{1}{2}\theta_1^T B_t \theta_1 + \theta_2 N_t - (e^{\theta_2} - 1)t \qquad (3.6.5) \\ &\quad + \sum_{s \in S_t} g(d_s(X_{s-})^{-1} \Delta X_s)^T \theta_3 - (\kappa(\theta_3) - \kappa(0))\Bigg], \end{aligned}$$

where $S_t = \{s \leq t : \Delta X_s \neq 0\}$.

Particular examples of diffusion processes with jumps are

$$dX_t = (\theta_1 + \theta_2 X_t)dt + dW_t + dZ_t, \tag{3.6.6}$$

$$dX_t = \theta_1 X_t dt + X_{t-}dW_t + X_{t-}dZ_t, \tag{3.6.7}$$

and

$$dX_t = \theta_1 X_t + \theta_2 X_t^2 + X_{t-}dW_t + X_{t-}dZ_t. \tag{3.6.8}$$

The first equation has been used to model the dynamics of soil moisture. The second has been used as a model for stock prices with the assumption that the jumps of Z are strictly larger than minus one. The last process is a model for the dynamics of a population that grows logistically between disasters. Here it is assumed that the jumps of Z belong to the interval $(0, 1)$.

3.7 Random fields

Let X^θ be a random field on $\mathbb{R}_+^2$ that solves the stochastic differential equation

$$dX_z^\theta = [a_z(X) + \theta b_z(X)]dz + dW_z, \quad z \in \mathbb{R}_+^2, \tag{3.7.1}$$

with initial conditions $X_{(x,0)} = X_{(0,y)} = 0$ for all x, $y \in \mathbb{R}_+$. In (3.7.1) W is a standard Wiener random field, while $a_z(\cdot)$ and $b_z(\cdot)$ are real functionals that for $z = (x, y)$ depend only on the behaviour of X in the domain $[0, x] \times [0, y]$. The parameter set Θ is a subset of $\mathbb{R}$. It is no restriction to assume that $0 \in \Theta$. In the following we will consider observation of X^θ in windows of the form $K_t = [0, t] \times [0, t]$.

Let C_t denote the set of continuous functions $x : K_t \to \mathbb{R}$ satisfying $x(s_1, 0) = x(0, s_2) = 0$ for all s_1, $s_2 \in \mathbb{R}$, and let $\mathcal{A}_t$ be the σ-algebra generated by the sets $M_{B,z} = \{x \in C_t : x(z) \in B\}$, where $z \in K_t$, and B is a Borel subset of $\mathbb{R}$. Moreover, let C be the set of continuous functions $\mathbb{R}_+^2 \to \mathbb{R}$ satisfying the same initial conditions as for C_t, and set $\mathcal{A} = \sigma(\mathcal{A}_t : t \geq 0)$. The processes $\{X^\theta : \theta \in \Theta\}$ induce a class of probability measures $\{P_\theta : \theta \in \Theta\}$ on $(C, \mathcal{A})$. Let P_θ^t denote the restriction on P_θ to $\mathcal{A}_t$. Then for all $t \geq 0$ and all $\theta \in \Theta$,

$$\begin{aligned} \frac{dP_\theta^t}{dP_0^t} &= \exp\left\{\theta\left[\int_{K_t} b_z(X)dX_z - \int_{K_t} b_z(X)a_z(X)dz\right]\right. \\ &\qquad \left. - \tfrac{1}{2}\theta^2 \int_{K_t} b_z(X)dz\right\}, \end{aligned} \tag{3.7.2}$$

provided that

$$P_\theta\left(\int_{K_t} b_z(X)^2 dz < \infty\right) = 1, \quad \theta \in \Theta, \quad t \geq 0. \tag{3.7.3}$$

In fact, under condition (3.7.3) the class $\{P_\theta : \theta \in \Theta\}$ is an exponential family relative to the filtration corresponding to observation in any family of windows of the form $G_t = [0, \eta_1(t)] \times [0, \eta_2(t)]$ where η_1 and η_2 are non-decreasing functions $[0,\infty) \to [0,\infty)$ satisfying $\eta_1(0) = \eta_2(0) = 0$.

3.8 The significance of the filtration

In this section we consider two examples that in different ways illustrate how important the filtration is for an exponential family. In short, to obtain an exponential family it is not enough to have the right class of processes, it is also necessary to observe the processes in the right way. The way of observing the processes is given by the filtration.

Let $(\Omega, \mathcal{F}, P)$ be a probability space. Suppose there exists a stochastic process X which is a standard Wiener process under P. Let $\{\mathcal{F}_t\}$ be the filtration generated by X, as defined earlier, and assume as usual that $\mathcal{F} = \sigma(\mathcal{F}_t : t \geq 0)$. As discussed in Chapter 2, we can define a class of probability measures on $\mathcal{F}$ by

$$\frac{dP_\theta^t}{dP^t} = L_t(\theta) = \exp[\theta X_t - \tfrac{1}{2}\theta^2 t], \quad \theta \in \mathbb{R}, \quad t \geq 0. \tag{3.8.1}$$

Under P_θ the process X is a Wiener process with drift θ.

Now define the family of first hitting times

$$\tau_u = \inf\{t : X_t \geq u\}, \quad u \geq 0,$$

with the usual convention $\inf \emptyset = \infty$. It is known that $P_\theta(\tau_u < \infty) = 1$ for all $\theta \geq 0$ and that $P_\theta(\tau_u = \infty) > 0$ for all $\theta < 0$. We will now consider the filtration $\{\mathcal{G}_u\}$ given by

$$\mathcal{G}_u = \mathcal{F}_{\tau_u},$$

with the definition $\mathcal{F}_\infty = \mathcal{F}$ and the restriction of P_θ to $\mathcal{G}_u$, which we denote by $\tilde{P}_\theta^u$. It holds that

$$\frac{d\tilde{P}_\theta^u}{d\tilde{P}^u} = L_{\tau_u}(\theta) = \exp(\theta u - \tfrac{1}{2}\theta^2 \tau_u), \quad \theta \geq 0, \tag{3.8.2}$$

with $L_t(\theta)$ given by (3.8.1); see Corollary B.1.2. We see that $\{P_\theta : \theta \geq 0\}$ is an exponential family relative to $\{\mathcal{G}_u\}$ as well. The whole class $\{P_\theta : \theta \in \mathbb{R}\}$, however, is not an exponential family relative to $\{\mathcal{G}_u\}$. This is because the restrictions of the elements of $\{P_\theta : \theta < 0\}$ to the set $\{\tau_u = \infty\}$ are singular for all $u > 0$.

Denote by $\bar{P}_\theta^u$ the restriction of P_θ to the σ-algebra $\mathcal{H}_u = \sigma(\tau_s : s \leq u)$. The Radon-Nikodym derivatives $d\bar{P}_\theta^u / d\bar{P}^u$ are given by (3.8.2), since $\mathcal{H}_u \subseteq$

$\mathcal{G}_u$ and (3.8.2) is $\mathcal{H}_u$-measurable. By the re-parametrization $\alpha = -\frac{1}{2}\theta^2$, we can write

$$\frac{d\bar{P}_\theta^u}{d\bar{P}^u} = \exp(\alpha\tau_u + \sqrt{-2\alpha}t), \quad \alpha \leq 0. \tag{3.8.3}$$

Under P, the stochastic process $\{\tau_u\}$ has independent increments, and τ_1 has cumulant transform $-\sqrt{-2s}$ with domain $s \in (-\infty, 0]$. Hence the restriction of the elements in $\{P_\theta : \theta \geq 0\}$ to $\sigma(\mathcal{H}_u : u \geq 0)$ equals the exponential family of processes with independent increments obtained from $\{\tau_u\}$ and P as discussed in Chapter 2.

Another example illustrating the significance of the filtration is the following. Suppose again that the process X is a Wiener process under P. Assume, moreover, that under P there exists a counting process N independent of X. Let $\{\mathcal{F}_t\}$ be the filtration generated by X and N, and assume that $\mathcal{F} = \sigma(\mathcal{F}_t : t \geq 0)$. We can then define new probability measures P_θ on $(\Omega, \mathcal{F})$ by

$$\frac{dP_\theta^t}{dP^t} = \exp\left\{\theta \int_0^t (-1)^{N_s} dX_s - \tfrac{1}{2}\theta^2 t\right\}, \quad \theta \in \mathbb{R}, \ t \geq 0. \tag{3.8.4}$$

That this defines a class of probability measures follows from the fact that

$$\int_0^t (-1)^{N_s} dX_s$$

is a standard Wiener process under P and that hence the expression in (3.8.4) is a P-martingale with respect to $\{\mathcal{F}_t\}$. Under P_θ there exists a Wiener process W^θ such that X solves the stochastic differential equation

$$dX_t = \theta(-1)^{N_t} dt + dW_t^\theta, \quad t \geq 0; \tag{3.8.5}$$

i.e., X_t has drift $-\theta$ or θ depending on whether N_t is odd or even. The process N is, under P_θ, the same counting process as under P. To see this, note that if $\mathcal{H}_t = \sigma(N_s : s \leq t)$ and $A = \{N_{t_1} = n_1, \ldots, N_{t_k} = n_k\}$, then for $t \geq t_k$,

$$\begin{aligned} P_\theta(A) &= E_P\left(1_A \frac{dP_\theta^t}{dP^t}\right) \\ &= E_P\left[1_A E_P\left(\frac{dP_\theta^t}{dP^t}\middle| \mathcal{H}_t\right)\right] \\ &= E_P[1_A] \\ &= P(A). \end{aligned}$$

Obviously, $\{P_\theta : \theta \in \mathbb{R}\}$ is an exponential family with respect to $\{\mathcal{F}_t\}$. However, with respect to the filtration generated by X it is not an exponential family. Let $\tilde{P}_\theta^t$ denote the restriction of P_θ to $\mathcal{G}_t = \sigma(X_s : s \leq t)$.

Then

$$\begin{aligned}\frac{d\tilde{P}_\theta^t}{d\tilde{P}^t} &= e^{-\frac{1}{2}\theta^2 t} E_P\left[\exp\left(\theta\int_0^t(-1)^{N_s}dX_s\right)\Big|\mathcal{G}_t\right] \\ &= e^{-\frac{1}{2}\theta^2 t} E_P\left[\exp\left\{\theta\left(\sum_{i=1}^{N_t}(-1)^{i-1}(X_{\tau_i}-X_{\tau_{i-1}})\right.\right.\right. \\ &\qquad\left.\left.\left. +(-1)^{N_t}(X_t - X_{\tau_{N_t}})\right)\right\}\Big|\mathcal{G}_t\right),\end{aligned}$$

where τ_i is the ith jump time of N. This is not of exponential form.

Note, that the situation is completely different if N is adapted to the filtration generated by X, because then $d\tilde{P}_\theta^t/d\tilde{P}^t = dP_\theta^t/dP^t$. An example is a process N the jumps of which are defined recursively by $\tau_i = \inf\{t > \tau_{i-1} : X_s = (-1)^{i-1}\}$; $\tau_0 = 0$.

3.9 Exercises

3.1 Consider the general type of discrete time models introduced in Section 3.3. Let the notation be as in that section, but to simplify the notation let U_t and $\gamma(\theta)$ be one-dimensional, and suppose that the distribution of X_0 does not depend on θ and that κ is differentiable. We assume further that

$$\Theta \subseteq \text{int}\{\theta : \int b_t(y; x_0, \cdots, x_{t-1})\exp[\theta^T T_t(y; x_0, \cdots, x_{t-1})]\mu(dy) < \infty\}$$

for all $x_0, \cdots, x_{t-1}$ and all $t \geq 1$, where μ denotes the dominating measure with respect to which (3.3.1) is a density. The dependence of $b_t(y)$ and $T_t(y)$ on the past has been expressed explicitly here.

Prove under these assumptions that for all $i \geq 1$,

$$E_\theta(T_i(X_i)|\mathcal{F}_{i-1}) = \dot{\gamma}(\theta)U_i$$

and

$$\text{Var}_\theta(T_i(X_i)|\mathcal{F}_{i-1}) = \ddot{\gamma}(\theta)U_i,$$

for all $\theta \in \Theta$, where $\dot{\gamma}$ denotes the vector of partial derivatives of γ with respect to the coordinates of θ. Analogously, $\ddot{\gamma}$ is the $k \times k$-matrix of second partial derivatives. Conclude from this that we can choose a representation (3.3.1) with $U_i \geq 0$ almost surely for all $i \geq 1$. Hint: Use classical exponential family results.

Suppose, moreover, that $E_\theta(|U_i|) < \infty$ for all $i \geq 1$, and prove that under P_θ the score function

$$\dot{l}_t(\theta) = \sum_{i=1}^t T_i(X_i) - \dot{\gamma}(\theta)\sum_{i=1}^t U_i$$

is a k-dimensional square integrable zero-mean martingale with quadratic characteristic $\langle \dot{l}(\theta) \rangle_t = \ddot{\gamma}(\theta) \sum_{i=1}^{t} U_i$. The score function is the vector of partial derivatives with respect to the coordinates of θ of the log-likelihood function $l_t(\theta) = \log L_t(\theta)$. Note that $-\ddot{l}_t(\theta) = \langle \dot{l}(\theta) \rangle_t$.

3.2 Consider the counting process model with intensity (3.4.1). Prove that the score function $\dot{l}_t(\alpha) = \frac{d}{d\alpha} \log L_t(\alpha)$ is a zero-mean square integrable martingale with quadratic variation $e^{\alpha} \int_0^t H_s ds$ under P_α.

3.3 Consider the class of diffusion-type processes given as solutions to (3.5.1). Let the notation be as in Section 3.5 and prove that the score function $\dot{l}_t(\theta) = A_t - B_t\theta$ is a local martingale under P_θ. Show that if $E_\theta(B_t) < \infty$ for all $t > 0$, then $U_t(\theta)$ is a zero-mean square integrable martingale under P_θ.

3.10 Bibliographic notes

The general definition (3.1.1) of exponential families of stochastic processes was first given in Küchler and Sørensen (1989) and was studied further in a series of papers by the same authors; see the list of references. Various sub-classes have been studied by several authors. References will be given in the following chapters. Markov processes with transition densities of the form (3.2.1) were introduced by Heyde and Feigin (1975) and studied further in Feigin (1981), Hudson (1982), and Hwang and Basawa (1994); see also Küchler and Sørensen (1997). The more general processes in Section 3.3 were considered by Bhat (1988a) and Sørensen (1996). The families of counting processes and marked counting processes in Section 3.4 were considered in Sørensen (1986). The special case of compound Poisson processes was studied by Stefanov (1982). The class of linear birth-and-death processes was treated from the point of view of exponential families by Franz (1982). Exponential families of diffusion-type processes have been studied by several authors, including Novikov (1972), Taraskin (1974), Brown and Hewitt (1975a), Feigin (1976), Sørensen (1983), and Küchler and Sørensen (1994b). The non-Markovian diffusion given by (3.5.8) was studied by Küchler and Mensch (1992) and Gushchin and Küchler (1996). Exponential families of diffusions with jumps were investigated by Sørensen (1991). The diffusion random field given by (3.7.1) was considered by Rózánski (1989). Further examples of exponential families of stochastic processes can be found in Basawa and Prakasa Rao (1980).

Readers who want to know more about the types of stochastic processes considered in this chapter can consult the following monographs. An introduction to Markov processes with countable state space can be found in Karlin and Taylor (1975). Counting processes and marked counting processes are treated in Liptser and Shiryaev (1977), Bremaud (1981), and

Karr (1991). Concerning diffusions and diffusion-type processes see Liptser and Shiryaev (1977) or Ikeda and Watanabe (1981). Results about first hitting times for diffusions can be found in Ito and McKean (1965). A good introduction to the general theory of stochastic processes can be found in the first chapters of Jacod and Shiryaev (1987).

4

First Properties

4.1 Exponential representations

In this chapter we consider a general exponential family on a filtered space $(\Omega, \mathcal{F}, \{\mathcal{F}_t\}_{t\geq 0})$ with an exponential representation of the form (3.1.1). For fixed $t \geq 0$ the class $\mathcal{P}_t = \{P_\theta^t : \theta \in \Theta\}$ is an exponential family in the classical sense; see, e.g., Barndorff-Nielsen (1978) or Brown (1986). We shall call these exponential families the *finite sample exponential families.* The set-up considered in this book generalizes the classical situation of repeated observation of independent random variables X_i, $i = 1, 2, \ldots$, with a common distribution Q_θ belonging to an exponential family. The product measure $Q_\theta^n = Q_\theta \times \cdots \times Q_\theta$ (n times) belongs to an exponential family of the same order as the original family, and Q_θ^n is the restriction of Q_θ^∞ to the σ-algebra generated by $\{X_1, \ldots, X_n\}$. Here, however, we consider general filtrations and measures P_θ that need not have a product structure. This more general setting gives rise to new problems, but also has interesting regularities, which we shall study in the rest of this book. An obvious problem is that we cannot simply define some exponential family that we might like on every $\mathcal{F}_t$ ($t \geq 0$). This does not necessarily imply that a family of probability measures exists on $\mathcal{F}$. At least the measures $\{P_\theta^t : t \geq 0\}$ must be consistent for all θ. We shall return to this problem in Chapter 7.

For every fixed t we can find a minimal representation of the finite sample exponential family $\{P_\theta^t : \theta \in \Theta\}$. Let $\bar{B}_t$ and $\bar{\gamma}_t(\theta)$ be the canonical statistic and the parameter function of such a representation, and let m_t denote the dimension of $\bar{B}_t$ and $\bar{\gamma}_t(\theta)$. That the representation is minimal means that the components of $\bar{B}_t$ as well as those of $\bar{\gamma}_t(\theta)$ are affinely independent, i.e.,

$$\sum_{i=1}^{m_t} a_i \bar{B}_t^{(i)} = a_0 \; P-\text{a.s} \;\Rightarrow\; a_0 = a_1 = \cdots = a_{m_t} = 0$$

and

$$\sum_{i=1}^{m_t} b_i \bar{\gamma}_t^{(i)}(\theta) = b_0 \;\; \forall \theta \in \Theta \;\Rightarrow\; b_0 = b_1 = \cdots = b_{m_t} = 0.$$

The dimension m_t is the same for all minimal representations of $\mathcal{P}_t$ and is called the order of the finite sample family. The order m_t of $\mathcal{P}_t$ may vary with t (for an example see Section 7.4), but we can nonetheless define a minimal representation of an exponential family on a filtered space. Since by assumption we have the representation (3.1.1), m_t is bounded by m. We can therefore define the *order* of $\mathcal{P}$ as $m^* = \max\{m_t : t \geq 0\}$.

For any representation where the canonical process is m^*-dimensional there exists at least one $t \geq 0$ for which the representation is minimal for $\mathcal{P}_t$. We call a representation

$$\frac{dP_\theta^t}{dP^t} = a_t^*(\theta) q_t^* \exp(\gamma_t^*(\theta)^T B_t^*), \;\; t \geq 0, \, \theta \in \Theta, \tag{4.1.1}$$

minimal if $\gamma_t^*(\theta)$ and B_t^* are m^*-dimensional for all $t \geq 0$ and if for every $t \geq 0$ there exist non-random $m^* \times m_t$-matrices F_t and G_t of rank m_t such that

$$B_t^* = F_t \bar{B}_t, \; \gamma_t^*(\theta) = G_t \bar{\gamma}_t(\theta), \text{ and } G_t^T F_t = I_{m_t}, \tag{4.1.2}$$

where $\bar{B}_t$ and $\bar{\gamma}_t(\theta)$ are, respectively, the canonical statistic and the parameter function of a minimal representation of $\mathcal{P}_t$, and where I_{m_t} is the m_t-dimensional identity matrix. For example, B_t^* could be obtained from $\bar{B}_t$ by adding $m^* - m_t$ coordinates with the value zero. The function $\gamma_t^*(\theta)$ could be obtained similarly from $\bar{\gamma}_t(\theta)$.

If a representation (4.1.1) is minimal, the canonical process B_t^* is minimal sufficient for $\mathcal{P}_t$ for all $t \geq 0$. This is a classical result for minimal representations of the finite sample families $\mathcal{P}_t$, see, e.g., Johansen (1979, p. 3).

The following lemma relates a minimal representation to another exponential representation.

Lemma 4.1.1 *Suppose (4.1.1) is a minimal representation of the exponential family $\mathcal{P}$ of order m^*, and let m_t denote the order of the finite sample family $\mathcal{P}_t$. Suppose (3.1.1) is another representation of $\mathcal{P}$. Then $m^* \leq m$, and there exist two $m^* \times m$-matrices of non-random functions D_t and $\bar{D}_t$ of rank m_t and two m^*-dimensional vectors of non-random functions E_t and $\bar{E}_t$ such that*

$$\begin{aligned} D_t B_t + E_t &= B_t^* \;\; P-a.s., \\ \bar{D}_t \gamma_t(\theta) + \bar{E}_t &= \gamma_t^*(\theta) \;\; \textit{for all } \theta \in \Theta, \end{aligned}$$

where $D_t\bar{D}_t^T$ *has rank* m_t. *In fact,* $D_t\bar{D}_t^T = F_tG_t^T$ *for the* F_t *and* G_t *appearing in (4.1.2).*

Proof. The result follows immediately from Lemma 8.1 in Barndorff-Nielsen (1978) (see also Johansen, 1979, p. 3) combined with (4.1.2). □

In the important case of a *time-homogeneous* exponential family with a representation (3.1.1) where γ is independent of t, we can parametrize the family by the set

$$\Gamma = \{\gamma(\theta) : \theta \in \Theta\}. \tag{4.1.3}$$

In analogy with the classical terminology we call this parametrization a *canonical parametrization*, and an exponential representation with this parametrization a *canonical representation.* If the exponential representation (3.1.1) is minimal, then $\gamma(\theta)$ is maximally identifiable, i.e.,

$$P_{\theta_1} = P_{\theta_2} \Leftrightarrow \gamma(\theta_1) = \gamma(\theta_2).$$

To prove the non-trivial part of this statement, assume that $P_{\theta_1} = P_{\theta_2}$. Then for all $t \geq 0$,

$$(\gamma(\theta_1) - \gamma(\theta_2))^T B_t = \phi_t(\theta_1) - \phi_t(\theta_2).$$

Since there exists at least one t for which the components of B_t are affinely independent stochastic variables, it follows that $\gamma(\theta_1) = \gamma(\theta_2)$.

For exponential families that are not time-homogeneous there is, in general, no sensible canonical parametrization. The set

$$\Gamma_t = \{\gamma_t(\theta) : \theta \in \Theta\} \tag{4.1.4}$$

will typically depend on t. Even when it is independent of t, i.e., when $\Gamma_t = \Gamma$ and when $\gamma_t : \Theta \to \Gamma$ is one-to-one, (4.1.4) and γ_t do not define a parametrization of the exponential family. This is because for $\varphi \in \Gamma$, the measure $P_{\gamma_t^{-1}(\varphi)}$ varies with t, except when γ_t is independent of t. We will, quite generally, call γ_t the canonical parameter function.

4.2 Exponential families of stochastic processes with a non-empty kernel

For most time-homogeneous exponential families of stochastic processes the canonical parameter set Γ given by (4.1.3) is a submanifold of $\mathbb{R}^m$ of dimension lower than m. In this section we shall see that if this is not the case, the exponential family has a structure similar to classical repeated sampling models. This follows from the next theorem. If int Θ is non-empty, we say that the exponential family has a *non-empty kernel.*

Theorem 4.2.1 *Consider a time-homogeneous exponential family $\mathcal{P}$ with canonical representation*

$$\frac{dP_\varphi^t}{dP_0^t} = \exp[\varphi^T B_t - \phi_t(\varphi)], \quad t \geq 0, \ \varphi \in \Gamma. \tag{4.2.1}$$

If $\operatorname{int}\Gamma \neq \emptyset$, *the canonical process B has independent increments under P_φ for all $\varphi \in \Gamma$. The Laplace transform of the increment $B_t - B_s$ under P_φ is for $\lambda \in \Gamma - \varphi$ given by*

$$E_\theta(e^{\lambda^T(B_t - B_s)}) = \exp\{[\phi_t(\lambda+\varphi) - \phi_t(\varphi)] - [\phi_s(\lambda+\varphi) - \phi_s(\varphi)]\}. \tag{4.2.2}$$

In particular, B_t has stationary increments if and only if $\phi_t(\varphi)$ is of the form $tf(\varphi)$.

Proof: Let φ_0 be an arbitrary element of Γ. Then

$$\frac{dP_\varphi^t}{dP_{\varphi_0}^t} = \exp[(\varphi - \varphi_0)^T B_t - \phi_t(\varphi) + \phi_t(\varphi_0)] \tag{4.2.3}$$

is a P_{φ_0}-martingale. Therefore,

$$\begin{aligned} &E_{\varphi_0}(\exp[\lambda^T(B_t - B_s)]|\mathcal{F}_s) \\ &\quad = \exp\{\phi_t(\lambda+\varphi_0) - \phi_s(\lambda+\varphi_0) - \phi_t(\varphi_0) - \phi_s(\varphi_0)\} \end{aligned} \tag{4.2.4}$$

for all $\lambda \in \Gamma - \varphi_0$. We see that the conditional Laplace transform of $B_t - B_s$ under P_{φ_0} equals a non-random function for γ in the open set $\operatorname{int}\Gamma - \varphi_0$. By analytic continuation the conditional Laplace transform is non-random in its entire domain, and hence $B_t - B_s$ is independent of $\mathcal{F}_s$ under P_{φ_0}. The expression (4.2.2) follows immediately from (4.2.4). □

When $\phi_t(\varphi) = tf(\varphi)$, the conclusion of Theorem 4.2.1 is that $\mathcal{P}$ is the natural exponential family of Lévy processes generated by B under P_0 (see Chapter 2), or a sub-family thereof. We can also think of the family as being generated by $f(\theta)$, which is necessarily an infinitely divisible cumulant transform.

Suppose the filtration is defined by observation of a stochastic process, as discussed in Chapter 3. Then the conditions of Theorem 4.2.1 do not imply that the basic observed process has independent increments. The following example illustrates this point.

Example 4.2.2 Let W_t be a Wiener process. Then a process of the form $X_t = \exp(W_t + \theta t)$, $\theta \in \mathbb{R}$, is called a geometric Brownian motion. The likelihood function corresponding to observation of X_t in $[0,t]$ is

$$L_t(\theta) = \exp[(\theta + \tfrac{1}{2})\log(X_t) + \tfrac{1}{2}(\theta + \tfrac{1}{2})^2 t], \quad \theta \in \mathbb{R}. \tag{4.2.5}$$

We see that int $\Gamma = \mathbb{R} \neq \emptyset$, and indeed, $B_t = \log(X_t) = W_t + \theta t$ has independent increments for all $\theta \in \mathbb{R}$. The process X solves the stochastic differential equation

$$dX_t = (\theta + \tfrac{1}{2})X_t dt + X_t dW_t, \quad t > 0, \tag{4.2.6}$$

with $X_0 = 1$, as follows by a straightforward application of Ito's formula. We see that X_t does not have independent increments. □

A converse result to Theorem 4.2.1 holds only if the parameter space is large enough. Therefore, we consider the parameter set

$$\tilde{\Gamma} = \cap_{t \geq 0}\{\gamma \in \mathbb{R}^m : E_0(\exp(\gamma^T B_t) < \infty)\}. \tag{4.2.7}$$

Theorem 4.2.3 *Consider a time-homogeneous exponential family with representation (4.2.1) where the canonical process B has independent increments under P_{φ_0} for some $\varphi_0 \in \Gamma$, and assume that $\mathcal{F} = \sigma(\mathcal{F}_t : t \geq 0)$. Then for every $\gamma \in \tilde{\Gamma}$ there exists a probability measure P_γ on $(\Omega, \mathcal{F})$ such that*

$$\frac{dP_\gamma^t}{dP_{\varphi_0}^t} = \exp[(\gamma - \varphi_0)^T B_t - \psi_t(\gamma) + \phi_t(\varphi_0)], \quad t \geq 0, \tag{4.2.8}$$

where

$$\psi_t(\gamma) = \log[E_0(\exp(\gamma^T B_t))].$$

The canonical process B has independent increments under all P_γ, $\gamma \in \tilde{\Gamma}$. If the affine span of Γ is $\mathbb{R}^m$, then int $\tilde{\Gamma} \neq \emptyset$.

Proof: The cumulant transform of B_t under P_{φ_0} is $\psi_t(s + \varphi_0) - \psi_t(\varphi_0)$ for $s \in \tilde{\Gamma} - \varphi_0$. Note that $\psi_t(\gamma) = \phi_t(\gamma)$ for $\gamma \in \Gamma$. Since B_t has independent increments under P_{φ_0}, it follows that $\Lambda_t(\gamma) = dP_\gamma^t / dP_{\varphi_0}^t$ given by (4.2.8) is a P_{φ_0}-martingale for every $\gamma \in \tilde{\Gamma}$. This, in turn, implies that the measures $\{P_\gamma^t : t \geq 0\}$ are consistent and hence (provided that $(\Omega, \mathcal{F})$ is natural measurable) that the measure P_γ exists, as discussed in Chapter 2. For $B \in \mathcal{F}_s$ and $A \in \mathcal{B}(\mathbb{R}^m)$ we have

$$\begin{aligned} E_\gamma[1_B 1_A(B_t - B_s)] &= E_{\theta_0}[1_B \Lambda_s(\gamma) E_{\theta_0}\{1_A(B_t - B_s)\Lambda_t(\gamma)\Lambda_s(\gamma)^{-1} | \mathcal{F}_s\}] \\ &= E_\gamma[1_B E_{\theta_0}\{1_A(B_t - B_s)\Lambda_t(\gamma)\Lambda_s(\gamma)^{-1}\}]. \end{aligned}$$

Therefore,

$$P_\gamma(B_t - B_s \in A | \mathcal{F}_s) = E_{\theta_0}(1_A(B_t - B_s)\Lambda_t(\gamma)\Lambda_s(\gamma)^{-1})$$

is independent of $\mathcal{F}_s$. The last statement of the theorem follows because $\tilde{\Gamma}$ is convex and $\Gamma \subset \tilde{\Gamma}$. □

Theorem 4.2.1 can often be used to show that a given time-homogeneous exponential family cannot be embedded in a larger time-homogeneous exponential family for which the canonical parameter set is, for instance, a manifold of dimension m. This is illustrated in the following example.

Example 4.2.4 For the class of pure birth processes the canonical process is $(X_t - x_0, S_t)$, where X is the observed birth process, x_0 is the initial population size, and

$$S_t = \int_0^t X_s ds;$$

cf. (3.4.3). This process does not have independent increments. Therefore, there does not exist a larger time-homogeneous exponential family on $(\Omega, \mathcal{F})$ with non-empty kernel containing the class of pure birth processes. □

The following Example 4.2.5 shows that Theorem 4.2.1 does not hold without the condition that $\gamma_t(\theta)$ does not depend on t.

Example 4.2.5 Consider the class of counting processes with intensity

$$\lambda_t(\alpha, \theta) = h_t(\alpha, \theta) X_{t-}, \tag{4.2.9}$$

where X is the observed process and where

$$h_t(\alpha, \theta) = \{(e^{(\theta-1)t} - 1)/(\theta - 1) + \alpha^{-1} e^{(\theta-1)t}\}^{-1}, \tag{4.2.10}$$

with $\theta \in \mathbb{R}$ and $\alpha > 0$. For $\theta = 1$ we have $h_t(\alpha, 1) = (\alpha^{-1} + t)^{-1}$. For $\alpha = 1 - \theta$ the process is a pure birth process. The likelihood function corresponding to observing X in $[0, t]$ is

$$L_t(\alpha, \theta) = \exp\left\{\theta \int_0^t X_s ds + \log(h_t(\alpha, \theta)) X_t - x_0 \log\ \alpha\right\}, \tag{4.2.11}$$

where x_0 is the initial value of X. We see that the model is an exponential family and that

$$\begin{aligned} &\{\gamma_t(\alpha, \theta) : (\theta, \alpha) \in \mathbb{R} \times \mathbb{R}_+\} \\ &= \{(\theta, \log h_t(\alpha, \theta) : (\theta, \alpha) \in \mathbb{R} \times \mathbb{R}_+\} \\ &= \{(\gamma_1, \gamma_2) : \gamma_2 < -\log[(\exp((\gamma_1 - 1)t) - 1)/(\gamma_1 - 1)]\}, \end{aligned}$$

so that the kernel is non-empty. The exponential family is, however, not time-homogeneous, and the increments of the canonical process

$$B_t = \left(\int_0^t X_s ds, X_t\right)$$

are not independent. The model is studied in more detail in Chapter 7. □

4.3 Exercises

4.1 Prove Lemma 4.1.1 in detail.

4.2 Prove (4.2.11) by means of the general expression (A.10.16) for the likelihood function for counting process models and Ito's formula (A.6.1).

4.3 Suppose X solves $dX_t = -X_t dt + dW_t$, $X_0 = 0$, under the probability measure P on $(\Omega, \mathcal{F})$, i.e., X in an Ornstein-Uhlenbeck process under P. Let $\{\mathcal{F}_t\}$ denote the filtration generated by X, and let, as usual, P^t denote the restriction of P to $\mathcal{F}_t$. Define for each $t > 0$ a class of probability measures $\{Q_{\theta,t} : \theta \in \mathbb{R}\}$ on $\mathcal{F}_t$ by

$$\frac{dQ_{\theta,t}}{dP^t} = \exp(\theta X_t - \phi_t(\theta)), \; \theta \in \mathbb{R},$$

where $\phi_t(\theta) = \frac{1}{4}\theta^2(1 - e^{-2t})$ is the cumulant transform of X_t under P. Show that there does not exist a class of probability measures $\mathcal{P} = \{P_\theta : \theta \in \mathbb{R}\}$ such that $P_\theta^t = Q_{\theta,t}$.

4.4 Bibliographic notes

The material in Section 4.1 is a translation of classical results for exponential families of distribution; see, e.g., Barndorff-Nielsen (1978) and Johansen (1979). The results on exponential families with a non-empty kernel were first given in Küchler and Sørensen (1994a); see also Jacod and Shiryaev (1987, Proposition X.2.3).

5

Random Time Transformation

5.1 An important type of continuous time model

A central theme in this chapter is the study of exponential families on a filtered space $(\Omega, \mathcal{F}, \{\mathcal{F}_t\})$ with a likelihood function of the form

$$\frac{dP_\theta^t}{dP_0^t} = \exp(\theta^T A_t - \kappa(\theta) S_t) \tag{5.1.1}$$

where $\theta \in \Theta \subseteq \mathbb{R}^k$ and int $\Theta \neq \emptyset$. Here $\kappa(\theta)$ and S_t are one-dimensional, while A_t is a k-dimensional vector process that is right-continuous with limits from the left. Moreover, we assume that S_t is a non-decreasing continuous random function of t and that $S_0 = 0$ and $S_t \to \infty$ P_θ-almost surely as $t \to \infty$ for all $\theta \in \Theta$. This type of exponential model that is curved, except when S_t is non-random, covers several important classes of stochastic processes; cf. Chapter 3. Note, for instance, that the natural exponential families of Lévy processes considered in Chapter 2 are a particular (non-curved) case. Note also that the representation (5.1.1) is of the form (3.1.2) with $\gamma_t(\theta)_i = \theta_i$ for $i = 1, \cdots, k$, $\gamma_t(\theta)_{k+1} = -\kappa(\theta)$, $B_t = (A_t^T, S_t)^T$, and $\phi_t(\theta) = 0$. Exponential families of the form (5.1.1) are investigated in Sections 5.1, 5.2, and 5.4. In Section 5.3 we shall consider a more general type of continuous time model, and in Section 5.5 we shall study a closely related type of discrete time model.

As explained in Chapter 3, when we consider an exponential family of stochastic processes, the filtration is always generated by the observed process. We will not state this explicitly in all examples.

Example 5.1.1 *Exponential families of diffusions.* Consider the class of stochastic differential equations

$$dX_t = [a(t, X_t) + \theta b(t, X_t)]dt + c(t, X_t)dW_t, \quad t > 0, \; X_0 = x_0, \tag{5.1.2}$$

where $c > 0$, W is a standard Wiener process, and all processes and functions are one-dimensional. We assume that $\theta \in \Theta$, where Θ is a real interval, and that (5.1.2) has a unique (strong) solution for all $\theta \in \Theta$. It is no restriction to assume that $0 \in \Theta$. This type of diffusion processes was discussed in Section 3.5.

The class of solutions of (5.1.2) induces a family of probability measures $\{P_\theta : \theta \in \Theta\}$ on the space of continuous functions from $[0, \infty)$ into $\mathbb{R}$ with the filtration generated by the cylinder sets. Under conditions given in Section 3.5 the measures $\{P_\theta^t : \theta \in \Theta\}$ are equivalent for every fixed $t > 0$, and the likelihood function is of the form (5.1.1) with $\kappa(\theta) = \frac{1}{2}\theta^2$,

$$S_t = \int_0^t b(s, X_s)^2 c(s, X_s)^{-2} ds, \tag{5.1.3}$$

and

$$A_t = \int_0^t b(s, X_s) c(s, X_s)^{-2} d\tilde{X}_s, \tag{5.1.4}$$

where

$$\tilde{X}_t = X_t - \int_0^t a(s, X_s) ds. \tag{5.1.5}$$

Some concrete examples of diffusion models of the type (5.1.2) were given in Section 3.5. □

Example 5.1.2 *Exponential families of counting processes.* Consider the class of counting processes with intensities of the form

$$\lambda_t = \nu H_t, \quad \nu > 0, \; t > 0, \tag{5.1.6}$$

where H is a positive predictable stochastic process with finite expectation for all $t < \infty$. Then the observed counting process X is finite for $t < \infty$, and the likelihood function is given by (5.1.1) with $\theta = \log \nu$, $\kappa(\theta) = e^\theta - 1$, $A_t = X_t$, and

$$S_t = \int_0^t H_s ds. \tag{5.1.7}$$

This type of counting process model was discussed in Section 3.4, where some concrete examples were given. □

Define a class of stopping times by

$$\tau_u = \inf\{t : S_t > u\}, \quad u \geq 0. \tag{5.1.8}$$

From the assumptions that S_t tends almost surely to infinity as $t \to \infty$ and is finite for $t < \infty$, it follows that $\tau_u < \infty$ and $\tau_u \to \infty$ P_θ-almost surely as $u \to \infty$ for all $\theta \in \Theta$. Since S is non-decreasing and continuous, the function $u \to \tau_u$ is strictly increasing and has an inverse function, namely S. Specifically, $S_{\tau_u} = u$. Further, τ_u is right-continuous with limits from the left. At points where S is strictly increasing, τ is a continuous function.

Consider the filtration $\{\mathcal{F}_{\tau_u}\}_{u \geq 0}$, where $\mathcal{F}_{\tau_u}$ is the σ-algebra of events happening up to the stopping time τ_u. The formal definition of this σ-algebra is $A \in \mathcal{F}_{\tau_u}$ if and only if $A \cap \{\tau_u \leq t\} \in \mathcal{F}_t$ for all $t \in [0, \infty)$ and $A \in \sigma\{\mathcal{F}_t : t \geq 0\}$. If $\mathcal{F}_t$ is defined by observing a stochastic process in $[0, t]$, then $\mathcal{F}_{\tau_u}$ corresponds to observing the process in $[0, \tau_u]$. Let $P_\theta^{\tau_u}$ denote the restriction of P_θ to $\mathcal{F}_{\tau_u}$. By the fundamental identity of sequential analysis (Corollary B.1.2),

$$\frac{dP_\theta^{\tau_u}}{dP_0^{\tau_u}} = \exp(\theta^T A_{\tau_u} - \kappa(\theta)u), \tag{5.1.9}$$

so $\{P_\theta : \theta \in \Theta\}$ is also an exponential family with respect to the filtration $\{\mathcal{F}_{\tau_u}\}$. Moreover, with respect to this filtration it has a non-empty kernel, and hence the canonical process

$$B_u = A_{\tau_u} \tag{5.1.10}$$

is, by Theorem 4.2.1, a Lévy process under every P_θ. Note that κ is the cumulant transform of B_1 under P_0, so we have shown that κ must be a cumulant transform. In particular, κ is a convex function. Since B is a Lévy process, $\kappa(\theta)$ must in fact be an infinitely divisible cumulant transform. With respect to $\{\mathcal{F}_{\tau_u}\}$, the class $\{P_\theta : \theta \in \Theta\}$ is the natural exponential family of Lévy processes generated by $\kappa(\theta)$ as defined in Chapter 2 or a subset of it.

Example 5.1.1 (continued). Here the cumulant transform $\kappa(\theta) = \frac{1}{2}\theta^2$ is Gaussian, and as in Example 2.1.2, we find that under P_θ the process A_{τ_u} is a Wiener process with drift θ. This result is known because under P_θ the process $A_t - \theta S_t$ is a continuous local martingale (and hence a locally square integrable martingale) with quadratic characteristic S_t. □

Example 5.1.2 (continued). For exponential families of this type, $\kappa(\theta) = e^\theta - 1$ is Poissonian, and as in Example 2.1.1 we see that A_{τ_u} is a Poisson process with intensity e^θ under P_θ. This is also a known result: After a stochastic time-transformation by the inverse of the integrated intensity of a counting process, the counting process becomes a Poisson process. □

If S is strictly increasing, then by the definition of τ_t, $\tau_{S_t} = t$, so that the process A_t in (5.1.1) is a stochastic time transformation of a Lévy process:

$$A_t = B_{S_t}. \tag{5.1.11}$$

Note that

$$\{S_t \leq u\} = \{\tau_u \geq t\} \in \mathcal{F}_{\tau_u},$$

so that S_t is a stopping time with respect to the filtration $\{\mathcal{F}_{\tau_u}\}$.

Under weak conditions on the distribution of S_t, see Section 8.1 and Corollary 7.3.3, the score vector, i.e., the vector of partial derivatives with respect to the coordinates of θ of the logarithm of the likelihood function

$$\dot{l}_t(\theta) = A_t - \dot{\kappa}(\theta) S_t, \tag{5.1.12}$$

is a square integrable P_θ-martingale with quadratic characteristic $\ddot{\kappa}(\theta)S_t$. Therefore, if S_t is constant in a time interval, then so is $\dot{l}_t(\theta)$ and hence A_t. From this consideration it follows that (5.1.11) also holds when S is not strictly increasing. A sufficient condition is that zero is an interior point in the domain of the Laplace transform of S_t for all $t > 0$ and under all P_θ; see Corollary 8.1.3 and Corollary 7.3.3. This is the case if the right tail of the density of S_t tends at least exponentially fast to zero. For many concrete models it is easy to verify directly that (5.1.12) is a square integrable P_θ-martingale. In the following we assume that (5.1.11) holds.

5.2 Statistical results

In this section we shall exploit the representation (5.1.11) to obtain results about statistical properties of models of the form (5.1.1). We assume throughout this section that (5.1.11) holds. Let C denote the closed convex support of the random variable B_1. Then C is the closed convex support for B_t/t for every $t > 0$. This is obvious when κ is steep; see below. A proof that also covers the non-steep case can be found in Casalis and Letac (1994). It follows that $A_t/S_t = B_{S_t}/S_t \in C$ for all $t > 0$.

The model (5.1.1) can also be parametrized by $\mu = \dot{\kappa}(\theta)$ when Θ is open. For the non-curved exponential family (5.1.9) this is the classical mean value parameter. The mean value of A_{τ_u} is μu. When S_t is random, the exponential family (5.1.1) is curved, so the parameter $\mu = \dot{\kappa}(\theta)$ does not have this interpretation. However, we see from (5.1.12) that A is a submartingale or a supermartingale (depending on the sign of μ) and that μS_t is the increasing (or decreasing) predictable process in the Doob-Meyer decomposition (see Section A.3) of A_t. Therefore, we call μ the *compensator parameter.*

In the rest of this section we impose the same regularity conditions on the cumulant transform κ as in Section 2.2. Specifically, we assume that κ is strictly convex and that $\operatorname{int}\operatorname{dom}\kappa \neq \emptyset$. The first assumption is equivalent to assuming that the distribution of A_t/S_t is not concentrated on an affine subspace of $\mathbb{R}^k$. We assume that $\Theta = \operatorname{dom}\kappa$. If κ is steep, $C = \operatorname{cl}\dot{\kappa}(\operatorname{int}\Theta)$. The definition of a steep cumulant transform was given in Section 2.2.

Theorem 5.2.1 *Suppose κ is steep and that $S_t > 0$ for $t > 0$ P_θ-a.s. for all $\theta \in \Theta$. Then the maximum likelihood estimator $\hat{\theta}_t$ based on observation in the time interval $[0,t]$ exists and is uniquely given by*

$$\hat{\theta}_t = \dot{\kappa}^{-1}(A_t/S_t) \tag{5.2.1}$$

if and only if $A_t/S_t \in \operatorname{int} C$.

Suppose $\theta \in \operatorname{int}\Theta$. Then under P_θ the maximum likelihood estimator exists and is unique for t sufficiently large, and

$$\hat{\theta}_t \to \theta \text{ almost surely} \tag{5.2.2}$$

as $t \to \infty$.

Proof. Set $u = S_t$. Then we can formally think of A_t as an observation of $B_u = A_{\tau_u}$ and of A_t/S_t as an observation of B_u/u. Therefore, the result about existence follows from Theorem 2.2.1. The strong consistency follows since

$$S_t^{-1}A_t = S_t^{-1}B_{S_t} \to \mu(\theta),$$

because B has independent stationary increments, $S_t \to \infty$, and $\mu(\theta)$ is the mean value of B_1 under P_θ. □

Theorem 5.2.2 *Suppose $\theta \in \operatorname{int}\Theta$ and that there exists an increasing positive non-random function $\varphi_\theta(t)$ such that under P_θ*

$$S_t/\varphi_\theta(t) \to \eta^2(\theta) \tag{5.2.3}$$

in probability as $t \to \infty$, where $\eta^2(\theta)$ is a finite non-negative random variable for which $P_\theta(\eta^2(\theta) > 0) > 0$. Then under P_θ

$$[(A_t - \dot{\kappa}(\theta)S_t)/\sqrt{S_t},\ S_t/\varphi_\theta(t)] \to N(0, \ddot{\kappa}(\theta)) \times F_\theta \tag{5.2.4}$$

weakly as $t \to \infty$ conditionally on $\{\eta^2(\theta) > 0\}$, where F_θ is the conditional distribution of $\eta^2(\theta)$ given $\{\eta^2(\theta) > 0\}$. Moreover, under P_θ,

$$[\sqrt{S}_t(\hat{\theta}_t - \theta),\ S_t/\varphi_\theta(t)] \to N(0, \ddot{\kappa}(\theta)^{-1}) \times F_\theta \tag{5.2.5}$$

and

$$-2\log Q_t \to \chi^2(k-l) \tag{5.2.6}$$

weakly as $t \to \infty$ conditionally on $\{\eta^2(\theta) > 0\}$. Here Q_t is the likelihood ratio test statistic for the hypothesis that the true parameter value belongs to an l-dimensional ($l < k$) subset of Θ of the type defined in Theorem 2.2.1.

Proof: The results follow from results for Lévy processes (Theorem 2.2.1) by stochastic time transformation. First, suppose that $P_\theta(\eta^2(\theta) > 0) = 1$. By the central limit theorem for Lévy processes, $Z_u = \sqrt{u}(B_u/u -$

$\dot{\kappa}(\theta))$ converges weakly to $N(0, \ddot{\kappa}(\theta))$, and by a continuous time version of Theorem 17.2 in Billingsley (1968), it follows that also $Z_{S_t} = (A_t - \dot{\kappa}(\theta)S_t)/\sqrt{S_t}$ converges weakly to $N(0, \ddot{\kappa}(\theta))$. Given this result, the proof of (5.2.6) is classical. To prove the simultaneous convergence (5.2.4), note that the convergence in the central limit theorem for Lévy processes is mixing, and that Anscombe's uniform continuity condition is satisfied because of the independent, stationary increments of a Lévy process; cf. Anscombe (1952). Therefore, Theorem 8 in Csörgö and Fischler (1973) ensures that Z_{S_t} converges mixing to $N(0, \ddot{\kappa}(\theta))$, which implies that $(Z_{S_t}, S_t/\varphi_\theta(t)) = [(A_t - \dot{\kappa}(\theta))S_t)/\sqrt{S_t}, S_t/\varphi_\theta(t)]$ converges weakly to $N(0, \ddot{\kappa}(\theta)) \times F_\theta$; see Theorem 1' in Aldous and Eagleson (1978). The result (5.2.5) follows from (5.2.4) by the delta-method (Taylor expansion). When $P_\theta(\eta^2(\theta) > 0) < 1$, the convergence results conditionally on $\{\eta^2(\theta) > 0\}$ follow because the convergence is mixing; see Aldous and Eagleson (1978). □

Note that the two proofs above also prove the following results about the compensator parameter μ:

$$\hat{\mu}_t = A_t/S_t \rightarrow \mu(\theta) \tag{5.2.7}$$

P_θ-almost surely, and

$$\sqrt{S}_t(\hat{\mu}_t - \mu(\theta)) \rightarrow N(0, \ddot{\kappa}(\theta))) \tag{5.2.8}$$

weakly under P_θ as $t \rightarrow \infty$ conditionally on $\{\eta^2(\theta) > 0\}$.

The following result about the asymptotic distributions of $\hat{\theta}_t$ and $\hat{\mu}_t$ when normalization is done with the deterministic function $\varphi_\theta(t)$ rather than with the random process S_t follows immediately from Theorem 5.2.2.

Corollary 5.2.3 *Under the conditions of Theorem 5.2.2,*

$$\sqrt{\varphi_\theta(t)}(\hat{\theta}_t - \theta) \rightarrow N(0, (\ddot{\kappa}(\theta)\eta^2(\theta))^{-1}) \tag{5.2.9}$$

and

$$\sqrt{\varphi_\theta(t)}(\hat{\mu}_t - \mu(\theta)) \rightarrow N(0, \ddot{\kappa}(\theta)\eta^{-2}(\theta)) \tag{5.2.10}$$

weakly as $t \rightarrow \infty$ under P_θ conditional on $\{\eta^2(\theta) > 0\}$. Here $N(0, M\eta^{-2}(\theta))$, where M is a $k \times k$-matrix, denotes the normal variance-mixture with characteristic function $u \mapsto E_\theta(\exp[-\frac{1}{2}\eta^{-2}(\theta)u^T Mu] | \eta^2(\theta) > 0)$.

Theorems 5.2.1 and 5.2.2 are examples of the technique by which statistical results for families of the form (5.1.1) can be derived from results for natural exponential families of Lévy processes. A different example of an application of this stochastic time change method is provided by the following theorem about conjugate priors to (5.1.1). As above, let C denote the closed convex support of the random variable $B_1 = A_{\tau_1}$, and let $\mathcal{M}_t \subseteq \mathbb{R}^{k+1}$ denote the support of (A_t, S_t). We define $\mathcal{K}$ as the convex cone generated by $\mathcal{M} = \cup_{t>0}\mathcal{M}_t$; i.e., $x \in \mathcal{K}$ if and only if x is a linear combination with non-negative coefficients of elements from $\mathcal{M}$. Of course, C and $\mathcal{K}$ do not depend on θ.

Theorem 5.2.4 *Suppose Θ is open and that $a \in \mathbb{R}^k$ and $s > 0$ satisfy $a/s \in \operatorname{int} C$. Then*

$$d\pi_{a,s}(\theta) = k(a,s)\exp(\theta^T a - \kappa(\theta)s)d\theta \tag{5.2.11}$$

defines a probability distribution on Θ, provided that the normalizing constant $k(a,s)$ is chosen suitably. Moreover,

$$\int_\Theta \dot{\kappa}(\theta)d\pi_{a,s}(\theta) = a/s. \tag{5.2.12}$$

These results hold, in particular, if $(a,s) \in \operatorname{int}\mathcal{K}$.

Proof: The first part of the theorem follows immediately from Theorem 1 and Theorem 2 in Diaconis and Ylvisaker (1979) because the exponential family (5.1.9) is regular under the conditions imposed. To prove the last statement, note that the set $\mathcal{C} = \{(sc, s) : c \in C,\ s \geq 0\}$ is a cone, the interior of which is $\{(sc, s) : c \in \operatorname{int} C,\ s > 0\}$. Since $\mathcal{M}_t \subseteq \mathcal{C}$ for all $t > 0$, it follows that $\operatorname{int}\mathcal{K} \subseteq \operatorname{int}\mathcal{C}$. □

Note that the integral in (5.2.12) is the mean value of the compensator parameter when θ has the distribution given by (5.2.11). Theorem 5.2.4 can be used to prove admissibility of estimators of θ; see Magiera (1991). The result that the process $B_u = A_{\tau_u}$ has independent stationary increments also has obvious applications to graphical control of the statistical model and to bootstrap.

Example 5.2.5 Consider the class of *Ornstein-Uhlenbeck processes* given as solutions of

$$dX_t = \theta X_t dt + dW_t,\ X_0 = x, \tag{5.2.13}$$

where W is a standard Wiener process and $\theta \in \mathbb{R}$. This is a model of the type considered in Example 5.1.1. The likelihood function has the form (5.1.1) with $\kappa(\theta) = \frac{1}{2}\theta^2$,

$$A_t = \int_0^t X_s dX_s = \tfrac{1}{2}(X_t^2 - x^2 - t),$$

and

$$S_t = \int_0^t X_s^2 ds.$$

The expression for A is easily found by means of Ito's formula. It can also be checked by Ito's formula that the solution of (5.2.13) is

$$X_t = xe^{\theta t} + \int_0^t e^{\theta(t-s)}dW_s. \tag{5.2.14}$$

The conditions of Theorem 5.2.1 are satisfied, so the maximum likelihood estimator $\hat{\theta}_t = A_t/S_t$ is consistent. In order to find the asymptotic distribution of $\hat{\theta}_t$, we need to verify (5.2.3). When $\theta < 0$, $X_t \to N(0, -\frac{1}{2}\theta^{-1})$ weakly as $t \to \infty$, which for instance follows from (5.2.14). By the ergodic theorem, $t^{-1}S_t \to -\frac{1}{2}\theta^{-1}$ P_θ-almost surely, so from Theorem 5.2.2 it follows that under P_θ,

$$\sqrt{S_t}(\hat{\theta}_t - \theta) \to N(0,1)$$

and

$$\sqrt{t}(\hat{\theta}_t - \theta) \to N(0, -2\theta)$$

weakly as $t \to \infty$.

When $\theta > 0$, the process $|X_t|$ tends to infinity. Indeed, by (5.2.14)

$$H_t = e^{-\theta t} X_t - x = \int_0^t e^{-\theta s} dW_s$$

is a zero-mean martingale. Since $E_\theta(H_t^2) = \frac{1}{2}\theta^{-1}(1 - e^{-2\theta t}) \leq \frac{1}{2}\theta^{-1}$, H is uniformly integrable, so by the martingale convergence theorem, $e^{-\theta t}X_t$ converges P_θ-almost surely to a random variable $H_\infty + x$. Hence $e^{-2\theta t}X_t^2$ converges P_θ-almost surely to $(H_\infty + x)^2$, and by the integral version of the Toeplitz lemma (see Appendix B), $e^{-2\theta t}S_t \to \frac{1}{2}\theta^{-1}(H_\infty + x)^2$ P_θ-almost surely. Now it follows by Theorem 5.2.2 that under P_θ,

$$\sqrt{S_t}(\hat{\theta}_t - \theta) \to N(0,1)$$

weakly as $t \to \infty$. By Corollary 5.2.3,

$$2\theta e^{-\theta t}(\hat{\theta}_t - \theta) \to N(0,\ \tfrac{1}{2}\theta^{-1}(H_\infty + x)^{-2})$$

weakly as $t \to \infty$. Here the righthand side denotes the normal variance mixture with mixing distribution $\frac{1}{2}\theta^{-1}(H_\infty + x)^2$. Since $H_t \sim N(0, \frac{1}{2}\theta^{-1}(1 - e^{-2\theta t}))$, it follows that the distribution of $2\theta(H_\infty + x)^2$ is a non-central $\chi^2(1)$-distribution with non-centrality parameter $x\sqrt{2\theta}$. If $x = 0$, the random variable $2\theta H_\infty^2$ is $\chi^2(1)$-distributed, and

$$2\theta e^{-\theta t}(\hat{\theta}_t - \theta) \to t(1)$$

weakly as $t \to \infty$, where $t(1)$ denotes a t-distribution with one degree of freedom.

Consider finally the case $\theta = 0$. For simplicity we assume that $x = 0$, so that $X_t = W_t$. Then $S_t = \int_0^t W_s^2 ds \to \infty$ P_θ-almost surely, so by Theorem 5.2.1 the maximum likelihood estimator $\hat{\theta}_t$ is consistent. However,

$$S_t = t^2 \int_0^1 (\tilde{W}_s^{(t)})^2 ds, \tag{5.2.15}$$

where $\tilde{W}_s^{(t)} = W_{st}/\sqrt{t}$. It is not difficult to see that $\tilde{W}^{(t)}$ is a standard Wiener process for all $t > 0$. Therefore, the integral on the righthand side of (5.2.15) has the same distribution for all $t > 0$. We see that $t^{-2}S_t$ converges in distribution, but not in probability, as $t \to \infty$. We can therefore not use the results in this section to derive the asymptotic distribution of the maximum likelihood estimator. Let us consider the situation for $\theta = 0$ in more detail. Since

$$A_t = t\int_0^1 \tilde{W}_s^{(t)} d\tilde{W}_s^{(t)} = \tfrac{1}{2}t[(\tilde{W}_1^{(t)})^2 - 1],$$

it follows that

$$t\hat{\theta}_t = \frac{\int_0^1 \tilde{W}_s^{(t)} d\tilde{W}_s^{(t)}}{\int_0^1 (\tilde{W}_s^{(t)})^2 ds} = \frac{(\tilde{W}_1^{(t)})^2 - 1}{2\int_0^1 (\tilde{W}_s^{(t)})^2 ds}$$

has the same distribution for all $t > 0$. This distribution is not a normal distribution with mean zero. It is not even symmetric around zero. To see this, note that $P(t\hat{\theta}_t > 0) = P((\tilde{W}_1^{(t)})^2 > 1) < \frac{1}{2}$, since $(\tilde{W}_1^{(t)})^2 \sim \chi^2(1)$. Normalization of $\hat{\theta}_t$ with $\sqrt{S_t}$ obviously does not produce a normal distribution either. Note also that if we consider the parameter $\alpha = \theta t$, then the likelihood function

$$L_t(\alpha) = \alpha\int_0^1 \tilde{W}_s^{(t)} d\tilde{W}_s^{(t)} - \frac{1}{2}\alpha^2\int_0^1 (\tilde{W}_s^{(t)})^2 ds$$

has the same distribution for all t. □

Example 5.2.6 The *pure birth process* is a counting process X with intensity μX_{t-}, $\mu > 0, t > 0$. We set $X_0 = x$. The class of pure birth processes is a model of the type considered in Example 5.1.2 and has a likelihood function of the type (5.1.1) with $\theta = \log\mu$, $\kappa(\theta) = e^\theta - 1$, $A_t = X_t$, and

$$S_t = \int_0^t X_s ds.$$

It is well-known that the linearity of the compensator implies that $E_\mu(X_t) < \infty$ (see, e.g., Jacobsen, 1982) and that

$$X_t - \mu\int_0^t X_s ds$$

is a P_μ-martingale, see Example 8.1.4. or Section 3 in Appendix 1. Therefore, $m_t = E_\mu(X_t|X_0 = x)$ satisfies $\dot{m}_t = \mu m_t$, so $E_\mu(X_t|X_0 = x) = xe^{\mu t}$. From this and from the fact that X is a Markov process it follows that

$$Z_t = e^{-\mu t}X_t$$

is a positive P_μ-martingale, so the martingale convergence theorem implies that there exists a random variable Z_∞ such that under P_μ,

$$e^{-\mu t} X_t \to Z_\infty \tag{5.2.16}$$

almost surely as $t \to \infty$. From the known fact that the distribution of $X_t - x$ is a negative binomial distribution with Laplace transform $[e^z - e^{\mu t}(e^z - 1)]^{-x}$, it follows that the Laplace transform of Z_∞ is $[1-z]^{-x}$, i.e., $2Z_\infty \sim \chi^2(2x)$.

From (5.2.16) and the integral version of the Toeplitz lemma, it follows that

$$\mu e^{-\mu t} S_t \to Z_\infty$$

almost surely as $t \to \infty$. Therefore, by Theorem 5.2.1, Theorem 5.2.2, and Corollary 5.2.3 we have the following results about the maximum likelihood estimator of the compensator parameter $\hat{\mu}_t = A_t/S_t$. It is consistent,

$$\sqrt{S_t}(\hat{\mu}_t - \mu) \to N(0, \mu),$$

and

$$\mu^{-1} e^{-\frac{1}{2}\mu t}(\hat{\mu}_t - \mu) \to N(0, Z_\infty^{-1})$$

weakly as $t \to \infty$ under P_μ. Here $N(0, Z_\infty^{-1})$ denotes the normal variance mixture with mixture distribution Z_∞^{-1}. Since $Z_\infty/x \sim \chi^2/f$ with $f = 2x$, we see that

$$\sqrt{x}\mu^{-1} e^{-\frac{1}{2}\mu t}(\hat{\mu}_t - \mu) \to t(2x)$$

weakly as $t \to \infty$. □

5.3 More general models

Some interesting exponential families of stochastic processes are not of the form (5.1.1), but have likelihood functions that are products of factors of this type. Examples are continuous time Markov processes with finite state space, the class of pure birth-and-death processes with immigration, and some classes of marked counting processes (Section 3.4). Specifically, the models considered in this section have likelihood functions

$$\frac{dP_\theta^t}{dP_0^t} = \exp\left(\sum_{i=1}^n \left[\theta_{(i)}^T A_t^{(i)} - \kappa_i(\theta_{(i)}) S_t^{(i)}\right]\right), \tag{5.3.1}$$

where $\theta^T = (\theta_{(1)}^T, \ldots, \theta_{(n)}^T) \in \Theta_1 \times \cdots \times \Theta_n$, $\Theta_i \subseteq \mathbb{R}^{k_i}$, and $\operatorname{int} \Theta_i \neq \emptyset$. For every i we assume that κ_i and $S_t^{(i)}$ are one-dimensional and that $A^{(i)}$ is a

k_i-dimensional process which is right-continuous with limits from the left. We also assume that for each i, $S^{(i)}$ is a non-decreasing continuous process for which $S_0 = 0$ and $S_t \to \infty$ as $t \to \infty$.

The product structure of the likelihood function, which is usually referred to as L-independence of the parameter components $\theta_{(1)}, \ldots, \theta_{(n)}$, implies that the statistical results for models of the type (5.1.1) proved above generalize immediately to the more general models (5.3.1). This is because likelihood inference about $\theta_{(i)}$ does not depend on the values of the other parameter components. In particular, the maximum likelihood estimator of $\theta_{(i)}$ depends on $A_t^{(i)}$ and $S_t^{(i)}$ only. We can therefore for each i consider the class of stopping times

$$\tau_u^{(i)} = \inf\{t : S_t^{(i)} > u\}, \quad u \geq 0,$$

and proceed exactly as in the previous sections. Thus the following result is proved.

Theorem 5.3.1 *Suppose (5.1.11) holds for the ith component of (5.3.1). Then the results of Theorem 5.2.1, Theorem 5.2.2, Corollary 5.2.3, and Theorem 5.2.4 hold for this component. If the conditions hold for all* $i = 1, \cdots, n$, *then (5.2.6) holds for the hypothesis that* $\theta^{(i)}$, $i \in J$, *is the true value of these components while the other components are unspecified. Here* J *is an arbitrary subset of* $\{1, \cdots, n\}$, $k = \sum_{i \in J} k_i$ *in (5.2.6).* □

Note that if (5.1.11) and the conditions of Theorem 5.2.4 hold for all components of (5.3.1), then a proper conjugate prior is given on Θ by the product of the measures (5.2.11) on each Θ_i.

5.4 Inverse families

Let X be a one-dimensional Lévy process with cumulant transform κ and satisfying (2.1.16). Let $\{P_\theta : \theta \in \Theta\}$ be the natural exponential family generated by X, and define an increasing family of stopping times by

$$\tau_u = \inf\{t > 0 : X_t > u\}. \tag{5.4.1}$$

These stopping times are not in general finite. Define

$$\Theta^+ = \{\theta \in \operatorname{int} \Theta : \dot{\kappa}(\theta) \geq 0\}.$$

Because

$$t^{-1} X_t \to \dot{\kappa}(\theta)$$

P_θ-almost surely as $t \to \infty$ for $\theta \in \operatorname{int} \Theta$, it follows that τ_u is almost surely finite under P_θ if $\dot{\kappa}(\theta) > 0$. For $\dot{\kappa}(\theta) = 0$, in which case X behaves like a random walk, it is also known that τ_u is finite. Thus (by Corollary B.1.2)

$\{P_\theta : \theta \in \Theta^+\}$ is an exponential family relative to the filtration $\{\mathcal{F}_{\tau_u}\}$, provided that $\Theta^+ \neq \emptyset$. Since κ is strictly convex, Θ^+ is an interval, and $\kappa(\theta) > 0$ except at the left end-point. If $\operatorname{int}\Theta \neq \Theta^+$, the whole class $\{P_\theta : \theta \in \Theta\}$ is not an exponential family relative to $\{\mathcal{F}_{\tau_u}\}$. To see this, note that $P_\theta(\tau_u = \infty) > 0$ for $\theta \in \operatorname{int}\Theta\backslash\Theta^+$, so that the measures $\{P_\theta : \theta \in \operatorname{int}\Theta\}$ are not equivalent on $\mathcal{F}_{\tau_u}$. Specifically, the elements of $\{P_\theta : \theta \in \operatorname{int}\Theta\backslash\Theta^+\}$ are singular on $\{\tau_u = \infty\}$. We have here used the definition $\mathcal{F}_\infty = \sigma(\mathcal{F}_t : t \geq 0)$.

Next assume that under P_0 the process X has no jumps upwards, i.e., that $X_t - X_{t-} \leq 0$. Then this holds under all P_θ, $\theta \in \Theta$, and $X_{\tau_u} = u$. A necessary and sufficient condition ensuring this property is that the Lévy measure ν has no mass on $(0, \infty)$. The likelihood function with respect to $\mathcal{F}_{\tau_u}$ is by the fundamental identity of sequential analysis (Corollary B.1.2)

$$L_{\tau_u}(\theta) = \exp(\theta u - \kappa(\theta)\tau_u), \quad \theta \in \Theta^+.$$

By the re-parametrization $\alpha = -\kappa(\theta)$, we get

$$L_{\tau_u}(\alpha) = \exp(\alpha\tau_u + \kappa^{-1}(-\alpha)), \quad \alpha \in -\kappa(\Theta^+). \tag{5.4.2}$$

The inverse κ^{-1} exists on Θ^+ because $\dot{\kappa}(\theta) > 0$ for $\theta \in \operatorname{int}\Theta^+$. The family with respect to $\{\mathcal{F}_{\tau_u}\}$ has a non-empty kernel, provided that we assume that Θ^+ contains more than one point, so by Theorem 4.2.1 the process $\{\tau_u\}$ is a Lévy process, and the cumulant transform of τ_1 under P_0 is $-\kappa^{-1}(s)$ with domain $-\kappa(\Theta^+)$. The class $\{P_\theta : \theta \in \Theta^+\}$ is also an exponential family relative to the filtration generated by the process $\{\tau_u\}$. This can be thought of as the *inverse family* of the original exponential family with respect to the filtration generated by X. We cannot by the same procedure come from the exponential family generated by $\{\tau_u\}$ to that generated by X, because the process $\{\tau_u\}$ has strictly positive jumps. Specifically, the pieces $(X_{\tau_{u-}}, X_{\tau_u})$, $u > 0$, of the sample path are lost.

The particular example where X is a Wiener process with drift $\theta \geq 0$ under P_θ was considered in Section 3.8. Under P_0 the cumulant transform of τ_1 is $-\sqrt{-2s}$ with domain $(-\infty, 0]$. Hence the cumulant transform of τ_1 under P_θ is $\sqrt{-2\theta}(1 - \sqrt{1 + s/\theta})$, $s \leq -\theta$, so the increments of τ follow an inverse Gaussian distribution.

Another example is obtained when $X_t = t - U_t$ where U is a standard Gamma-process under P_0. Then $U_t \sim \Gamma(t, 1 + \theta)$ under P_θ, $\theta \in (-1, \infty)$, and $\kappa(\theta) = \theta - \log(1 + \theta)$. There is no explicit expression for the cumulant transform of τ_1, but it is known (Letac, 1986) that the density function of $\tau_u - u$ under P_0 is

$$t \mapsto \frac{ut^{t+u-1}e^{-t}}{\Gamma(t+u+1)}, \quad t > 0. \tag{5.4.3}$$

5.5 Discrete time models

Consider a d-dimensional discrete-time stochastic process for which the conditional distribution of X_t given $\mathcal{F}_{t-1} = \sigma(X_0, X_1, \ldots, X_{t-1})$ has a density with respect to some dominating measure μ of the form

$$f_t(y;\theta) = b_t(y)\exp[\theta^T T_t(y) - \kappa(\theta)U_t] \tag{5.5.1}$$

for $t = 1, 2, \cdots$ and $\theta \in \Theta$. For given y and t, the random variables $b_t(y), T_t(y)$, and U_t are $\mathcal{F}_{t-1}$-measurable. Sometimes we make this explicit in the notation by writing, e.g., $T_t(y; X_0, X_1, \ldots, X_{t-1})$. The dimension of θ and T_t is d, while $\kappa(\theta)$ and U_t are one-dimensional. To simplify matters we assume that $X_0 = x$. This type of model is a particular case of the models introduced in Section 3.3, and the likelihood function is proportional to

$$L_t(\theta) = \exp[\theta^T A_t - \kappa(\theta)S_t], \quad t = 1, 2, \ldots, \tag{5.5.2}$$

where

$$A_t = \sum_{i=1}^{t} T_i(X_i) \tag{5.5.3}$$

and

$$S_t = \sum_{i=1}^{t} U_i. \tag{5.5.4}$$

From Exercise 3.1 it follows that without loss of generality we can assume that $U_i \geq 0$ almost surely for all $i \geq 1$, so that S is an increasing process.

The conditional Laplace transform of $A_t - A_{t-1}$ given $\mathcal{F}_{t-1}$ is

$$E_\theta(\exp[u(A_t - A_{t-1})]|\mathcal{F}_{t-1}) = \exp[U_t\{\kappa(\theta + u) - \kappa(\theta)\}],$$

which has the form of the Laplace transform of the increment of a Lévy process over a random time interval of length $U_t = S_t - S_{t-1}$. One might therefore expect that the canonical process is distributed like a randomly time transformed Lévy process, where the random time-transformation (the intrinsic time) is given by the increasing process S_t. This is in fact the result of the following representation theorem for the canonical process.

We will make the following assumption about U_i.

Condition 5.5.1 *For all $i \geq 1$ the set of possible values of U_i is either (a) a subset of the positive integers containing 1 or (b) a subset of $(0, \infty)$ containing an interval $(0, \delta)$ for some $\delta > 0$.*

Let $\mathcal{L}(Z|P_\theta)$ denote the distribution of the stochastic process Z under P_θ.

Theorem 5.5.2 *Suppose the function $y \mapsto T_t(y)$ is P_θ-almost surely invertible for all $t \in \mathbb{N}$, that $\operatorname{int}\Theta \neq \emptyset$, and that Condition 5.5.1. holds.*

Then there exists, on some probability space $(\tilde{\Omega}, \tilde{\mathcal{F}}, \tilde{P}_\theta)$, *a k-dimensional process Y that under Condition 5.5.1 (a) is a discrete time process with independent stationary increments and under Condition 5.5.1 (b) is a Lévy process. In both cases*

$$\mathcal{L}(\{A_t, S_t) : t \in \mathbb{N}\} | P_\theta) = \mathcal{L}(\{(Y_{\tau_t}, \tau_t) : t \in \mathbb{N}\} | \tilde{P}_\theta). \tag{5.5.5}$$

Here $\{\tau_t\}$ *is an increasing sequence of stopping times with respect to the filtration* $\{\mathcal{G}_t\}$, *where* $\mathcal{G}_t = \sigma(Y_s : s \leq t)$. *The distribution of* Y *is given by*

$$\tilde{E}_\theta\{\exp(uY_t)\} = \exp[t\{\kappa(\theta + u) - \kappa(\theta)\}] \tag{5.5.6}$$

for all $t > 0$.

Proof: The Laplace transform of the conditional distribution under P_θ of $T_t(X_t)$ given $\mathcal{F}_{i-1}$ is

$$u \mapsto \exp[\{\kappa(\theta + u) - \kappa(\theta)\}U_t], \tag{5.5.7}$$

at least for $u \in \Theta - \theta$, which under the conditions of the theorem contains an open set. Under Condition 5.5.1 (b) we see that $\exp[s\{\kappa(\theta + u) - \kappa(\theta)\}]$ is a Laplace transform for all $s > 0$ (i.e., $\exp[\kappa(\theta + u) - \kappa(\theta)]$ is infinitely divisible), so we can define a Lévy process Y on some sufficiently rich probability space $(\tilde{\Omega}, \tilde{\mathcal{F}}, \tilde{P}_\theta)$ such that (5.5.6) holds for all $t > 0$. In case (a), a discrete time process Y is defined similarly.

Now define a new discrete time process $\tilde{X}$ on $(\tilde{\Omega}, \tilde{\mathcal{F}}, \tilde{P}_\theta)$ by $\tilde{X}_0 = x$,

$$\tilde{X}_1 = T_1^{-1}(Y_{\tau_1}; \tilde{X}_0),$$

where

$$\tau_1 = U_1(x),$$

and then for $t \geq 2$ continuing by

$$\tau_t = \sum_{i=1}^{t} U_i(\tilde{X}_0, \cdots, \tilde{X}_{i-1})$$

and

$$\tilde{X}_t = T_t^{-1}(Y_{\tau_t} - Y_{\tau_{t-1}}; \tilde{X}_0, \cdots, \tilde{X}_{t-1}).$$

Since τ_t is $\mathcal{G}_{\tau_{t-1}}$-measurable, it is clear that τ_t is a stopping time with respect to $\{\mathcal{G}_t\}$. Indeed, $\{\tau_t \leq u\} = \{\tau_t \leq u\} \cap \{\tau_{t-1} \leq u\} \in \mathcal{G}_u$.

The Laplace transform of the conditional distribution under $\tilde{P}_\theta$ of $T_t(\tilde{X}_t)$ given $\tilde{X}_0, \cdots, \tilde{X}_{t-1}$ equals the Laplace transform of $Y_{\tau_t} - Y_{\tau_{t-1}}$ given $\tilde{X}_0, \cdots, \tilde{X}_{t-1}$, which is $\exp[\{\kappa(\theta + u) - \kappa(\theta)\}U_t(\tilde{X}_0, \cdots, \tilde{X}_{t-1})]$. Thus the conditional distribution under $\tilde{P}_\theta$ of $T_t(\tilde{X}_t)$ given $\tilde{X}_0, \cdots, \tilde{X}_{t-1}$ equals the conditional distribution under P_θ of $T_t(X_t)$ given $X_0, \cdots, X_{t-1}$ (cf. (5.5.7)) and since T_t is invertible, it follows that

$$\mathcal{L}(X_0, X_1, \cdots, X_t | P_\theta) = \mathcal{L}(\tilde{X}_0, \tilde{X}_1, \cdots, \tilde{X}_t | \tilde{P}_\theta)$$

for all $t \geq 0$. From this, (5.5.5) follows immediately. □

Now let us apply Theorem 5.5.2 to derive statistical results for the models considered in this section from results for Lévy processes in analogy with what was done in Section 5.2. In the following we assume that the conditions of Theorem 5.5.2 are satisfied.

Without loss of generality we can assume that $0 \in \Theta$ and that $\kappa(0) = 0$, which implies that κ is a cumulant transform. As in Section 5.2 we suppose that κ is strictly convex and that $\operatorname{int}\operatorname{dom}\kappa \neq \emptyset$. Let C denote the closed convex support of the random variable Y_1. Then C is also equal to the closed convex support of Y_t/t and A_t/S_t for all $t \in \mathbb{N}$. As was noted in Section 2.2, the function κ is differentiable on $\operatorname{int}\operatorname{dom}\kappa$, and $\dot{\kappa} : \operatorname{int}\operatorname{dom}\kappa \to \operatorname{int} C$ is a homeomorphism.

The score function is

$$\dot{l}_t(\theta) = A_t - \dot{\kappa}(\theta)S_t. \tag{5.5.8}$$

Theorem 5.5.3 *Suppose κ is steep and that $\Theta = \operatorname{int}\operatorname{dom}\kappa$. Then the maximum likelihood estimator $\hat{\theta}_t$ based on the observations $X_0, \cdots, X_t$ exists and is uniquely given by*

$$\hat{\theta}_t = \dot{\kappa}^{-1}(A_t/S_t) \tag{5.5.9}$$

if and only if $A_t/S_t \in \operatorname{int} C$

If $S_t \to \infty$ P_θ-almost surely, then under P_θ the maximum likelihood estimator exists for t sufficiently large, and $\hat{\theta}_t \to \theta$ almost surely as $t \to \infty$.

Suppose further that there exists an increasing positive non-random sequence $\varphi_\theta(t)$ such that

$$S_t/\varphi_\theta(t) \to \eta^2(\theta) \tag{5.5.10}$$

in probability under P_θ, where $\eta^2(\theta)$ is a finite non-negative random variable for which $P_\theta(\eta^2(\theta) > 0) > 0$. Then under P_θ,

$$(\dot{l}_t(\theta)/\sqrt{S_t}, S_t/\varphi_\theta(t)) \to N(0, \ddot{\kappa}(\theta)) \times F_\theta \tag{5.5.11}$$

weakly as $t \to \infty$ conditionally on $\{\eta^2(\theta) > 0\}$, where F_θ is the conditional distribution of $\eta^2(\theta)$ given $\{\eta^2(\theta) > 0\}$. Moreover, under P_θ,

$$(\sqrt{S_t}(\hat{\theta}_t - \theta), S_t/\varphi_\theta(t)) \to N(0, \ddot{\kappa}(\theta)^{-1}) \times F_\theta \tag{5.5.12}$$

and

$$\sqrt{\varphi_\theta(t)}(\hat{\theta}_t - \theta) \to Z \tag{5.5.13}$$

weakly as $t \to \infty$ conditionally on $\{\eta^2(\theta) > 0\}$, where Z is the normal variance mixture with characteristic function $E_\theta(\exp[-\frac{1}{2}\eta(\theta)^{-2}u^T\ddot{\kappa}(\theta)^{-1}u]| \eta^2(\theta) > 0)$. Finally,

$$-2\log Q_t \to \chi^2(d-l)$$

weakly as $t \to \infty$ conditionally on $\{\eta^2(\theta) > 0\}$, where Q_t is the likelihood ratio test statistic for the hypothesis that the true parameter belongs to an l-dimensional ($l < d$) subset of Θ of the type defined in Theorem 2.2.1.

Proof: The results follow from results for processes with independent, stationary increments by stochastic time-transformation in a way analogous to the proofs of Theorems 5.2.1 and 5.2.2 and Corollary 5.2.3. □

Also, an analogue of Theorem 5.2.4 holds and can be proved in exactly the same way.

Example 5.5.4 Define a stochastic process X by $X_0 = x > 0$ and by assuming that the conditional distribution of X_t given $X_0, \cdots, X_{t-1}$ is the Gamma-distribution with density

$$f_t(y;\theta) = \theta^{M_{t-1}} y^{M_{t-1}-1} e^{-\theta y} / \Gamma(M_{t-1}),$$

for all $t \geq 1$, where $\theta > 0$ and

$$M_{t-1} = t^{-1} \sum_{i=0}^{t-1} X_i.$$

Since $E_\theta(X_t | X_0, \cdots, X_{t-1}) = \theta^{-1} M_{t-1}$, we can think of X as an autoregression of the form

$$X_t = \theta^{-1} M_{t-1} + Z_t,$$

where Z_t is an error term with mean zero. The error terms are not independent.

The likelihood function is of the form (5.5.2) with $\kappa(\theta) = -\log(\theta)$,

$$A_t = -\sum_{i=1}^{t} X_i,$$

and

$$S_t = \sum_{i=1}^{t} M_{i-1}.$$

Condition 5.5.1 (b) is satisfied, so we have the representation result (5.5.5), where $-Y$ is the gamma process with Laplace transform $(1-u/\theta)^{-t}$ at time $t > 0$.

To conclude that the maximum likelihood estimator $\hat{\theta}_t = -S_t/A_t$ is consistent, we need to show that $S_t \to \infty$ almost surely. Note that

$$E_\theta(M_t | \mathcal{F}_{t-1}) = \left(\frac{\theta^{-1} - 1}{t+1} + 1 \right) M_{t-1}, \tag{5.5.14}$$

so

$$Z_t = M_t / \psi_t(\theta)$$

is a positive P_θ-martingale if $\psi_t(\theta)$ is defined by

$$\psi_t(\theta) = \left(\frac{\theta^{-1} - 1}{t+1} + 1 \right) \psi_{t-1}(\theta), \quad t = 1, 2, \cdots,$$

with $\psi_0(\theta) = 1$. By the martingale convergence theorem, $Z_t \to Z_\infty(\theta)$ almost surely under P_θ as $t \to \infty$. From (5.5.14) it follows that $E_\theta(Z_t) = E_\theta(M_t) = x$ if $\theta = 1$, and that

$$E_\theta(M_t) = \psi_t(\theta)x \to \begin{cases} 0 & \text{if } \theta > 1 \\ \infty & \text{if } \theta < 1. \end{cases}$$

Now,

$$\log \psi_n(\theta) \sim (\theta^{-1} - 1)(\sum_{i=1}^{n}(i+1)^{-1} + K) \sim (\theta^{-1} - 1)[\log(n+1) + \tilde{K}],$$

so

$$\psi_n(\theta) \sim (n+1)^{\theta^{-1}-1}.$$

Since

$$\sum_{i=1}^{n} \psi_i(\theta) \sim n^{\theta^{-1}}$$

as $n \to \infty$, it follows by the Toeplitz lemma that

$$S_t/t^{\theta^{-1}} \to c(\theta)Z_\infty(\theta)$$

almost surely under P_θ, where $c(\theta) = \lim_{t\to\infty} t^{-\theta^{-1}} \sum_{i=1}^{t} \psi_i(\theta)$. Hence $\hat{\theta}_t$ is consistent on $\{Z_\infty(\theta) > 0\}$, and the asymptotic distribution results hold conditionally on $Z_\infty(\theta) > 0$. In particular, the speed of convergence of $\hat{\theta}_t$ is $t^{\frac{1}{2}\theta^{-1}}$. For $\theta \leq 1$ it is easy to see that $Z_\infty(\theta) > 0$ almost surely. □

5.6 Exercises

5.1 Consider the class of Ornstein-Uhlenbeck processes given by

$$dX_t = \theta X_t dt + dW_t, \quad X_0 = 0,$$

for which the likelihood function is given by (3.5.7). Suppose $\theta > -\beta$, where $\beta > 0$. Define a family of stopping times by

$$\tau_u = \inf\{t : \int_0^t X_s^2 ds \geq u\}.$$

Prove that the process

$$Y_u = X_{\tau_u}^2 - \tau_u$$

is a Lévy process, that the expectation of Y_1 is θ, and that the process

$$Z_t = X_t^2 - t + \beta \int_0^t X_s^2 ds$$

tends to infinity almost surely as $t \to \infty$. Next, define a family of stopping times by

$$\sigma_u = \inf\{t : Z_t \geq u\},$$

and show that the process $X^2_{\sigma_u} - \sigma_u$ is a Lévy process.

5.2 Prove Theorem 5.3.1 (or parts of it) as outlined above the statement of the theorem.

5.3 Fill in the details of the proof of Theorem 5.5.3.

5.4 Prove that (5.2.14) solves (5.2.13).

5.5 Prove that $\tilde{W}^{(t)}$ in (5.2.15) is a standard Wiener process for all $t > 0$.

5.6 Consider a birth-and-death process X with immigration, where $X_0 = x$ and

$$P(X_{t+h} = j | X_t = i) = \begin{cases} i\lambda h + o(h) & \text{for } j = i+1, i \geq 1 \\ 1 - i(\lambda + \mu)h + o(h) & \text{for } j = i \geq 1 \\ i\mu h + o(h) & \text{for } j = i-1, i \geq 1 \\ \nu h + o(h) & \text{for } i = 0, j = 1 \\ 1 - \nu h + o(h) & \text{for } i = j = 0, \end{cases}$$

where $\lambda > 0, \mu > 0$, and $\nu > 0$. Show that this is a model of the type considered in Section 5.3. Find the maximum likelihood estimators $\hat{\lambda}_t$, $\hat{\mu}_t$, and $\hat{\nu}_t$; show that they are consistent; and find their asymptotic distribution.

5.7 Bibliographic notes

Various aspects of general exponential families of stochastic processes of the type (5.1.1) have been studied by several authors: Basawa (1991), Barndorff-Nielsen (1984), Barndorff-Nielsen and Cox (1984), Sørensen (1986), Stefanov (1986a,b), and Küchler and Sørensen (1994a). The more general models (5.3.1) were considered by Sørensen (1986) and Stefanov (1995).

The fact that if $\mathcal{F}_t$ is generated by observation of a stochastic process in $[0, t]$, then $\mathcal{F}_\tau$ where τ is a stopping time is generated by observation of the process in $[0, \tau]$ was proved by Courrège and Priouret (1965). The importance of this result to sequential estimation was pointed out by Keiding (1978). The result that (5.1.10) is a Lévy process was obtained by Stefanov (1986a) in the case where (A, S) is a strong Markov process. The general result was given in Küchler and Sørensen (1994a). Stochastic time-change methods similar to the one used in Section 5.2 were also used by Feigin (1976) to study the asymptotic behaviour of maximum likelihood estimators for some stochastic process models. That the results of Diaconis and Ylvisaker (1979) can be used in the context of exponential families of processes was first noticed by Magiera and Wilczýnski (1991). The result of Theorem 5.2.4 was given in Küchler and Sørensen (1994a). This result can be used to prove the admissibility of estimators of θ. For results in this direction see Magiera (1990). The statistical properties of the Ornstein-Uhlenbeck process have been studied by numerous authors; see, e.g., Bawasa and Prakasa Rao (1980). In particular, the case $\theta = 0$ was first studied by Feigin (1979). The statistical properties of the pure birth process and the linear birth-and-death process were studied by Keiding (1974, 1975).

The inverse families discussed in Section 5.4 were introduced in Franz (1977) and in Küchler and Sørensen (1994a) and are closely related to the concept of reciprocal exponential families introduced by Letac (1986) and studied further in Letac and Mora (1990). The two particular examples considered in that section were also studied by Letac (1986), who called the exponential family generated by (5.4.3) the Ressel family.

The results in Section 5.5 were first given for Markov processes by Feigin (1981). The generalization of the results to exponential families of general adapted discrete time processes was given in Sørensen (1996).

6

Exponential Families of Markov Processes

We shall now consider time-homogeneous exponential families under which the observed process is a Markov process. This turns out to be the case if and only if an exponential representation exists where the canonical process is an additive functional.

6.1 Conditional exponential families

In all of this chapter, $(\Omega, \mathcal{F})$ is a measurable space, and $\{\mathcal{F}_t : t \in \mathcal{T}\}$ is a filtration therein with $\mathcal{F} = \sigma(\mathcal{F}_t : t \in \mathcal{T})$, where $\mathcal{T} = [0, \infty)$ or $\mathcal{T} = \mathbb{N} = \{0, 1, \cdots\}$; i.e., the time is continuous or discrete. We assume that the σ-algebra $\mathcal{F}_t$ is generated by $\{X_s : s \leq t\}$, where $X = \{X_t : t \in \mathcal{T}\}$ is a stochastic process on $(\Omega, \mathcal{F})$ with state space $(E, \mathcal{E})$. Suppose that for every $\theta \in \Theta$, where Θ is some non-void subset of $\mathbb{R}^k$, and for every $x \in E$, probabilities $P_{\theta,x}$ are given on $(\Omega, \mathcal{F})$ with $P_{\theta,x}(X_0 = x) \equiv 1$ and such that $x \mapsto P_{\theta,x}(A)$ is $\mathcal{E}$-measurable for every $A \in \mathcal{F}$ and every $\theta \in \Theta$. By $E_x^\theta(\cdot\,; A)$ we denote the expectation over the set A with respect to $P_{\theta,x}$. For every family $\{\pi_\theta, \theta \in \Theta\}$ of probabilities π_θ on $\mathcal{E}$ define

$$P_{\theta,\pi_\theta}(A) = \int_E P_{\theta,x}(A)\pi_\theta(dx), \quad A \in \mathcal{F},\, \theta \in \Theta.$$

Obviously, we have

$$P_{\theta,\pi_\theta}(X_0 \in B) = \pi_\theta(B), \quad B \in \mathcal{E}.$$

If there is no doubt about the "initial" probability π_θ, we shall write P_θ instead of $P_{\theta\pi_\theta}$. As usual, denote by P_θ^t ($P_{\theta,x}^t$) the restriction of P_θ ($P_{\theta,x}$) to $\mathcal{F}_t$, and define $\mathcal{P} = \{P_\theta : \theta \in \Theta\}$ and $\mathcal{P}_x = \{P_{\theta,x} : \theta \in \Theta\}$. Moreover, we put $I\!P_F^\theta = \{P_{\theta,x} : x \in F\}$ and $I\!P_F^{\theta,t}\{P_{\theta,x}^t : x \in F\}$ for $F \subseteq E$ and define $I\!P^\theta = I\!P_E^\theta$ and $I\!P^{\theta,t} = I\!P_E^{\theta,t}$. If we want to express that a certain equation holds $P_{\theta,x}$-almost surely ($P_{\theta,x}^t$-almost surely) for every $x \in E$, we shall write $I\!P^\theta$-a.s. ($I\!P^{\theta,t}$-a.s., respectively). We say that $\mathcal{P}_x$ is locally dominated by $P_{\theta_0,x}$ if $P_{\theta,x}^t \ll P_{\theta_0,x}^t$ for all $\theta \in \Theta$ and all $t \in \mathcal{T}$.

Definition 6.1.1 *Assume $F \subseteq E$ to be fixed. Then $\{I\!P_F^\theta : \theta \in \Theta\}$ is called* an F-conditional exponential family *if there exists an m-dimensional $\{\mathcal{F}_t\}$-adapted process $B = \{B_t : t \in \mathcal{T}\}$ on $(\Omega, \mathcal{F})$, non-random functions $\gamma(\theta) : \Theta \mapsto I\!R^m$ and $\phi_t(\theta) : \mathcal{T} \times \Theta \mapsto I\!R$, and a fixed $\theta_0 \in \Theta$ such that for every $\theta \in \Theta$ and all $t \in \mathcal{T}$ the measures $P_{\theta,x}^t$ and $P_{\theta_0,x}^t$ are equivalent and*

$$L_t(\theta) = \frac{dP_{\theta,x}^t}{dP_{\theta_0,x}^t} = \exp[\gamma(\theta)^T B_t - \phi_t(\theta)] \quad \text{for all } x \in F \tag{6.1.1}$$

with $B_0 = 0$ and $\phi_0(\theta) \equiv 0$. If $\{I\!P_F^\theta : \theta \in \Theta\}$ is an F-conditional exponential family for some $F \subseteq E$ and if $\pi(E \setminus F) = 0$ for some measure π on $(E, \mathcal{E})$, then we call $\{I\!P^\theta : \theta \in \Theta\}$ a π-a.s. conditional exponential family, *and if $F = E$, it is simply called a* conditional exponential family.

Note that a conditional exponential family is time-homogeneous. The connection between an exponential family and a conditional exponential family is given by the following proposition.

Proposition 6.1.2 *The class $\mathcal{P}$ is a time-homogeneous exponential family if and only if $(\pi_\theta : \theta \in \Theta)$ is an exponential family of distributions and $\{I\!P^\theta : \theta \in \Theta\}$ is a π_{θ_0}-a.s. conditional exponential family for some $\theta_0 \in \Theta$.*

Proof: Let $\mathcal{P}$ be a time-homogeneous exponential family, i.e., $dP_\theta^t / dP_{\theta_0}^t = \exp[\gamma(\theta)^T B_t - \phi_t(\theta)]$, and assume $C \in \mathcal{E}$. Then we have that

$$\begin{aligned} \pi_\theta(C) &= P_\theta(X_0 \in C) \\ &= \int_{\{X_0 \in C\}} \exp[\gamma(\theta)^T B_0(\omega) - \phi_0(\theta)] P_{\theta_0}(d\omega) \\ &= \int_C \exp[\gamma(\theta)^T b_0(x) - \phi_0(\theta)] \pi_{\theta_0}(dx), \quad \theta \in \Theta, \end{aligned}$$

where b_0 is an $\mathcal{E}$-measurable function with $b_0(X_0(\omega)) = B_0(\omega)$, P_{θ_0}-a.s. Moreover, we have for every $A \in \mathcal{F}_t$ and $C \in \mathcal{E}$ that

$$\int_C P_{\theta,x}(A) \exp[\gamma(\theta)^T B_0 - \phi_0(\theta)] \pi_{\theta_0}(dx)$$

$$= P_\theta(A\cap\{X_0\in C\}) = \int\limits_{A\cap\{X_0\in C\}} \frac{dP^t_\theta}{dP^t_{\theta_0}} P_{\theta_0}(d\omega)$$

$$= \int\limits_C\int\limits_A \exp[\gamma(\theta)^T B_t - \phi_t(\theta)] P_{\theta_0,x}(d\omega)\pi_{\theta_0}(dx).$$

Thus, we obtain that π_{θ_0}-a.s.

$$\frac{dP^t_{\theta,x}}{dP^t_{\theta_0,x}} = \exp[\gamma(\theta)^T(B_t - B_0) - \{\phi_t(\theta) - \phi_0(\theta)\}].$$

Hence, $\{I\!P^\theta : \theta\in\Theta\}$ is a π_{θ_0}-a.s. conditional exponential family. The converse implication is obvious. □

As usual, we can without loss of generality assume that $\theta_0 = 0$

6.2 Markov processes

Let us recall some definitions to fix the notation. For further properties of Markov processes, see, e.g., Chung (1982). For every $x\in E$, let $P_x(\cdot)$ be a probability on $(\Omega,\mathcal{F})$ such that $P_x(X_0 = x) = 1$, $P_x(X_t\in E)\equiv 1$, and $x\mapsto P_x(A)$ is measurable with respect to $\mathcal{E}$ for every $A\in\mathcal{F}$. Then X is called a *Markov process* with respect to $\{\mathcal{F}_t\}$ under $(P_x : x\in E)$ if

$$P_x(X_t\in C|\mathcal{F}_s) = P_{X_s}(X_{t-s}\in C), \qquad P_x\text{-a.s.} \tag{6.2.1}$$

for every s, $t\in\mathcal{T}$ with $0\le s<t$, all $C\in\mathcal{E}$, and all $x\in E$.

A function $P(t,x,C)$ on $\mathcal{T}\times E\times\mathcal{E}$ with values in $[0,1]$ that is $\mathcal{E}$-measurable for every $(t,C)\in\mathcal{T}\times\mathcal{E}$ and satisfies $P(t,x,E)\equiv 1$ is called a *transition function* if

$$\int\limits_E P(s,x,dy)P(t,y,C) = P(s+t,x,C), \quad s,t\in\mathcal{T},\ x\in E,\ C\in\mathcal{E}. \tag{6.2.2}$$

It is called a *transition function for* X if additionally

$$P(t,x,C) = P_x(X_t\in C), \qquad t\in\mathcal{T},\ x\in E,\ C\in\mathcal{E}. \tag{6.2.3}$$

If $\mathcal{T} = \{0,1,2,\ldots\}$, we define $P(x,C) = P(1,x,C)$ and call it a *transition kernel* on $(E,\mathcal{E})$.

When in the rest of this book we write that X is a Markov process under $(P_x : x\in E)$, we will always implicitly suppose that the probability measures $\{P_x\}$ satisfy the conditions given above and that X has a transition function. If we want to stress the fact that $P_x(X_t\in E) = 1$ for all $t\ge 0$, we will talk about a *conservative Markov process.*

Now let us give some examples of conditional exponential families of Markov processes.

Example 6.2.1 First assume $\mathcal{T} = \{0, 1, 2, \cdots\}$. For every $\theta \in \Theta \subseteq \mathbb{R}^k$ let $P_\theta(x, dy)$ be a transition kernel on $(E, \mathcal{E})$ with

$$P_\theta(x, dy) = \exp[\gamma(\theta)^T m(x, y) - \psi(\theta)] P_{\theta_0}(x, dy),$$

where $\gamma : \Theta \mapsto \mathbb{R}^d$, $m : E \times E \mapsto \mathbb{R}^d$, $\psi : \Theta \mapsto \mathbb{R}$, m being measurable. Thus for every $x \in E$ the measures $\{P_\theta(x, \cdot) : \theta \in \Theta\}$ form an exponential family of distributions.

For every $x \in E$, $\theta \in \Theta$, and $n \in \mathbb{N}$ define a probability measure $P^n_{\theta,x}$ on $(E^n, \mathcal{E}^n)$ by

$$P^n_{\theta,x}(dx_1, \ldots, dx_n) = \prod_{k=1}^{n} P_\theta(x_{k-1}, dx_k) \quad \text{with} \quad x_0 = x.$$

For each x and θ these measures can be extended to a measure $P_{\theta,x}$ on $(\Omega, \mathcal{F}) = (E^{\mathbb{N}}, \mathcal{E}^{\mathbb{N}})$. Thus, there exists a family $\{P_{\theta,x} : x \in E, \theta \in \Theta\}$ of probabilities such that under $P_{\theta,x}$ the process $X = \{X_n : n \in \mathbb{N}\}$, with $X_n(\omega) = \omega_n$ for $\omega \in \Omega$, is a Markov chain with transition kernel $P_\theta(x, dy)$. Moreover,

$$\frac{dP^n_{\theta,x}}{dP^n_{\theta_0,x}} = \exp\{\gamma(\theta)^T M_n - n\psi(\theta)\}$$

with

$$M_n = \sum_{k=1}^{n} m(X_{k-1}, X_k),$$

so $\{P_{\theta,x} : x \in E, \theta \in \Theta\}$ is a conditional exponential family.

This is the time-homogeneous case of the exponential families considered in Section 3.2. Particular examples covered by this general set-up are the class of all discrete time Markov processes with state space $E = \{1, \ldots, l\}$, the Gaussian autoregressions of order one, and classes of branching processes where the offspring distribution belongs to an exponential family, see Section 3.2. □

The next example presents conditional exponential families of continuous time processes.

Example 6.2.2 Suppose $\mathcal{T} = [0, \infty)$ and $\Omega = C(\mathcal{T})$, the set of continuous functions $\mathcal{T} \mapsto \mathbb{R}$, and let $\mathcal{F}$ be generated by the cylindric sets in $C(\mathcal{T})$. In Section 3.5 conditions are given ensuring the existence of a class of probability measures on $\mathcal{F}$, here denoted by $\{P_{\theta,x} : \theta \in \Theta, x \in E\}$, where $\Theta \subseteq \mathbb{R}$ and $E \subseteq \mathbb{R}$, such that under $P_{\theta,x}$ the process $X_t(\omega) = \omega_t$, $t \geq 0$, is a solution of the stochastic differential equation

$$dX_t = \theta b(X_t)dt + \sigma(X_t)dW_t, \quad X_0 = x, \tag{6.2.4}$$

where W is a standard Wiener process under $P_{\theta,x}$, and where $\sigma(x) > 0$ for all $x \in E$ (see also Liptser and Shiryaev, 1977). We assume that $0 \in \Theta$. The state space of X under $P_{\theta,x}$ is E. Moreover,

$$\frac{dP^t_{\theta,x}}{dP^t_{0,x}} = \exp\left\{\theta \int_0^t b(X_s)\sigma^{-2}(X_s)dX_s - \tfrac{1}{2}\theta^2 \int_0^t b^2(X_s)\sigma^{-2}(X_s)ds\right\}.$$

Since the coefficients b and σ depend on X_t only, a solution of (6.2.4) is a Markov process. Hence $\{P_{\theta,x} : \theta \in \Theta, x \in E\}$ is a conditional exponential family.

If $\{\mathbb{P}^\theta, \theta \in \Theta\}$ is a conditional exponential family of Markov processes, then the corresponding transition functions $\{P^\theta(t, x, \cdot), \theta \in \Theta\}$ for fixed t and x may, but need not, form exponential families. This is shown by the following two particular examples. We shall return to this question in Section 6.3 below.

For $b(y) = -y$ and $\sigma(y) = 1$, the diffusion X is an Ornstein-Uhlenbeck process for all $\theta \in \Theta = \mathbb{R}$. Under $P_{\theta,x}$, the distribution of X_t is Gaussian with mean value $xe^{-\theta t}$ and variance $(e^{2\theta t} - 1)(2\theta)^{-1}$. Thus, the transition functions for fixed x and t form an exponential family of distributions.

For $b(y) = y^{-1}$ and $\sigma(y) = 1$ with $y > 0$ and for $\theta \geq \frac{1}{2}$, we obtain as the solution of (6.2.4) the Bessel process, which is a Markov process on $E = (0, \infty)$ with transition densities

$$P_\theta(t, x, dy) = t^{-1} \exp\left[-\frac{x^2 + y^2}{2t}\right](xy)^{\frac{1}{2}-\theta} y^{2\theta} I_{\theta-\frac{1}{2}}\left(\frac{xy}{t}\right)dy; \qquad (6.2.5)$$

see, e.g., Karlin and Taylor (1981). Here I_α denotes the modified Bessel function of the first kind of order α. These transition densities obviously do not form an exponential family. In this case we have excluded 0 from Θ, but the above expression for the likelihood function still holds if $P_{0,x}$ denotes the measure corresponding to the Wiener process. □

Finally, we give an example of a conditional exponential family of point processes.

Example 6.2.3 Let $N = \{N_t : t \geq 0\}$ be a counting process on $(\Omega, \mathcal{F}, P)$ with $N_0 = x$. Now, for every $x \in \mathbb{N}$ and $\theta \in \mathbb{R}$ consider the measure $P_{\theta,x}$ on $\mathcal{F}$ under which N is a counting process with predictable intensity

$$H^\theta_t(\omega) = e^\theta h(N_{t-}(\omega)), \qquad t > 0,$$

with respect to $\{\mathcal{F}_t\}$, where h is a non-random function from $\{0, 1, 2, \cdots\}$ into $(0, \infty)$. Then $\{P_{\theta,x} : \theta \in \mathbb{R}, x \in \mathbb{N}\}$ forms a conditional exponential family of counting processes with

$$\frac{dP^t_{\theta,x}}{dP^t_{0,x}} = \exp\left\{\theta(N_t - N_0) - (e^\theta - 1)\int_0^t h(N_s)ds\right\}, \quad t > 0.$$

The process N is a Markov process under $\{P_{\theta,x} : x \in \mathbb{N}\}$ for every $\theta \in \mathbb{R}$, so $\{P_{\theta,x} : x \in \mathbb{N}, \theta \in \mathbb{R}\}$ is a conditional exponential family. This is a particular case of the counting processes in Section 3.4. Some simple examples are $h \equiv 1$ (Poisson process), $h(x) = x$ (pure birth process), or $h(x) = x(K - x)$ for some $K \in \mathbb{N}$ (logistic birth process with carrying capacity K). A slightly different example is obtained for $H_t^\theta(\omega) = e^\theta 1_{[0,T_1]}(t)$ (one-jump-process with exponentially distributed jump time T_1). □

In the rest of this chapter we shall assume that there exists a family of so-called *shift operators* $\eta = \{\eta_t : t \in \mathcal{T}\}$ on Ω associated with X. These are mappings $\eta_t : \Omega \mapsto \Omega$ with

$$X_s(\eta_t\omega) = X_{s+t}(\omega), \qquad s, t \in \mathcal{T}, \quad \omega \in \Omega.$$

The assumption of the existence of η is no restriction because we can always assume that $\Omega = E^{\mathcal{T}}$ and define $(\eta_t\omega)_s = \omega_{s+t}$ for $\omega = (\omega_u : u \in \mathcal{T}) \in \Omega$ (see, e.g., Dynkin, 1965).

Definition 6.2.4 *An $\mathbb{R}^d$-valued $\{\mathcal{F}_t\}$-adapted process $V = \{V_t : t \in \mathcal{T}\}$ on $(\Omega, \mathcal{F})$ is called an* additive functional *with respect to η, X, and $\mathbb{P}^\theta$ if*

$$V_{t+s}(\omega) = V_s(\omega) + V_t(\eta_s\omega), \quad s, t \in \mathcal{T},\ \mathbb{P}^\theta\text{-a.s.}$$

When X and $\mathbb{P}^\theta$ are given in advance, we just write that V is an additive functional (with respect to η). Examples of additive functionals are M_n in Example 6.2.1, $V_t = \int_0^t b(X_s)\sigma^{-2}(X_s)dX_s$, and $\tilde{V}_t = \int_0^t b^2(X_s)\sigma^{-2}(X_s)ds$ in Example 6.2.2 and $\Lambda_t = \int_0^t h(N_s)ds$ in Example 6.2.3. Indeed, we have, for example,

$$M_{n+l} = \sum_{k=1}^{l+n} m(X_{k-1}, X_k) = \sum_{k=1}^{l} m(X_{k-1}, X_k) + \sum_{k=1}^{n} m(X_{k-1} \circ \eta_l, X_k \circ \eta_l)$$

and

$$V_{t+s} = \int_0^s b(X_u)\sigma^{-2}(X_u)dX_u + \int_0^t b(X_u \circ \eta_s)\sigma^{-2}(X_u \circ \eta_s)d(X_u \circ \eta_s).$$

For $\tilde{V}_t$ and Λ_t the proofs are analogous.

It should be noted that in the case of discrete time, $T = \mathbb{N}$, the additive functionals can be described very easily. Indeed, in this case $\eta_n = (\eta_1)^n$, and if $\{U_n : n \geq 0\}$ is an additive functional, we obtain $U_n = U_{n-1} + U_1 \circ \eta_{n-1} = \cdots = \sum_{k=1}^n U_1 \circ \eta_{k-1}$. Thus, U_n is determined by U_1, and because of the $\mathcal{F}_1$-measurability of U_1, there exists a measurable function $m : E \times E \mapsto \mathbb{R}^d$ such that $U_1(\omega) = m(X_0(\omega), X_1(\omega))$. This implies that

$$U_n = \sum_{k=1}^n m(X_{k-1}, X_k). \tag{6.2.6}$$

Thus, the canonical process in Example 6.2.1 is the most general additive functional. Of course, some of the components of m may be functions of either X_0 or X_1 only.

The Markov property (6.2.1) can now be formulated in a slightly more general form:

$$E_x(V \circ \eta_t | \mathcal{F}_s) = E_{X_s}(V), \quad P_x\text{-a.s.} \tag{6.2.7}$$

for every $s, t \in \mathcal{T}$ with $0 \leq s < t$, all $C \in \mathcal{E}$, and all $x \in E$, where V is any bounded random variable.

Definition 6.2.5 *A real-valued non-negative $\{\mathcal{F}_t\}$-adapted process $M = \{M_t : t \in \mathcal{T}\}$ on $(\Omega, \mathcal{F})$ is called a* multiplicative functional *with respect to η, X, and $\mathbb{P}^\theta$ if*

$$M_{t+s}(\omega) = M_s(\omega) \cdot M_t(\eta_s \omega), \quad s, t \in \mathcal{T}, \quad \mathbb{P}^\theta\text{-a.s.}$$

If X and $\mathbb{P}^\theta$ are given, we just write that M is a multiplicative functional (with respect to η). Obviously, if V is a real-valued additive functional, then $M_t = \exp[V_t]$ is a multiplicative functional, and conversely, if M is a strictly positive multiplicative functional, then $V_t = \log M_t$ is an additive functional.

6.3 The structure of exponential families of Markov processes

Let a conditional exponential family $\{\mathbb{P}^\theta : \theta \in \Theta\}$ in the sense of Definition 6.1.1 and a stochastic process X be given, and suppose in the sequel that the following condition holds.

Condition 6.3.1 *There exists a $\theta_0 \in \Theta$ such that X is a right-continuous Markov process with respect to $\{\mathcal{F}_t\}$ under $\mathbb{P}^{\theta_0} = \{P_{\theta_0,x} : x \in E\}$, the state space E of X is a separable complete metric space, and X has the transition function $\{P_{\theta_0}(t,x,B) : (t,x,B) \in \mathcal{T} \times E \times \mathcal{E}\}$.*

Under this assumption we have

Theorem 6.3.2 *The process X is a Markov process under $\mathbb{P}^\theta = \{P_{\theta,x} : x \in E\}$ if and only if*

$$L_t(\theta) = \frac{dP^t_{\theta,x}}{dP^t_{\theta_0,x}} = \exp[\gamma(\theta)^T B_t - \phi_t(\theta)] \tag{6.3.1}$$

is a multiplicative functional with respect to $\mathbb{P}^{\theta_0}$. If this is the case, the transition function of X under $\mathbb{P}^\theta$ is given by

$$P_\theta(t,x,B) = P_{\theta,x}(X_t \in B) = E^{\theta_0}_x[1_B(X_t) L_t(\theta)]. \tag{6.3.2}$$

Proof: Assume that $L_t(\theta)$ is a multiplicative functional. Then, for every $A \in \mathcal{F}_s$, $x \in E$, and for all bounded $\mathcal{E}$-measurable functions $f : E \mapsto \mathbb{R}$, the Markov property of X under $\mathbb{P}^{\theta_0}$ yields

$$\begin{aligned} E_x^\theta[f(X_{s+t}); A] &= E_x^{\theta_0}[f(X_{s+t})L_{s+t}(\theta); A] \\ &= E_x^{\theta_0}[f(X_t \circ \eta_s)L_s(\theta) \cdot \{L_t(\theta) \circ \eta_s\}; A] \\ &= E_x^{\theta_0}[L_s(\theta)E_{X_s}^{\theta_0}[f(X_t)L_t(\theta)]; A] \\ &= E_x^\theta[E_{X_s}^\theta(f(X_t)); A]. \end{aligned}$$

Thus, X is Markovian under $\mathbb{P}^\theta$, and (6.3.2) is obvious by choosing $f = 1_B$.

Conversely, assume that X is Markovian under $\mathbb{P}^\theta$. Then it is clear that (6.3.2) holds. Define the Radon-Nikodym derivative

$$q_t(x,y) = \frac{P_\theta(t,x,dy)}{P_{\theta_0}(t,x,dy)}, \qquad t \in \mathcal{T};\ x, y \in E.$$

Let $\delta = \{t_i : i = 1, \ldots, n\}$ be a partition of $[0,t]$ with $0 = t_0 < t_1 < \ldots < t_n = t$ and define

$$q_t^\delta = \prod_{i=0}^{n-1} q_{t_{i+1}-t_i}(X_{t_i}, X_{t_{i+1}}). \tag{6.3.3}$$

Let $x \in E$. Then it holds that

$$q_t^\delta = E_x^{\theta_0}(L_t(\theta)|X_{t_0}, \ldots, X_{t_n}), \qquad P_{\theta_0,x}\text{-a.s.} \tag{6.3.4}$$

If $\delta' = \{s_i : i = 1, \ldots, m\}$ is a partition of $[0,s]$ with $0 = s_0 < s_1 < \ldots < s_m = s$, then define $\delta \cup \delta'$ to be the partition of $[0, s+t]$ given by $\{t_i : i = 1, \ldots, n\} \cup \{t + s_i : i = 1, \ldots, m\}$. Now we have

$$q_{t+s}^{\delta \cup \delta'} = q_t^\delta q_s^{\delta'} \circ \eta_t. \tag{6.3.5}$$

If $\delta_k = \{t_i^{(k)} : i = 0, \ldots, n_k\}$ is an increasing sequence of partitions of $[0,t]$ such that $\lambda(\delta_k) = \max\{t_{i+1} - t_i : i \le n_k - 1\}$ tends to zero, then by (6.3.4) $(q_t^{\delta_k} : k \ge 1)$ is a uniformly integrable martingale with respect to $\{\mathcal{H}_k^t\}$, where $\mathcal{H}_k^t := \sigma(X_{t_i^{(k)}} : i = 0, \ldots, n_k)$ and

$$q_t^{\delta_k} \to E_x^{\theta_0}(L_t(\theta)|\mathcal{H}^t), \qquad P_{\theta_0,x}\text{-a.s.} \tag{6.3.6}$$

as $k \to \infty$. Here $\mathcal{H}^t = \sigma(\mathcal{H}_k^t : k \ge 1)$. Under the imposed condition that X is right-continuous, one can show that $\mathcal{H}^t$ is equal to $\mathcal{F}_t$, and because of the adaptedness of $L_t(\theta)$, the expression (6.3.6) is equal to $L_t(\theta)$ almost surely with respect to $P_{\theta_0,x}$. Now (6.3.5) implies that

$$L_{t+s}(\theta) = L_t(\theta) \cdot L_s(\theta) \circ \eta_t, \qquad P_{\theta_0,x}\text{-a.s.,} \tag{6.3.7}$$

and because of the arbitrariness of $x \in E$, this is what we wanted to show. □

Note that the exponential structure of $L_t(\theta)$ was not used in the proof above.

Under our assumption that $\{I\!P^\theta : \theta \in \Theta\}$ is a conditional exponential family of Markov processes, it is not true, in general, that the corresponding transition functions given by (6.3.2) also form an exponential family of probabilities on $(E, \mathcal{E})$ for every fixed $(t, x) \in \mathcal{T} \times E$. Concerning this question we have the following result.

Proposition 6.3.3 *If $\{I\!P^\theta, \theta \in \Theta\}$ is a conditional exponential family of Markov processes, then for fixed $(t, x) \in \mathcal{T} \times E$ their transition functions $\{P_\theta(t, x, \cdot), \theta \in \Theta\}$ form an exponential family of distributions if and only if*

$$E_x^{\theta_0}[\exp(\gamma(\theta)^T B_t)|X_t] = \exp[\beta_{t,x}(\theta)^T \varphi(t, x, X_t) - \psi_{t,x}(\theta)] \qquad (6.3.8)$$

for some measurable φ and some functions $\beta_{t,x}$ and $\psi_{t,x}$.

Proof: The result follows because

$$\begin{aligned} P_\theta(t, x, B) &= P_{\theta,x}(X_t \in B) = E_x^{\theta_0}(1_B(X_t)L_t(\theta)) \\ &= E_x^{\theta_0}(1_B(X_t)E_x^{\theta_0}(L_t(\theta)|X_t)) \\ &= \int_B E_x^{\theta_0}(L_t(\theta)|X_t = y)P_{\theta_0}(t, x, dy) \end{aligned}$$

for all $B \in \mathcal{E}$. □

In Example 6.2.2 it was illustrated how (6.3.8) may, but need not, hold. Now consider the case $\mathcal{T} = I\!N$. Then the condition (6.3.8) holds for $t = 1$. Indeed, $L_1(\theta) = \exp[\gamma(\theta)^T B_1 - \phi_1(\theta)]$, where B_1 is $\mathcal{F}_1$-measurable. Hence, there is a measurable function m on $E \times E$ such that $B_1(\omega) = m(X_0(\omega), X_1(\omega))$. Thus we have the following result.

Corollary 6.3.4 *Let $\mathcal{T} = I\!N$ and let $\{I\!P^\theta : \theta \in \Theta\}$ be a conditional exponential family of Markov processes. Then the one-step transition kernels*

$$P_\theta(x, dy) = P_\theta(1, x, dy) = P_{\theta,x}(X_1 \in dy)$$

form an exponential family of probabilities on $(E, \mathcal{E})$. In particular, if $L_t(\theta)$ is given by (6.1.1), we have

$$\frac{P_\theta(x, dy)}{P_{\theta_0}(x, dy)} = \exp[\gamma(\theta)^T m(x, y) - \phi_1(\theta)], \qquad (6.3.9)$$

where m is a measurable function on $E \times E$ such that $B_1(\omega) = m(X_0(\omega), X_1(\omega))$.

Conversely, if (6.3.9) holds, then one can construct a conditional exponential family of Markov chains with a representation (6.1.1), where the γ is the same as in (6.3.9), $B_t = \sum_{k=1}^t m(X_{k-1}, X_k)$, and $\phi_t(\theta) = t \cdot \phi_1(\theta)$; see Example 6.2.1.

Note that Corollary 6.3.4 also applies to discrete observation of Markov processes with $\mathcal{T} = [0, \infty)$. If for fixed $t \in [0, \infty)$ the class $\{P_\theta(t, x, \cdot) : \theta \in \Theta\}$ is an exponential family of distributions with representation of the same form as (6.3.9) for all x, then the distributions of the random vector $(X_t, X_{2t}, \cdots, X_{Nt})$ form an exponential family for all $N \in \mathbb{N}$.

For $\mathcal{T} = [0, \infty)$ the fact that the transition kernels form an exponential family of distributions does not imply that the corresponding class of Markov chains is a conditional exponential family. Consider, for instance, the family of Brownian motions with zero drift but with different diffusion coefficients. In this case the processes are not even locally absolutely continuous, so the likelihood function $L_t(\theta)$ does not exist. To ensure the existence of $L_t(\theta)$ it is necessary and sufficient that the family$\{q_t^{\delta_k} : \delta_k$ is a partition of $[0, t]$, $k = 1, 2, \cdots\}$, defined in the proof of Theorem 6.3.2, be a uniformly integrable martingale. If this holds, it is an open question whether it may happen that the transition functions form exponential families, while the processes as measures on $\mathcal{F}_t$ do not.

We will now continue our study of the structure of conditional exponential families of Markov processes.

Proposition 6.3.5 *Let $\{\mathbb{P}^\theta : \theta \in \Theta\}$ be a conditional exponential family. If there exists an exponential representation (6.1.1) where the canonical process $B = \{B_t : t \in \mathcal{T}\}$ is an additive functional with respect to X, η and $\mathbb{P}^{\theta_0}$, and if (in case $\mathcal{T} = [0, \infty)$) B is right-continuous, then for every $\theta \in \Theta$ the likelihood process*

$$L_t(\theta) = \exp[\gamma(\theta)^T B_t - \phi_t(\theta)], \quad t \in \mathcal{T},$$

is a multiplicative functional with respect to X, η, and $\mathbb{P}^{\theta_0}$, and

$$\phi_t(\theta) = t \cdot \phi_1(\theta), \quad t \in \mathcal{T}, \; \theta \in \Theta. \tag{6.3.10}$$

Proof: Note that

$$\begin{aligned} \exp(\phi_{t+s}(\theta)) &= E_x^{\theta_0}[\exp\{\gamma(\theta)^T B_{t+s}\}] \qquad (6.3.11) \\ &= E_x^{\theta_0}[\exp\{\gamma(\theta)^T (B_s + B_t \circ \eta_s)\}] \\ &= E_x^{\theta_0}\{\exp(\gamma(\theta)^T B_s) E_{X_s}^{\theta_0}[\exp(\gamma(\theta)^T B_t)]\} \\ &= [\exp \phi_t(\theta)] \cdot [\exp \phi_s(\theta)], \end{aligned}$$

for all $s, t \in \mathcal{T}$ and all $\theta \in \Theta$. If $\mathcal{T} = \mathbb{N}$, this proves the proposition. Assume $\mathcal{T} = [0, \infty)$ and use the right-continuity of B to conclude from Fatou's lemma that

$$\begin{aligned} \limsup_{h \downarrow 0} \phi_{t+h}(\theta) &= \limsup_{h \downarrow 0} \log E_x^{\theta_0}[\exp(\gamma(\theta)^T B_{t+h})] \\ &\leq \log E_x^{\theta_0}[\lim_{h \downarrow 0} \exp(\gamma(\theta)^T B_{t+h})] = \phi_t(\theta). \end{aligned}$$

Thus, the function $s \mapsto \phi_s(\theta)$ is bounded from above on $[t, t+\varepsilon]$ for some $\varepsilon > 0$. Therefore, (6.3.11) implies (6.3.10) also for $\mathcal{T} = [0, \infty)$ (see, e.g., Dynkin and Juschkevič (1969), p. 228 f.) In particular, $L(\theta)$ is a multiplicative functional with respect to $I\!P^{\theta_0}$ because B is an additive functional by assumption. □

Note that in Proposition 6.3.5 we did not need the usual regularity condition that $\phi_t(\theta)$ is right-continuous. Right-continuity of B is enough. By combining Proposition 6.3.5 and Theorem 6.3.2 we immediately have the following result.

Corollary 6.3.6 *Under the conditions of Proposition 6.3.5 the process X is a Markov process under $I\!P^\theta$ for all $\theta \in \Theta$.*

Proposition 6.3.7 *Let $\{I\!P^\theta : \theta \in \Theta\}$ be a conditional exponential family with a representation (6.1.1). Suppose there exist d_1 values of θ, which we denote by $\theta_1, \cdots, \theta_{d_1}$, such that $L_t(\theta_i)$ is a multiplicative functional with respect to X, η, and $I\!P^{\theta_0}$ for $i = 1, \cdots, d_1$, and such that $\gamma(\theta_1), \cdots, \gamma(\theta_{d_1})$ are linearly independent. Then there exists an exponential representation with canonical process $\tilde{B}$ such that the first d_1 coordinates of $\tilde{B}$ form an additive functional with respect to X, η, and $I\!P^{\theta_0}$. If the first d_1 coordinates of $\tilde{B}$ are right-continuous, it follows that X is a Markov process for all probability measures in the subfamily parametrized by $\Theta_M = \{\theta \in \Theta : \gamma(\theta) \in M\}$, where $M = \mathrm{span}(\gamma(\theta_1), \ldots, \gamma(\theta_{d_1}))$. If, in particular, $d_1 = d$, the process $\tilde{B}$ is an additive functional with respect to $I\!P^{\theta_0}$, and X is Markovian under P^θ for all $\theta \in \Theta$.*

Proof: Define

$$U_t(\theta) = \log L_t(\theta) = \gamma(\theta)^T B_t - \phi_t(\theta).$$

Next, extend $\gamma_i = \gamma(\theta_i)$, $i = 1, \ldots, d_1$, to a basis $\gamma_1, \ldots, \gamma_d$ for $I\!R^d$, define the $d \times d$ matrix $\Gamma = \{\gamma_1, \ldots, \gamma_d\}$, and define $\tilde{\gamma}(\theta) = \Gamma^{-1}\gamma(\theta)$ and $\bar{B}_t = \Gamma^T B_t$. Then $\tilde{\gamma}(\theta)^T \bar{B}_t = \gamma(\theta)^T B_t$, and $\tilde{\gamma}(\theta_i)$ is the unit vector with the ith coordinate equal to one. In particular,

$$\begin{bmatrix} U_t(\theta_1) \\ \vdots \\ U_t(\theta_{d_1}) \end{bmatrix} = \bar{B}_t^{(1)} - \psi_t,$$

where $\bar{B}_t^{(1)}$ denotes the first d_1 components of $\bar{B}_t$ and where $\psi_t = (\phi_t(\theta_1), \ldots, \phi_t(\theta_{d_1}))^T$. Since $U_t(\theta_i)$ is an additive functional for $i = 1, \ldots, d_1$, so is $\bar{B}_t^{(1)} - \psi_t$. Hence, an exponential representation of the type required is given by

$$U_t(\theta) = \tilde{\gamma}(\theta)^T \tilde{B}_t - \tilde{\phi}_t(\theta),$$

where

$$\tilde{B}_t = \begin{bmatrix} \bar{B}_t^{(1)} - \psi_t \\ \bar{B}_t^{(2)} \end{bmatrix}, \quad \tilde{\phi}_t(\theta) = \phi_t(\theta) - \tilde{\gamma}(\theta)^T \begin{bmatrix} \psi_t \\ 0 \end{bmatrix},$$

and where $\bar{B}_t^{(2)}$ denotes the last $d - d_1$ components of $\bar{B}_t$. Define $\tilde{\gamma}_{(1)}(\theta)$ and $\tilde{B}_t^{(1)}$ in analogy with $\bar{B}_t^{(1)}$. Then, for the subfamily parametrized by Θ_M we have

$$U_t(\theta) = \tilde{\gamma}_{(1)}(\theta)^T \tilde{B}_t^{(1)} - \tilde{\phi}_t(\theta),$$

i.e., the canonical process is an additve functional with respect to P_0. Hence, by Proposition 6.3.5, $U_t(\theta)$ is an additive functional for all $\theta \in \Theta_M$, and by Theorem 6.3.2 the process X is then Markovian for all $\theta \in \Theta_M$. The last statement of the proposition now follows trivially because $\Theta_M = \Theta$ if $d_1 = d$. □

Note that if $d = 1$, then X is either Markovian under all measures in the family or only under the dominating measure (the latter has been assumed, cf. Condition 6.3.1). Note also that Proposition 6.3.7 contains a converse result to Proposition 6.3.5.

Corollary 6.3.8 *If $L_t(\theta)$ is a multiplicative functional with respect to $\mathbb{P}^{\theta_0}$ for all $\theta \in \Theta$, then there exists an exponential representation where the canonical process is an additive functional with respect to $\mathbb{P}^{\theta_0}$.*

Proof: The condition of Proposition 6.3.7 is satisfied for $d_1 = d$ if we choose a minimal representation (6.1.1). □

To summarize, we essentially have that the process X is a Markov process under $\mathbb{P}^{\theta}$ for all $\theta \in \Theta$ if and only if there exists an exponential representation where the canonical process is an additive functional with respect to $\mathbb{P}^{\theta_0}$. Note, in particular, that we have proved that in the discrete time case all conditional exponential families of Markov processes are of the form considered in Example 6.2.1; cf. (6.2.6).

Let us conclude this section by illustrating the result of Proposition 6.3.7 by an example.

Example 6.3.9 The stochastic differential equation

$$dX_t = (\theta_1 X_t + \theta_2 X_{t-r})dt + dW_t, \quad t > 0,\ r > 0,$$

with the initial condition $X_s = f(s)$, $s \in [-r, 0]$, where f is a given function $[-r, 0] \mapsto \mathbb{R}$, has a unique (strong) solution for all $(\theta_1, \theta_2) \in \mathbb{R}^2$. This type of Langevin equation with a time-delayed term will be studied intensively in Chapter 9. Obviously, X is a Markov process if and only if $\theta_2 = 0$.

The likelihood function is given by

$$L_t(\theta_1,\theta_2) = \exp\left\{\theta_1\int_0^t X_s dX_s + \theta_2\int_0^t X_{s-r}dX_s \right.$$

$$\left. -\tfrac{1}{2}\theta_1^2\int_0^t X_s^2 ds - \tfrac{1}{2}\theta_2^2\int_0^t X_{s-r}^2 ds - \theta_1\theta_2\int_0^t X_s X_{s-r} ds\right\},$$

provided that we use $\theta_1 = \theta_2 = 0$ as dominating measure. For $\theta_1 = \theta_2 = 0$, X is a Wiener process and hence Markovian. The functionals

$$\int_0^t X_s dX_s \qquad \text{and} \qquad \int_0^t X_s^2 ds$$

are additive; the others are not. We see that the assumptions of Proposition 6.3.7 hold with $d_1 = 2$ and $\Theta_M = \{\theta \in \mathbb{R}^2 : \theta_2 = 0\}$. The subfamily parametrized by Θ_M is the class of Ornstein-Uhlenbeck processes. □

6.4 Exercises

6.1 Show in the case of an Ornstein-Uhlenbeck process (see Example 6.2.2) by direct calculations that the quantity q_t^δ defined in the proof of Theorem 6.3.2 converges to

$$\exp\left\{(\theta-\theta_0)\int_0^t X_s dX_s - \tfrac{1}{2}(\theta^2-\theta_0^2)\int_0^t X_s^2 ds\right\}$$

as $\lambda(\delta) \to 0$.

6.2 Show in the case of a Bessel process (see Example 6.2.2) by direct calculations that the quantity q_t^δ defined in the proof of Theorem 6.3.2 converges to

$$\exp\left\{(\theta-\theta_0)\int_0^t X_s^{-1} dX_s - \tfrac{1}{2}(\theta^2-\theta_0^2)\int_0^t X_s^{-2} ds\right\}$$

as $\lambda(\delta) \to 0$. Hint: Use the asymptotic expansion of the Bessel function $I_\nu(\cdot)$ in Abramowitz and Stegun (1970, p. 377) up to second-order terms.

6.3 Prove that every conditional exponential family of Markov processes with discrete time has a likelihood function of the form

$$L_t(\theta) = \exp\left[\gamma(\theta)^T \sum_{i=1}^t m(X_{k-1}, X_k) - t\phi(\theta)\right],$$

for some measurable function $m : E \times E \mapsto \mathbb{R}^d$, where E is the state space of the Markov processes in the family, and where γ is d-dimensional. Hint: Proposition 6.3.5, Corollary 6.3.8, and the remarks after Definition 6.2.4.

6.4 Let $X = \{X_t : t \in T\}$ be a real-valued Markov process with respect to $\{\mathcal{F}_t\}$ under $\{P_x : x \in E\}$. Suppose X has right-continuous sample paths with limits from the left. Show that for every $\epsilon > 0$ the process

$$A_t(\omega) = \sum_{s \leq t} 1_{\{|\Delta X_s| > \epsilon\}}(\omega)$$

is a real-valued additive functional for X.

6.5 Bibliographic notes

The material in this chapter is mainly drawn from Küchler and Sørensen (1997). The proof of Theorem 6.3.2 relies strongly on known facts from the general theory of Markov processes. For details see Dynkin (1965, Chapter IX) and Kunita (1976). Real-valued additive functionals of Markov processes have been studied in detail in the literature; see Blumenthal and Getoor (1968), Çinlar et al. (1980) and the references therein. These functionals are a powerful tool and are related to local times, random time transformations, and, generally put, to the potential theory of Markov processes. The corresponding theory has, up to now, only to a small extent been used to study exponential families of Markov processes. We have, in fact, only used elementary properties of additive and multiplicative functionals.

Exponential families of discrete time Markov processes have been studied by Heyde and Feigin (1975), Feigin (1981), Hudson (1982), Bhat (1988a), and Hwang and Basawa (1994). Feigin (1975) proved a representation theorem for asymptotic inference. It should be noted that Feigin's concept of a conditional exponential family is entirely different from our conditional exponential families. His families are, in fact, only exponential families in our sense in the special case of what he calls additive conditional exponential families. Hudson (1982) proved that under regularity conditions exponential families of discrete time Markov processes are locally asymptotically mixed normal. Hwang and Basawa (1994) showed local asymptotic normality of conditional exponential families in Feigin's sense and applied the result to non-linear time series models and threshold autoregressive models. Exponential families of Markov processes with finite state space where studied by Stefanov (1984, 1995), who considered sequential estimation for such models with discrete as well as continuous time. Exponential families of diffusion processes have been studied by several authors, including Novikov

(1972), Taraskin (1974), Brown and Hewitt (1975a), Sørensen (1983), and Küchler and Sørensen (1994b). The stochastic differential equation with time-delay discussed in Example 6.3.9 was studied by Küchler and Mensch (1992) and Gushchin and Küchler (1996).

7

The Envelope Families

7.1 General theory

We have seen that most exponential families of stochastic processes are curved exponential families in the sense that the canonical parameter space is a curved submanifold of a Euclidean space; cf. Theorem 4.2.1. It is therefore important to develop statistical theory for general curved exponential families of processes. Several modern statistical techniques for curved exponential families use properties of the full exponential family generated by the curved model. Examples are methods based on differential geometric considerations or on approximately ancillary statistics; see, e.g., Amari (1985), Barndorff-Nielsen and Cox (1994) and references therein. See also Sweeting (1992) and Jensen (1997).

In this chapter we study the full exponential families generated by the finite sample exponential families $\mathcal{P}_t = \{P_\theta^t : \theta \in \Theta\}$ and investigate their interpretation as stochastic process models. To emphasize the facts that in general it is not possible to enlarge the family $\mathcal{P}$ on $\mathcal{F}$ and that a stochastic process interpretation of the full families is not straightforward, we call these, in the stochastic process setting, envelope families. From a probabilistic point of view it is interesting that this statistical investigation provides a new way of deriving other stochastic processes from a given class of processes.

In the following we consider a family of probability measures $\mathcal{P}$ with a general exponential representation of the form (3.1.1). For fixed $t \geq 0$ we define the full exponential family generated by P_0^t and B_t in the classical way. Specifically, we let $\tilde{\Gamma}_t$ denote the domain of the Laplace transform of B_t under P_0

$$\tilde{\Gamma}_t = \{\gamma \in \mathbb{R}^m : E_0(\exp(\gamma^T B_t)) < \infty\} \tag{7.1.1}$$

and define a class of probability measures $\mathcal{Q}_t = \{Q_\gamma^{(t)} : \gamma \in \tilde{\Gamma}\}$ on $\mathcal{F}_t$ by

$$\frac{dQ_\gamma^{(t)}}{dP_0^t} = \exp(\gamma^T B_t - \Psi_t(\gamma)), \quad \gamma \in \tilde{\Gamma}_t, \tag{7.1.2}$$

with Ψ_t denoting the cumulant transform of B_t under P_0

$$\Psi_t(\gamma) = \log E_0(\exp(\gamma^T B_t)). \tag{7.1.3}$$

We call the class $\mathcal{Q}_t$ the *envelope exponential family on* $\mathcal{F}_t$. These families will play an important role in the following three chapters.

An obvious question is whether for fixed γ there exists a probability measure P_γ on $\mathcal{F}$ such that $Q_\gamma^{(t)}$ is the restriction P_γ to $\mathcal{F}_t$ for all $t \geq 0$. A first problem is that it is quite possible that the sets $\tilde{\Gamma}_t$ differ for varying t; see the examples in Sections 7.4, 7.5, and 7.6. This is only a minor problem: We can simply restrict the envelope families to $\tilde{\Gamma} = \cap_{t\geq 0}\tilde{\Gamma}_t$. For time-homogeneous families, $\Gamma \subseteq \tilde{\Gamma}$, where Γ is given by (4.1.3). Moreover, since $\tilde{\Gamma}_t$ for all $t \geq 0$ is a convex subset of $\mathbb{R}^m$, so is $\tilde{\Gamma}$. Therefore, if Γ is a curved sub-manifold of $\mathbb{R}^m$ (i.e., not a linear subspace), then necessarily $\tilde{\Gamma}$ is larger than Γ.

The crucial problem is that for fixed γ the measures $\{Q_\gamma^{(t)} : t \geq 0\}$ might not be consistent, i.e., the restriction of $Q_\gamma^{(t)}$ to $\mathcal{F}_s$, $s < t$, is not equal to $Q_\gamma^{(s)}$. The measures $\{Q_\gamma^{(t)} : t \geq 0\}$ are consistent if and only if the stochastic process $dQ_\gamma^{(t)}/dP_0^t$ is a P_0-martingale. Thus we must restrict the parameter set $\tilde{\Gamma}$ to

$$\Gamma^* = \{\gamma \in \tilde{\Gamma} : dQ_\gamma^{(t)}/dP_0^t \text{ is a } P_0\text{-martingale}\}. \tag{7.1.4}$$

If the space $(\Omega, \mathcal{F})$ is standard measurable, then there exists for each $\gamma \in \Gamma^*$ a probability measure P_γ on $\mathcal{F}$ such that $P_\gamma^t = Q_\gamma^{(t)}$; see Ikeda and Watanabe (1981, p. 176). Typically, Γ^* is contained in a sub-manifold of $\mathbb{R}^m$ of dimension smaller than m. This is necessarily so in case B_t is not a process with independent increments; see Theorem 4.2.1. If, on the other hand, B is a Lévy process, the model is similar to repeated sampling from a classical exponential family of distributions, and provided the representation is minimal, the dimension of Γ^* is m. Of course, the probability measures $\{P_\theta^t : t \geq 0\}$ are consistent for every $\theta \in \Theta$, and as is easily verified, the stochastic process dP_θ^t/dP_0^t is a martingale under P_0 for every fixed $\theta \in \Theta$. Thus $\Gamma \subseteq \Gamma^*$, and a natural definition of a full exponential family of processes would be to require that $\Gamma^* = \Gamma$. This holds for most examples considered in this book.

One conclusion of these considerations is that the requirement that the measures $\{Q_\gamma^{(t)} : t \geq 0\}$ be consistent is too strong if we want a stochastic process interpretation of the envelope family on $\mathcal{F}_t$. In the following we consider another approach, which, particularly in the case where a Markov process is observed, is much more fruitful.

For fixed $t > 0$ and $\gamma \in \tilde{\Gamma}_t$ we consider the restriction of $Q_\gamma^{(t)}$ to $\mathcal{F}_s$, $s \leq t$. We denote this measure by $Q_\gamma^{(t,s)}$. Quite generally, we have that

$$\frac{dQ_\gamma^{(t,s)}}{dP_0^s} = E_0\left(\frac{dQ_\gamma^{(t)}}{dP_0^t}|\mathcal{F}_s\right) = \exp(\gamma^T B_s + C_s^{(t)}(\gamma) - \Psi_t(\gamma)), \qquad (7.1.5)$$

where

$$C_s^{(t)}(\gamma) = \log\ E_0[\exp(\gamma^T(B_t - B_s))|\mathcal{F}_s]. \qquad (7.1.6)$$

We see that in general, $\{Q_\gamma^{(t,s)} : \gamma \in \tilde{\Gamma}_t\}$ is not an exponential family for $s < t$.

A very broad class of stochastic processes is the class of semimartingales, which we shall consider in detail in Chapter 11. The local behaviour of a semimartingale is determined by its local characteristics. Suppose B is a semimartingale and that $\{\mathcal{F}_s\}$ is generated by observing a semimartingale X. Then $\{X_s : s \leq t\}$ is also a semimartingale under $Q_\gamma^{(t)}$, and its local characteristics under $Q_\gamma^{(t)}$ can be determined from (7.1.5) by Theorem A.10.1. This gives an interpretation of the envelope family on $\mathcal{F}_t$ as a stochastic process model. In the next section we shall see that rather explicit results can be given for Markov processes. Whether $\{X_s : s \leq t\}$ can be extended to time points larger than t in a natural way cannot be determined in general.

Note, incidentally, that $Q_\gamma^{(t,s)} = Q_\gamma^{(s)}$ only when $C_s^{(t)}(\gamma) = \Psi_t(\gamma) - \Psi_s(\gamma)$, which happens precisely when B has independent increments under P_0, a situation that does not bring anything essentially new compared to repeated sampling from classical exponential families of distributions.

When the canonical parameter space Γ is a k-dimensional submanifold of $\mathbb{R}^m$ $(k < m)$ that is not an affine subspace, we say that $\mathcal{P}$ is a *curved* (m, k) *exponential model.* For most curved (m, k) exponential families it is possible to find a representation of the form

$$\frac{dP_\theta^t}{dP_0^t} = \exp(\theta^T A_t - \alpha_t(\theta)^T S_t - \phi_t(\theta)), \ \ t \geq 0, \qquad (7.1.7)$$

where $\alpha_t(\theta)$ is an $(m - k)$-dimensional vector with $\alpha_t(0) = 0$ and where A and S are vectors of $\{\mathcal{F}_t\}$-adapted processes of dimension k and $(m - k)$, respectively. We call a representation of the form (7.1.7) a *natural representation* if it is minimal and int $\Theta \neq \emptyset$. Such a representation is not unique. The parametrization defined by (7.1.7) will be called a *natural parametrization.* The natural exponential family generated by a Lévy process has a representation of this form, as have families obtained by stochastic time transformation from such models; see Chapter 5. In Chapter 11 we shall define the natural exponential family generated by a semimartingale. In this general case a representation of the form (7.1.7) is obtained too.

If $\mathcal{P}$ is a time-homogeneous curved exponential family with natural representation (7.1.7), it follows easily that for $\gamma = (\gamma_1, \gamma_2)$ with γ_1 k-dimensional

$$\begin{aligned} C_u^{(t)}(\gamma) &= \log(E_{\gamma_1}\{\exp[(\gamma_2 + \alpha(\gamma_1))^T(S_t - S_u)]|\mathcal{F}_u\}) \\ &\quad + \phi_t(\gamma_1) - \phi_u(\gamma_1). \end{aligned} \tag{7.1.8}$$

7.2 Markov processes

To obtain more manageable expressions, let us consider the case where $\{\mathcal{F}_t\}$ is generated by observation of a Markov process X with state space E. We will look at the conditional exponential family, where we condition on $X_0 = x (x \in E)$; see Chapter 6. As in that chapter, it is useful to make the initial condition x explicit in the notation, so we replace P_θ by $P_{\theta,x}$, $Q_\gamma^{(t)}$ by $Q_{\gamma,x}^{(t)}$, etc. In (7.1.2) we replace $\Psi_t(\gamma)$ by $\Psi_t(\gamma, x)$. We assume that X is a Markov process under $\{P_{0,x} : x \in E\}$ and that B is a right-continuous additive functional with respect to X and $\{P_{0,x} : x \in E\}$. Then X is a Markov process under $\{P_{\theta,x} : x \in E\}$ for every $\theta \in \Theta$; see Chapter 6. Under these assumptions,

$$C_s^{(t)}(\gamma) = \log\ E_{0,X_s}(\exp\{\gamma^T B_{t-s}\}) = \Psi_{t-s}(\gamma, X_s). \tag{7.2.1}$$

By inserting this in (7.1.5) we find that

$$\frac{dQ_{\gamma,x}^{(t,s)}}{dP_{0,x}^s} = \exp(\gamma^T B_s + \Psi_{t-s}(\gamma, X_s) - \Psi_t(\gamma, x)). \tag{7.2.2}$$

This is only an exponential family in case $\Psi_u(\gamma, y)$ is a finite sum of the form

$$\Psi_u(\gamma, y) = \sum_i f_u^{(i)}(\gamma) g_u^{(i)}(y), \quad u \le t. \tag{7.2.3}$$

We see that for exponential families of Markov processes the key to studying the envelope families is the function $\psi_u(\gamma, y)$. Before giving results on how to determine $\psi_t(\gamma)$ for general curved exponential families, we shall first consider some important classes of Markov processes.

Example 7.2.1 Suppose we observe a diffusion process X which under P_θ solves the stochastic differential equation

$$dX_t = \theta\mu(X_t)dt + \sigma(X_t)dW_t, \quad X_0 = x, \quad \theta \in \Theta, \tag{7.2.4}$$

where W is a Wiener process, $\Theta \in \mathbb{R}$, and $\sigma > 0$; cf. Section 3.5. It is well known that provided that

$$P_\theta\left(\int_0^t \mu^2(X_s)\sigma^{-2}(X_s)ds < \infty\right) = 1,$$

this model is an exponential family of stochastic processes with likelihood function

$$\frac{dP_\theta^t}{dP_0^t} = \exp\left(\theta \int_0^t \frac{\mu(X_u)}{\sigma^2(X_u)}dX_u - \tfrac{1}{2}\theta^2 \int_0^t \frac{\mu^2(X_u)}{\sigma^2(X_u)}du\right), \tag{7.2.5}$$

which is of the form (7.1.7). By applying Ito's formula to the function $(y,s) \to \Psi_{t-s}(\gamma, y)$, we find that

$$\begin{aligned}\Psi_{t-s}(\gamma, X_s) - \Psi_t(\gamma, X_0) &= \int_0^s \frac{\partial}{\partial y}\Psi_{t-u}(\gamma, X_u)dX_u \\ &+ \int_0^s \frac{\partial}{\partial u}\Psi_{t-u}(\gamma, X_u)du \\ &+ \frac{1}{2}\int_0^s \frac{\partial^2}{\partial y^2}\Psi_{t-u}(\gamma, X_u)\sigma^2(X_u)du.\end{aligned} \tag{7.2.6}$$

By inserting this in (7.2.2) we see that the envelope family is given by

$$\begin{aligned}\log \frac{dQ_\gamma^{(t,s)}}{dP_0^s} &= \int_0^s \frac{\gamma_1\mu(X_u) + \frac{\partial}{\partial y}\Psi_{t-u}(\gamma, X_u)\sigma^2(X_u)}{\sigma^2(X_u)}dX_u \\ &+ \int_0^s \left(\frac{\gamma_2\mu^2(X_u)}{\sigma^2(X_u)} + \frac{\partial}{\partial u}\Psi_{t-u}(\gamma, X_u)\right. \\ &\qquad \left. + \frac{1}{2}\frac{\partial^2}{\partial y^2}\Psi_{t-u}(\gamma, X_u)\sigma^2(X_u)\right) du,\end{aligned} \tag{7.2.7}$$

for $\gamma = (\gamma_1, \gamma_2) \in \tilde{\Gamma}_t$.

Now, what kind of process is $(X_s : s \leq t)$ under $Q_\gamma^{(t)}$? We have made an absolute continuous change of measure, so it is still a continuous semimartingale, and the quadratic variation is unchanged; i.e., the diffusion coefficient is the same as under P_0; cf. Theorem A.10.1. By Theorem A.10.1 we also see that under $Q_\gamma^{(t)}$ the process X is a diffusion process with time-dependent drift coefficient given by

$$d_t(\gamma, y, s) = \gamma_1\mu(y) + \frac{\partial}{\partial y}\Psi_{t-s}(\gamma, y)\sigma^2(y), \quad s \leq t. \tag{7.2.8}$$

So under $Q_\gamma^{(t)}$, the process X solves the equation

$$dX_s = d_t(\gamma, X_s, s)ds + \sigma(X_s)dW_s, \quad X_0 = x, \quad s \leq t. \tag{7.2.9}$$

Whether this can be extended to a process on the whole real line in a natural way depends on whether $\frac{\partial}{\partial y}\psi_t(\gamma, y)$ has a well-behaved extension to $t < 0$. □

Example 7.2.2 Suppose the observed Markov process X is a counting process with intensity $\lambda_t(\theta) = (1-\theta)F(X_{t-})$, where $\theta \in (-\infty, 1)$ and F is a mapping $\mathbb{N} \to (0,\infty)$ satisfying $F(x) \leq a + bx$ for some $a \geq 0$ and $b \geq 0$. Then X is non-explosive for all θ (see Jacobsen, 1982, p. 115), and

$$\frac{dP_\theta^t}{dP_0^t} = \exp\left(\theta \int_0^t F(X_s)ds + \log(1-\theta)(X_t - X_0)\right), \tag{7.2.10}$$

where x is the initial value $X_0 = x$. Under $Q_\gamma^{(t)}$, $\gamma \in \tilde{\Gamma}_t$ given by (7.1.2), the process $\{X_s : s \leq t\}$ has almost surely sample paths like a counting process. In particular, jumps are only upwards of size one, and the jumps do not accumulate. This is because $Q_\gamma^{(t)}$ is dominated by P_0^t. In order to find the intensity of $\{X_s : s \leq t\}$ under $Q_\gamma^{(t)}$, note that from (8.2.1) it follows by Lemma B.2.1 that for $u < s \leq t$,

$$\begin{aligned} Q_\gamma^{(t)}(X_s = i|\mathcal{F}_u) \;=\; & E_0\Big[1_{\{i\}}(X_s)\exp\Big\{\gamma_1 \int_u^s F(X_v)dv \\ & + \gamma_2(X_s - X_u) + \Psi_{t-s}(\gamma, X_s) \\ & - \psi_{t-u}(\gamma, X_u)\Big\}\Big| X_u\Big]. \end{aligned} \tag{7.2.11}$$

Therefore,

$$\begin{aligned} &(s-u)^{-1}Q_\gamma^{(t)}(X_s = i+1|X_u = i) \\ &\quad = \exp(\gamma_2 + \psi_{t-s}(\gamma, i+1) - \psi_{t-u}(\gamma, i)) \\ &\qquad \times (s-u)^{-1}E_0\left(1_{\{i+1\}}(X_s)\exp\left[\gamma_1 \int_u^s F(X_v)dv\right]\Big| X_u = i\right) \\ &\quad \to \exp(\gamma_2 + \psi_{t-u}(\gamma, i+1) - \psi_{t-u}(\gamma, i))F(i), \end{aligned} \tag{7.2.12}$$

where the convergence is as $s \downarrow u$. Here we have used that the intensity of X under P_0 is $F(X_{t-})$. We have, for simplicity, assumed that $\psi_t(\gamma, x)$ is a left-continuous function of time. The intensity under $Q_\gamma^{(t)}$ is thus given by

$$\begin{aligned} \lambda_s^{(t)}(\gamma) \;=\; & \exp[\gamma_2 + \psi_{t-s}(\gamma, X_{s-} + 1) \\ & - \psi_{t-s}(\gamma, X_{s-})]F(X_{s-}), \quad s \leq t. \end{aligned} \tag{7.2.13}$$

Whether this can be extended to an intensity process for $s > t$ depends on whether ψ has a well-behaved extension to $t < 0$. □

Example 7.2.3 Now let the observed Markov process X be a d-dimensional discrete time process with one-step transition density

$$f_{X_t|X_{t-1}}(y;\theta|x) = b_t(x,y)\exp[\theta^T m_t(y,x) - \alpha(\theta)^T h_t(x)], \quad \theta \in \Theta, \tag{7.2.14}$$

where $\Theta \subseteq \mathbb{R}^k$, $m_t(y,x)$ is a k-dimensional vector, and $\alpha(\theta)$ and $h_t(x)$ are $(m-k)$-dimensional. We assume that $X_0 = x$. Then

$$\frac{dP_\theta^t}{dP_0^t} = \exp\left(\theta^T \sum_{i=1}^t m_i(X_i, X_{i-1}) - \alpha(\theta)^T \sum_{i=0}^t h_i(X_{i-1})\right). \tag{7.2.15}$$

Straightforward calculations show that the process $\{X_s : s \le t\}$ is also a Markov process under $Q_\gamma^{(t)}$ for $\gamma \in \tilde{\Gamma}_t$ and that the distribution of X_{s+1} given X_s under $Q_\gamma^{(t)}$ has Laplace transform given by

$$\begin{aligned} E_\gamma(e^{\beta^T X_{s+1}} | X_s) &= \exp\{(\alpha(\gamma_1) + \gamma_2)^T h_{s+1}(X_s) - \psi_{t-s}(\gamma, X_s)\} \\ &\quad \times E_{\gamma_1}(e^{\beta^T X_{s+1} + \psi_{t-s+1}(\gamma, X_{s+1})} | X_s) \end{aligned} \tag{7.2.16}$$

for $\gamma_1 \in \Theta$ and $s \in \{0, \cdots, t-1\}$. The last conditional expectation is under P_{γ_1}. □

7.3 Explicit calculations

In this section we give some results on how to calculate the function $\Psi_t(\gamma)$ explicitly for general curved exponential families. We assume that the family has a natural representation (7.1.7).

Let $P_{\theta,1}^t$ and $P_{\theta,2}^t$ denote the marginal distributions of A_t and S_t, respectively, under P_θ. Further, define for all $t \ge 0$ and $\theta \in \Theta$ the Laplace transforms

$$c_1(w; \theta, t) = E_\theta(e^{w^T A_t}) \tag{7.3.1}$$

and

$$c_2(w; \theta, t) = E_\theta(e^{w^T S_t}), \tag{7.3.2}$$

and denote by $D_i(\theta, t)$ the domain of $c_i(\cdot; \theta, t)$, $i = 1, 2$.

Proposition 7.3.1 *The envelope exponential family on $\mathcal{F}_t$ contains the measures given by*

$$\frac{dQ_{\theta,\varphi}^{(t)}}{dP_0^t} = \exp[\theta^T A_t + \varphi^T S_t - \Psi_t(\theta, \varphi)], \tag{7.3.3}$$

where

$$\Psi_t(\theta, \varphi) = \log\ c_2(\varphi + \alpha_t(\theta); \theta, t) + \phi_t(\theta) \tag{7.3.4}$$

and

$$(\theta, \varphi) \in \mathcal{M}_t = \{(\theta, \varphi) : \theta \in \Theta,\ \ \varphi + \alpha_t(\theta) \in D_2(\theta, t)\}. \tag{7.3.5}$$

Suppose int $\mathcal{M}_t \ne \emptyset$, *and let $\bar{\mathcal{M}}_t$ denote the largest subset of $\mathbb{R}^m$ to which $\Psi_t(\theta, \varphi)$ can be extended by analytic continuation. Then $\tilde{\Gamma}_t = \bar{\mathcal{M}}_t$, and the measures in the envelope family are given by (7.3.3) with $\Psi_t(\theta, \varphi)$ defined by analytic continuation.*

Remark. It is well-known that $\bar{\mathcal{M}}_t$ is a convex set and that the convex hull of $\{(\theta, -\alpha_t(\theta)) : \theta \in \Theta\}$ is contained in $\bar{\mathcal{M}}_t$. For a discussion of how to determine $\bar{\mathcal{M}}_t$, see Hoffmann-Jørgensen (1994).

Proof: Let $\bar{P}_\theta^t$ denote the conditional distribution under P_θ of A_t given S_t, and set

$$f_t(x;\theta) = \frac{dP_{\theta,2}^t}{dP_{0,2}^t}(x).$$

Then

$$\frac{d\bar{P}_\theta^t}{d\bar{P}_0^t} = \exp[\theta^T A_t - \alpha_t(\theta)^T S_t - \phi_t(\theta) - \log\ f_t(S_t;\theta)],$$

from which we see that

$$E_0(e^{\theta^T A_t}|S_t) = \exp[\alpha_t(\theta)^T S_t + \phi_t(\theta) + \log\ f_t(S_t;\theta)].$$

Therefore,

$$\begin{aligned} E_0(e^{\theta^T A_t + \varphi^T S_t}) &= E_0[e^{\varphi^T S_t} E_0(e^{\theta^T A_t}|S_t)] \\ &= E_0(e^{(\varphi+\alpha_t(\theta))^T S_t} f_t(S_t;\theta))e^{\phi_t(\theta)} \\ &= E_\theta(e^{(\varphi+\alpha_t(\theta))^T S_t})e^{\phi_t(\theta)} \\ &= c_2(\varphi + \alpha_t(\theta); \theta, t)e^{\phi_t(\theta)}, \end{aligned}$$

provided that $\theta \in \Theta$ and $\varphi + \alpha_t(\theta) \in D_2(\theta, t)$.

The extension to $\bar{\mathcal{M}}_t$ follows from well-known properties of the Laplace transform; see, e.g., Hoffmann-Jørgensen (1994). □

From Proposition 7.3.1 it follows by standard exponential family arguments that the simultaneous Laplace transform of (A_t, S_t) under $Q_{\theta,\varphi}^{(t)}$ is given by

$$\begin{aligned} &E_{\theta,\varphi}(e^{u^T A_t + v^T S_t}) \\ &\quad = \frac{c_2(v + \varphi + \alpha_t(u+\theta); u+\theta, t)}{c_2(\varphi + \alpha_t(\theta); \theta, t)} \exp[\phi_t(u+\theta) - \phi_t(\theta)] \end{aligned} \tag{7.3.6}$$

with domain $\tilde{\Gamma}_t - (\theta, \varphi)$. In particular, the simultaneous Laplace transform of (A_t, S_t) under P_θ is obtained for $\varphi = -\alpha_t(\theta)$. Hence

$$\begin{aligned} &E_\theta(e^{u^T A_t + v^T S_t}) \\ &\quad = c_2(\alpha_t(u+\theta) - \alpha_t(\theta) + v; u+\theta, t) \exp[\phi_t(u+\theta) - \phi_t(\theta)]. \end{aligned} \tag{7.3.7}$$

By arguments similar to those for Proposition 7.3.1 we can prove the following result.

Proposition 7.3.2 *Suppose the function $\theta \mapsto \alpha_t(\theta)$ is invertible on $\Theta_t^* \subseteq \Theta$ and set $\Lambda_t = -\alpha_t(\Theta_t^*)$. Then*

$$\mathcal{M}_t^* = \{(\theta, \varphi) : \varphi \in \Lambda_t,\ \theta - \alpha_t^{-1}(-\varphi) \in D_1(\alpha_t^{-1}(-\varphi), t)\} \subseteq \tilde{\Gamma}_t, \tag{7.3.8}$$

and for $(\theta,\varphi) \in \mathcal{M}_t^*$ *the function* $\Psi_t(\theta,\varphi)$ *in (7.3.3) can be expressed as*

$$\Psi_t(\theta,\varphi) = \log[c_1(\theta - \alpha_t^{-1}(-\varphi); \alpha_t^{-1}(-\varphi), t)] + \phi_t(\alpha_t^{-1}(-\varphi)). \quad (7.3.9)$$

If $\operatorname{int} \mathcal{M}_t^* \neq \emptyset$, *the whole envelope family can be obtained by analytic continuation as described in Proposition 7.3.1.*

By equating (7.3.4) and (7.3.9) for $\varphi = x - \alpha_t(\theta)$ we find that

$$\begin{aligned} c_2(x;\theta,t) &= c_1(\theta - \alpha_t^{-1}(\alpha_t(\theta) - x); \alpha_t^{-1}(\alpha_t(\theta) - x), t) \quad (7.3.10)\\ &\quad \times \exp[\phi_t(\alpha_t^{-1}(\alpha_t(\theta) - x)) - \phi_t(\theta)] \end{aligned}$$

for $x \in (\Lambda_t + \alpha_t(\theta)) \cap D_2(\theta,t)$. From Proposition 7.3.2 we also find an expression for the simultaneous Laplace transform of (A_t, S_t) under $Q_{\theta,\varphi}^{(t)}$ for $(u,v) \in \mathcal{M}_t - (\theta,\varphi)$:

$$\begin{aligned} E_{\theta,\varphi}(e^{u^T A_t + v^T S_t}) &= \exp[\phi_t(\alpha_t^{-1}(-v-\varphi)) - \phi_t(\alpha_t^{-1}(-\varphi))] \quad (7.3.11)\\ &\quad \times \frac{c_1(u + \theta - \alpha_t^{-1}(-v-\varphi); \alpha_t^{-1}(-v-\varphi), t)}{c_1(\theta - \alpha_t^{-1}(-\varphi); \alpha_t^{-1}(-\varphi), t)}. \end{aligned}$$

For the simultaneous Laplace transform of (A_t, S_t) under P_θ we find

$$\begin{aligned} E_\theta(e^{u^T A_t + v^T S_t}) &= \exp[\phi_t(\alpha_t^{-1}(\alpha_t(\theta) - v)) - \phi_t(\theta)] \quad (7.3.12)\\ &\quad \times c_1(u + \theta - \alpha_t^{-1}(\alpha_t(\theta) - v); \alpha_t^{-1}(\alpha_t(\theta) - v), t). \end{aligned}$$

Obviously, we could also derive (7.3.11) and (7.3.12) from (7.3.6) and (7.3.7) by means of (7.3.10).

The following corollary will be useful in Chapters 8 and 10.

Corollary 7.3.3 *It holds that*

$$[\theta \in \operatorname{int}\Theta \ \text{ and } \ 0 \in \operatorname{int} D_2(\theta,t)] \ \Rightarrow \ (\theta, -\alpha_t(\theta)) \in \operatorname{int}\tilde{\Gamma}_t$$

and that

$$[\theta \in \operatorname{int}\Theta_t^* \ \text{ and } \ 0 \in \operatorname{int} D_1(\theta,t)] \ \Rightarrow \ (\theta, -\alpha_t(\theta)) \in \operatorname{int}\tilde{\Gamma}_t,$$

where the set Θ_t^* *was defined in Proposition 7.3.2.*

Proof: The result follows from Proposition 7.3.1 and Proposition 7.3.2 because $\tilde{\Gamma}_t$ is a convex set. □

The sets $D_1(\theta,t)$ and $D_2(\theta,t)$ are the domains of the Laplace transforms of A_t and S_t, respectively, under P_θ. Therefore, to show that a member of our exponential family of processes belongs to the interior of the envelope exponential family, we need only know something about the tail behaviour of the distribution under P_θ of A_t or of S_t. If, for instance, S_t is one-dimensional and positive, it suffices that S_t have a density function that tends exponentially fast to zero at infinity.

7.4 The Gaussian autoregression

The real-valued Gaussian autoregression of order one is defined by

$$X_i = \theta X_{i-1} + Z_i, \quad i = 1, 2, \cdots, \tag{7.4.1}$$

where $\theta \in \mathbb{R}$, $X_0 = x_0$, and where the Z_i's are independent standard normal distributed random variables. This is a curved exponential family of processes with natural representation

$$\frac{dP_\theta^t}{dP_0^t} = \exp\left[\theta \sum_{i=1}^{t} X_i X_{i-1} - \tfrac{1}{2}\theta^2 \sum_{i=1}^{t} X_{i-1}^2\right]. \tag{7.4.2}$$

The envelope family on $\mathcal{F}_t$ has a representation of the form

$$\frac{dQ_{\theta,\varphi}^{(t)}}{dP_0^t} = \exp\left[\theta \sum_{i=1}^{t} X_i X_{i-1} + \varphi \sum_{i=1}^{t} X_{i-1}^2 - \Psi_t(\theta,\varphi;x_0)\right], \tag{7.4.3}$$

where $(\theta, \varphi) \in \tilde{\Gamma}_t$. In accordance with the notation for Markov processes in Chapter 6, we include x_0 among the arguments of the function Ψ.

For the Gaussian autoregression, the function $\Psi_t(\theta, \varphi; x_0)$ is easily found by direct calculation. Indeed,

$$\begin{aligned}
&\exp(\Psi_t(\theta,\varphi;x_0)) \\
&= \int_{-\infty}^{\infty} \cdots \int_{-\infty}^{\infty} \exp\left[\theta \sum_{i=1}^{t} x_i x_{i-1} + \varphi \sum_{i=1}^{t} x_{i-1}^2\right] \\
&\qquad \times (2\pi)^{-t/2} \exp\left[-\tfrac{1}{2}\sum_{i=1}^{t} x_i^2\right] dx_t \cdots dx_1 \\
&= \exp[x_0^2(\varphi - \theta^2/(4A_t))](2\pi)^{-t/2} \\
&\qquad \times \int_{-\infty}^{\infty} \cdots \int_{-\infty}^{\infty} \exp\left[\sum_{i=1}^{t} A_{t-i+1}(x_i + \theta x_{i-1}/(2A_{t-i+1}))^2\right] dx_t \cdots dx_1,
\end{aligned}$$

where the quantities $A_1, \cdots, A_t$ are functions of θ and φ defined iteratively by

$$A_1 = -\tfrac{1}{2} \text{ and } A_i = \varphi - \tfrac{1}{2} - \theta^2/(4A_{i-1}). \tag{7.4.4}$$

Clearly, $\Psi_t(\theta, \varphi; x_0)$ is finite if and only if $A_i(\theta, \varphi) < 0$ for $i = 1, \cdots, t$, and if this is the case,

$$\Psi_t(\theta,\varphi;x_0) = x_0^2[A_{t+1}(\theta,\varphi) + \frac{1}{2}] - \frac{1}{2}\sum_{i=1}^{t} \log(-2A_i(\theta,\varphi)). \tag{7.4.5}$$

Explicit, but complicated, expressions for the A_i's can be derived from results in White (1958).

The set $\tilde{\Gamma}_t = \{(\theta, \varphi) : A_i(\theta, \varphi) < 0,\ i = 1, \cdots, t\}$ is not easy to characterize in an explicit way. However, because $\tilde{\Gamma}_t$ is a convex set containing $\{(\theta, -\frac{1}{2}\theta^2) : \theta \in \mathbb{R}\}$, it follows that

$$\{(\theta, \varphi) : \varphi \leq -\tfrac{1}{2}\theta^2\} \subseteq \tilde{\Gamma}_t$$

for all $t \geq 1$. Moreover, from the inequality $A_2 = \varphi - \frac{1}{2} + \frac{1}{2}\theta^2 < 0$ we see that

$$\tilde{\Gamma}_t \subseteq \{(\theta, \varphi) : \varphi < -\tfrac{1}{2}\theta^2 + \tfrac{1}{2}\} \tag{7.4.6}$$

for $t \geq 2$. Even more can be said. Set $\eta = \varphi - \frac{1}{2}$, which by (7.4.6) is strictly negative. If $|\eta| > |\theta|$, the mapping

$$F_{\eta,\theta}(x) = \eta - \theta^2/(4x)$$

has two fixed points:

$$x_{\pm} = \tfrac{1}{2}(\eta \pm \sqrt{\eta^2 - \theta^2}) < 0. \tag{7.4.7}$$

By iterated application of $F_{\eta,\theta}$ starting from an x in $(-\infty, x_+)$, the sequence thus defined tends monotonically towards the lower fixed point x_-. If $|\theta| = |\eta|$, there is only one fixed point, $x_- = x_+ = \frac{1}{2}\eta < 0$. Also in this case, repeated application of $F_{\eta,\theta}$ defines for a starting point $x \leq x_+$ a sequence that tends monotonically towards x_+. From these considerations it follows that

$$\begin{aligned}
&\varphi \leq \tfrac{1}{2} - |\theta| \ \text{ and } \ -\tfrac{1}{2} = A_1(\theta, \varphi) \leq x_+(\theta, \varphi) \\
&\Downarrow \\
&A_i(\theta, \varphi) < 0, \ \ \forall i \geq 1.
\end{aligned}$$

If moreover $-\frac{1}{2} < x_+(\theta, \varphi)$, then

$$A_i(\theta, \varphi) \to x_-(\theta, \varphi) \ \text{as} \quad i \to \infty. \tag{7.4.8}$$

On the other hand, if $|\eta| \geq |\theta|$ and $x > x_+$, then iterated application of $F_{\eta,\theta}$ starting at x defines a sequence that grows monotonically with increasing increments until a strictly positive number is reached in a finite number of steps. In the case $|\eta| < |\theta|$ this is so for any negative starting point. Therefore,

$$\begin{aligned}
\tilde{\Gamma} &= \cap_{t>0}\tilde{\Gamma}_t & (7.4.9)\\
&= \{(\theta, \varphi) : \varphi \leq \tfrac{1}{2} - |\theta|,\ -\tfrac{1}{2} \leq x_+(\theta, \varphi)\} \backslash \{(0, \tfrac{1}{2})\} \\
&= \{(\theta, \varphi) : \varphi \leq \tfrac{1}{2} - |\theta|,\ \varphi < \tfrac{1}{2},\ |\theta| \leq 1\} \cup \{(\theta, \varphi) : \varphi \leq -\tfrac{1}{2}\theta^2\}.
\end{aligned}$$

We also see that the inclusion (7.4.6) is strict. For $\varphi > \frac{1}{2} - |\theta|$, the number of iterations needed to reach a positive value is bounded by $n(\varphi)$, defined as one plus the number of iterations needed when starting from $\varphi - \frac{1}{2}$. Hence, when t is larger than $n(-\frac{1}{2})$,

$$\tilde{\Gamma}_t \cap \{(\theta,\varphi) : |\theta| \le 1\} = \tilde{\Gamma} \cap \{(\theta,\varphi) : |\theta| \le 1\}.$$

Conversely, when $|\theta| > 1$, we can for any $(\theta, -\frac{1}{2}\theta^2) \in \mathrm{bd}\,\tilde{\Gamma}$ and any $t \ge 1$ find a neighbourhood $N_t(\theta)$ of $(\theta, -\frac{1}{2}\theta^2)$ such that for all $(\tilde{\theta}, \tilde{\varphi}) \in N_t(\theta)$ the upper fixed point $x_+(\tilde{\theta}, \tilde{\varphi})$ is sufficiently close to $-\frac{1}{2}$ that iterated application of $F_{\tilde{\eta},\tilde{\theta}}$ starting at $-\frac{1}{2}$ defines a sequence that remains negative at least in the first $t-1$ iterations, i.e., such that $N_t(\theta) \subset \tilde{\Gamma}_t$. Thus

$$\tilde{\Gamma} \cap \{(\theta,\varphi) : |\theta| > 1\} \subset (\mathrm{int}\,\tilde{\Gamma}_t) \cap \{(\theta,\varphi) : |\theta| > 1\}.$$

In particular,

$$(\theta, -\tfrac{1}{2}\theta^2) \in \mathrm{int}\,\tilde{\Gamma}_t \;\text{ for }\; |\theta| \neq 1 \tag{7.4.10}$$

for all $t \ge 1$. We see that the critical autoregression ($|\theta| = 1$) corresponds to a boundary point of the envelope family for t large enough.

Note, incidentally, that the lower fixed point x_- corresponds to a stable autoregression, while x_+ corresponds to an explosive autoregression. The critical case $|\theta| = 1$ corresponds to $x_+ = x_-$. Specifically, $x_-(\theta,\varphi) = -\frac{1}{2}$ if and only if $|\theta| \le 1$ and $\varphi = -\frac{1}{2}\theta^2$, whereas $x_+(\theta,\varphi) = -\frac{1}{2}$ if and only if $|\theta| \ge 1$ and $\varphi = -\frac{1}{2}\theta^2$. Obviously, $A_i = -\frac{1}{2}$ for all $i \ge 1$ in these cases.

The function $\Psi_t(\theta,\varphi;x_0)$ given by (7.4.5) has the form (7.2.3) for which the family $\{Q^{(t,s)}_{\theta,\varphi} : (\theta,\varphi) \in \tilde{\Gamma}_t\}$ is an exponential family for all $s \le t$. As in earlier sections, $Q^{(t,s)}_{\theta,\varphi}$ denotes the restriction of $Q^{(t)}_{\theta,\varphi}$ to $\mathcal{F}_s (s \le t)$. By (7.2.2) and (7.4.5) we see that

$$\begin{aligned} \frac{dQ^{(t,s)}_{\theta,\varphi}}{dP^s_0} &= \exp\Bigg[\theta \sum_{i=1}^{s} X_i X_{i-1} \\ &\quad + \varphi \sum_{i=1}^{s} X_{i-1}^2 + (A_{t-s+1}(\theta,\varphi) + \tfrac{1}{2})X_s^2 \\ &\quad - (A_{t+1}(\theta,\varphi) + \tfrac{1}{2})x_0^2 + \tfrac{1}{2} \sum_{i=t-s+1}^{t} \log(-2A_i(\theta,\varphi))\Bigg]. \end{aligned} \tag{7.4.11}$$

Note that $\mathcal{Q}_t = \{Q^{(t)}_{\theta,\varphi} : (\theta,\varphi) \in \tilde{\Gamma}_t\}$ is an exponential family on the filtered space $(\Omega, \mathcal{F}_t, \{\mathcal{F}^{(t)}_s\})$, where $\mathcal{F}^{(t)}_s = \mathcal{F}_s$ for $s \le t$ and $\mathcal{F}^{(t)}_s = \mathcal{F}_t$ for $s > t$. This exponential family is not time-homogeneous, it is curved (except for $s = t$), and it depends on t. In order to find out what class of stochastic processes corresponds to $Q^{(t)}_{\theta,\varphi}$, consider the simultaneous Laplace transform under $Q^{(t)}_{\theta,\varphi}$ of the random variables

$$W_i = X_i + \tfrac{1}{2}\theta A^{-1}_{t-i+1}(\theta,\varphi) X_{i-1}, \;\; i = 1, \cdots, t, \tag{7.4.12}$$

which can be found by direct calculation:

$$\begin{aligned}
&E_{\theta,\varphi}\left\{\exp\left[\sum_{i=1}^{t} s_i(X_i+\tfrac{1}{2}\theta A_{t-i}^{-1}X_{i-1})\right]\right\} \qquad (7.4.13)\\
&= \exp[\tfrac{1}{2}s_1\theta x_0 A_t^{-1}]E_{\theta,\varphi}\left\{\exp\left[s_tX_t+\sum_{i=1}^{t-1}(s_i+\tfrac{1}{2}s_{i+1}\theta A_{t-i}^{-1})X_i\right]\right\}\\
&= \exp[\tfrac{1}{2}s_1\theta x_0 A_t^{-1}]E_0\left\{\exp\left[s_tX_t+\sum_{i=1}^{t-1}(s_i+\tfrac{1}{2}s_{i+1}\theta A_{t-i}^{-1})X_i\right]\frac{dQ_{\theta,\varphi}^{(t)}}{dP_0^t}\right\}\\
&= \exp\left[-\sum_{i=1}^{t} s_i^2/(4A_{t-i+1})\right]\prod_{i=1}^{t}\sqrt{-A_i/\pi}\\
&\quad\times\int_{-\infty}^{\infty}\cdots\int_{-\infty}^{\infty}\exp\left[\sum_{i=1}^{t}A_{t-i+1}\{x_i+\tfrac{1}{2}(s_i+\theta x_{i-1})A_{t-i+1}^{-1}\}^2\right]dx_t\cdots dx_1\\
&= \prod_{i=1}^{t}\exp[\tfrac{1}{2}(2A_{t-i+1})^{-1}s_i^2].
\end{aligned}$$

We see that the random variables W_i, $i=1,\cdots,t$, are independent and that

$$W_i \sim N(0, -\tfrac{1}{2}A_{t-i+1}^{-1}),\quad i=1,\cdots,t.$$

This means that under $Q_{\theta,\varphi}^{(t)}$ the process $\{X_i : i=1,\cdots,t\}$ is the autoregression

$$X_i = -\tfrac{1}{2}\theta A_{t-i+1}^{-1}(\theta,\varphi)X_{i-1}+W_i, \qquad (7.4.14)$$

where the regression parameter $-\frac{1}{2}\theta A_{t-i+1}^{-1}(\theta,\varphi)$, as well as the variance of W_i, depends on time i. This result could also have been derived via (7.2.16).

We can easily extend this process to times after t by defining $A_{-i} = -\frac{1}{2}$, $i=0,1,2,\cdots$; i.e., after time t the process behaves like our original autoregression. This defines a family $\{P_{\theta,\varphi} : (\theta,\varphi)\in\tilde{\Gamma}_t\}$ of probability measures on $\mathcal{F}$. The likelihood function for $s\le t$ is given by (7.4.11) and for $s\ge t$ by

$$\frac{dP_{\theta,\varphi}^s}{dP_0^s} = \exp\left[\theta\sum_{i=1}^{s}X_iX_{i-1}+\varphi\sum_{i=1}^{t}X_{i-1}^2-\tfrac{1}{2}\theta^2\sum_{i=t+1}^{s}X_{i-1}^2-\psi_t(\theta,\psi;x_0)\right]$$

with $(\theta,\varphi)\in\tilde{\Gamma}_t$. This is a time-inhomogeneous exponential family on $(\Omega,\mathcal{F},\{\mathcal{F}_s\})$ that depends on t, the time of our original envelope family (7.4.3).

Consider the situation where t is large. From time t onwards the process behaves like the original autoregression, but when $t-i$ is large, we have from (7.4.8) that (except for $\varphi=-\frac{1}{2}\theta^2$ with $|\theta|>1$) the regression parameter is close $to - \frac{1}{2}\theta x_-(\theta,\varphi)^{-1}$ and the variance of W_i close to $-\frac{1}{2}x_-(\theta,\varphi)^{-1}$. Moreover,

$$x_-(\theta, -\frac{1}{2}\theta^2) = \begin{cases} -\frac{1}{2} & \text{for } |\theta| < 1 \\ -\frac{1}{2}\theta^2 & \text{for } |\theta| \geq 1. \end{cases}$$

This is not surprising since the fixed point x_- corresponds to the stable and critical case $|\theta| \leq 1$. However, it shows that there is an intriguing discontinuity for the explosive case. In the stable and critical case, $|\theta| \leq 1$, the process (7.4.14) resembles the original autoregression when φ is close to $-\frac{1}{2}\theta^2$. This is not so when $|\theta| > 1$. Here the process (7.4.14) has regression coefficient close to θ^{-1} and variance near θ^{-2} when φ approaches $-\frac{1}{2}\theta^2$. This process is not exploding, and for large values of θ it has weak correlations and small variances. These properties are very different from those of the original process (7.4.1) that is obtained when φ equals $-\frac{1}{2}\theta^2$ exactly.

If we restrict the parameter set to $\Gamma_t^* = \tilde{\Gamma}_t \backslash \{(\theta,\varphi) : |\theta| \leq 1, \varphi > -\frac{1}{2}\theta^2\}$, the process (7.4.14) can be extended beyond t in a more natural way. This is done by defining iteratively

$$A_{-i} = \frac{\theta^2}{4(\varphi - \frac{1}{2} - A_{-i+1}}), \quad i = 0, 1, 2, \cdots,$$

such that the A_i's are related by (7.4.4) also for non-positive indices. Considerations like those for $i \geq 1$ show that for $(\theta,\varphi) \in \Gamma_t^*$ we have $x_-(\theta,\varphi) \leq -\frac{1}{2}$ and $A_{-i}(\theta,\varphi) < 0$, $i = 0, 1, 2, \ldots$. If moreover $x_-(\theta,\varphi) < -\frac{1}{2}$, then

$$A_{-i}(\theta,\varphi) \to x_+(\theta,\varphi) \text{ as } \quad i \to \infty.$$

For $(\theta,\varphi) \in \tilde{\Gamma}_t \backslash \Gamma_t^*$ there exists an $i \geq 0$ such that $A_{-i}(\theta,\varphi) > 0$. The likelihood function for the second extended process is given by (7.4.11) for all $s \in \mathbb{N}$. When the time is much larger than t, the process has a nearly constant regression coefficient and disturbance-term variance. For φ near $-\frac{1}{2}\theta^2$ this asymptotic process is, in the explosive and critical case $|\theta| \geq 1$, very similar to the original autoregression obtained for $\varphi = -\frac{1}{2}\theta^2$. Conversely, in the stable case ($|\theta| < 1$), a process with $-\frac{1}{2}\theta^2 - \varphi$ close to but not equal to zero and the process where $\varphi = -\frac{1}{2}\theta^2$ are fundamentally different for s much larger than t (they are, respectively, unstable with large fluctuations and stable).

7.5 The pure birth process

The pure birth processes are counting processes with intensity λX_{t-}, where $\lambda > 0$. We assume that $X_0 = x_0$ is given. The likelihood function is

$$\frac{dP_\theta^t}{dP_0^t} = \exp\left[\theta \int_0^t X_s ds + \log(1-\theta)(X_t - x_0)\right], \tag{7.5.1}$$

where $\theta = 1 - \lambda < 1$. This is a curved exponential family, for which we can determine the envelope families by means of Proposition 7.3.1. To do this, we use that $S_t = X_t - x_0$ follows a negative binomial distribution with point probabilities

$$P_\theta(S_t = x) = \binom{x_0 + x - 1}{x} e^{(\theta-1)x_0 t}(1 - e^{(\theta-1)t})^x. \tag{7.5.2}$$

The Laplace transform is

$$E_\theta(e^{zS_t}) = \left[e^{(1-\theta)t} - e^z(e^{(1-\theta)t} - 1)\right]^{-x_0} \tag{7.5.3}$$

with domain $z < -\log[1 - \exp((\theta - 1)t)]$. From this it follows that the envelope family on $\mathcal{F}_t$ is given by

$$\frac{dQ_{\theta,\varphi}^{(t)}}{dP_0^t} = \exp\left[\theta \int_0^t X_s ds + \varphi(X_t - x_0) - x_0\beta_t(\theta,\varphi)\right], \tag{7.5.4}$$

where

$$\beta_t(\theta,\varphi) = -\log[e^{(1-\theta)t} - e^{\varphi}(1-\theta)^{-1}(e^{(1-\theta)t} - 1)] \tag{7.5.5}$$

and

$$\varphi < \log\left\{\frac{1-\theta}{1 - \exp((\theta-1)t)}\right\}. \tag{7.5.6}$$

Here we have used that $\beta_t(\theta,\varphi)$ is also defined for $\theta \geq 1$, provided that (7.5.6) holds. Note that the canonical parameter set of the class of linear birth processes $\Gamma = \{(\theta, \log(1-\theta)) : \theta < 1\}$ is contained in $\tilde{\Gamma}_t$, which is open for all $t > 0$. Note also that

$$\tilde{\Gamma} = \cap_{t \geq 0} \tilde{\Gamma}_t = \operatorname{conv}\Gamma,$$

where $\operatorname{conv}\Gamma$ denotes the convex hull of Γ.

Since the cumulant transform appearing in (7.5.4) is a product of x_0 and a function of the parameters θ and φ, the family $\{Q_{\theta,\varphi}^{(t,s)} : (\theta,\varphi) \in \tilde{\Gamma}_t\}$ obtained by restriction to $\mathcal{F}_s$ ($s < t$) is, as noted in Section 7.2, an exponential family. By (7.2.2) and (7.5.5) we see that

$$\frac{dQ_{\theta,\varphi}^{(t,s)}}{dP_0^s} = \exp\left(\theta \int_0^s X_u du + h_{t-s}(\theta,\varphi)X_s - x_0 h_t(\theta,\varphi)\right) \tag{7.5.7}$$

with

$$h_u(\theta,\varphi) = \varphi + \beta_u(\theta,\varphi). \tag{7.5.8}$$

We see that the exponential family of processes on $(\Omega, \mathcal{F}_t, (\mathcal{F}_{s\wedge t}))$ defined by $\{Q_{\theta,\varphi}^{(t)} : (\theta,\varphi) \in \tilde{\Gamma}_t\}$ is not time-homogeneous. In Example 7.2.2 we saw that under $Q^{(t)}_{\theta,\varphi}$ the process $\{X_s : s \leq t\}$ is a counting process with intensity

$$\lambda_s^{(t)}(\theta,\varphi) = \exp[h_{t-s}(\theta,\varphi)]X_{s-}. \tag{7.5.9}$$

For every $(\theta,\varphi) \in \tilde{\Gamma}$ the function $\beta_u(\theta,\varphi)$ is defined not only for $u \in [0,t]$, but also for $u < 0$. The function $\exp[h_{t-s}(\theta,\varphi)]$ is thus defined for all $s > 0$ and remains bounded for $s \to \infty$ for all $(\theta,\varphi) \in \tilde{\Gamma}_t$. Hence (7.5.9) defines a non-exploding counting process for all $s > 0$. For $\theta < 1$ the function $\exp[h_{t-s}(\theta,\varphi)]$ tends to $(1-\theta)$ monotonically as $s \to \infty$; for $\theta = 1$ it tends to zero as s^{-1}; and for $\theta > 1$ it tends exponentially to zero as $s \to \infty$. Thus a counting process with intensity (7.5.9) behaves almost like the counting process under P_θ for $\theta < 1$, whereas this is not the case for $\theta \geq 1$.

The fact that (7.5.9) defines a counting process implies that for each $(\theta,\varphi) \in \tilde{\Gamma}_t$ there exists a measure $P^{(t)}_{\theta,\varphi}$ on $\mathcal{F}$ the restriction of which to $\mathcal{F}_s$ is given by (7.5.7) for all $s > 0$. For $(\theta,\varphi) \in \tilde{\Gamma}_t$ it holds that $(\theta, h_{t-s}(\theta,\varphi)) \in \tilde{\Gamma}_s$ for all $s \geq 0$. Thus for all $s > 0$ the measure $Q^{(t,s)}_{\theta,\varphi}$ belongs to the exponential family $\{Q^{(s)}_{\theta,\varphi} : (\theta,\varphi) \in \tilde{\Gamma}_t\}$. The curve $(\theta, h_{t-s}(\theta,\varphi))$ tends monotonically to $(\theta, \log(1-\theta))$ for $\theta < 1$, i.e., to the curve Γ corresponding to the original counting process model. For $\theta \geq 1$ the function $h_{t-s}(\theta,\varphi)$ decreases *to* $-\infty$ for $s \to \infty$. For $s = t$ the curve passes through the point (θ,φ) for all $\theta \in \mathbb{R}$. Note that the exponential family $\{P^{(t)}_{\theta,\varphi} : (\theta,\varphi) \in \tilde{\Gamma}_t\}$ has a canonical parameter space with a non-empty interior, i.e., it is not a curved exponential family. Note also that a counting process with intensity (7.5.9) is well-defined for $t = 0$. In that case the parameter set is $(\theta,\varphi) \in \mathbb{R}^2$, and the likelihood function is given by (4.2.11). This was the model discussed in Example 4.2.5.

7.6 The Ornstein-Uhlenbeck process

Consider the class of solutions to the stochastic differential equations

$$dX_t = \theta X_t dt + dW_t, \quad X_0 = x_0, \tag{7.6.1}$$

for $\theta \in \mathbb{R}$. The likelihood function corresponding to observation of X in $[0,t]$ is

$$L_t(\theta) = \exp\left(\theta[X_t^2 - x_0^2]/2 - \tfrac{1}{2}\theta^2 \int_0^t X_s^2 ds - \tfrac{1}{2}\theta t\right). \tag{7.6.2}$$

The envelope families can be determined by Proposition 7.3.2 and are given by

$$\frac{dQ^{(t)}_{\theta,\varphi}}{dP_0^t} = \exp\left(\theta[X_t^2 - x_0^2]/2 + \varphi \int_0^t X_s^2 ds - \Psi_t(\theta,\varphi;x_0)\right) \tag{7.6.3}$$

with

$$\Psi_t(\theta,\varphi;x_0) = -\tfrac{1}{2}\log\{\cosh(t\sqrt{-2\varphi}) - \theta\sinh(t\sqrt{-2\varphi})/\sqrt{-2\varphi}\} + \frac{x_0^2(\frac{1}{2}\theta^2+\varphi)}{\sqrt{-2\varphi}\coth(t\sqrt{-2\varphi})-\theta} \qquad (7.6.4)$$

and with parameter space $\tilde{\Gamma}_t$ given by

$$\varphi < \tfrac{1}{2}\pi^2 t^{-2}, \;\; \theta < \sqrt{-2\varphi}\coth(t\sqrt{-2\varphi}); \qquad (7.6.5)$$

for details see Sørensen (1998).

The function $\Psi_t(\theta,\varphi;x_0)$ has the form (7.2.3), for which the family $\{Q_{\theta,\varphi}^{(t,s)} : (\theta,\varphi)\in\tilde{\Gamma}_t\}$ is an exponential family for all $s \le t$. By (7.2.2) and (7.6.4) we see that

$$\frac{dQ_{\theta,\varphi}^{(t,s)}}{dP_0^s} = \exp\left(h(\theta,\varphi;t-s)X_s^2 + \varphi\int_0^s X_u^2 du + m(\theta,\varphi;t,s) - h(\theta,\varphi;t)x_0^2\right), \qquad (7.6.6)$$

where

$$h(\theta,\varphi;u) = \tfrac{1}{2}\theta + \frac{\frac{1}{2}\theta^2+\varphi}{\sqrt{-2\varphi}\coth(u\sqrt{-2\varphi})-\theta} \qquad (7.6.7)$$

and

$$m(\theta,\varphi;t,s) = \tfrac{1}{2}\log\left(\frac{\cosh(t\sqrt{-2\varphi}) - \theta\sinh(t\sqrt{-2\varphi})/\sqrt{-2\varphi}}{\cosh((t-s)\sqrt{-2\varphi}) - \theta\sinh((t-s)\sqrt{-2\varphi})/\sqrt{-2\varphi}}\right). \qquad (7.6.8)$$

We see that $\{Q_{\theta,\varphi}^{(t)} : (\theta,\varphi)\in\tilde{\Gamma}_t\}$ defines an exponential family of stochastic processes that is not time-homogeneous.

By results in Example 7.2.1 we see that under $Q_{\theta,\varphi}^{(t)}$ the process X solves the stochastic differential equation

$$dX_s = 2h(\theta,\varphi,t-s)X_s ds + dW_s, \;\; X_0 = x_0, \;\; s \le t. \qquad (7.6.9)$$

For $\varphi \le 0$ and $\theta \ge -\sqrt{-2\varphi}$ the function h is well-defined and bounded for $u < 0$, so (7.6.9) has a solution for all $s \ge 0$. This is not the case if $\varphi \le 0$ and $\theta < -\sqrt{-2\varphi}$, or if $\varphi > 0$. In these cases the drift tends to infinity (or minus infinity) at a finite non-random time larger than t. The model (3.5.10) considered in Section 3.5 is the one obtained for $t = 1$ restricted to $\varphi \le 0$ and $\theta > -\sqrt{-2\varphi}$ and with the reparametrization $(\theta_1,\theta_2) = (\theta,\sqrt{-2\varphi})$.

7.7 A goodness-of-fit test

An obvious test of the appropriateness of a stochastic process model, which for observation in $[0, t]$ is a curved exponential family, is the likelihood ratio test of the curved model against the full envelope family on $\mathcal{F}_t$. For an interpretation of this test and an evaluation of its relevance, the results in this chapter are useful. For the Ornstein-Uhlenbeck process, for instance, the drift under the alternative model is not strictly proportional to the state of the process, but a certain temporal variation of the constant of proportionality is allowed. Similar remarks hold for the Gaussian autoregression and the pure birth process. The following simple, but interesting, example illustrates the main ideas.

Example 7.7.1 *Censored data.* Consider the following well-known model for censored observations of independent random variables with hazard function $(1-\theta)h$, $\theta < 1$, defined on $(0, \infty)$. We suppose that $h > 0$ and that it is integrable on $(0, t)$ for all $t > 0$. Let U and V be two independent random variables concentrated on $(0, \infty)$ such that the hazard function of U is $(1-\theta)h$, and denote the cumulative distribution function of V by G. Define two counting processes N and M by

$$N_t = 1_{\{U \leq t \wedge V\}} \text{ and } M_t = 1_{\{V \leq t \wedge U\}}. \tag{7.7.1}$$

The intensity of N with respect to the filtration generated by N and M is

$$\lambda_t(\theta) = (1-\theta)h(t)1_{\{t \leq V\}}1_{\{N_{t-}=0\}}, \quad \theta < 1. \tag{7.7.2}$$

Observation of N and M in the time interval $[0, t]$ is equivalent to observation of a random variable U with hazard function $(1-\theta)h$ censored at time $V \vee t$. The process N makes at most one jump. This happens at the time U if $U \leq V \wedge t$; otherwise, N does not jump. If $N_t = 0$, we know that U has been censored ($U > V \wedge t$), and from the sample path of M we know the time of censoring, $V \wedge t$.

The likelihood function for the model based on observation of N and M in $[0, t]$ is given by

$$\frac{dP_\theta^t}{dP_0^t} = \exp\left(\theta \int_0^{t \wedge V \wedge U} h(s)ds + \log(1-\theta)N_t\right). \tag{7.7.3}$$

We see that the model is a curved exponential family of stochastic processes. A possible test that the hazard function of the observed random variable belongs to the class $(1-\theta)h$, $\theta < 1$, is the likelihood ratio test for the curved family (7.7.3) against the envelope family on $\mathcal{F}_t$. In the following we obtain an interpretation of the alternative hypothesis by considering the envelope family as a stochastic process model.

In order to find the envelope family on $\mathcal{F}_t$ by Proposition 7.3.1, we need the Laplace transform of N_t under P_θ. It is easy to see that

$$E_\theta(e^{wN_t}) = \beta_t(\theta) + e^w(1 - \beta_t(\theta)),$$

where

$$\beta_t(\theta) = P_\theta(N_t = 0) = \int_0^\infty (1 - F_\theta(v \wedge t))dG(v)$$

is the probability of obtaining a censored observation, and

$$F_\theta(x) = 1 - \exp\{-(1-\theta)\int_0^x h(v)dv\}$$

is the cumulative distribution function corresponding to the hazard function $(1-\theta)h$. By Proposition 7.3.1 the envelope family is given by

$$\frac{dQ^{(t)}_{\theta,\varphi}}{dP^t_0} = \exp\left(\theta\int_0^{t\wedge V\wedge U} h(s)ds + \varphi N_t - \Psi_t(\theta,\varphi)\right), \tag{7.7.4}$$

with $\varphi \in \mathbb{R}$ and

$$\Psi_t(\theta,\varphi) = \log[\beta_t(\theta) + (1-\theta)^{-1}e^{\varphi}(1-\beta_t(\theta))].$$

The process N is not a Markov process (except for $h \equiv 1$), so the conclusions in Example 7.2.2 do not apply directly; but we can proceed in a very similar way. Under $Q^{(t)}_{\theta,\varphi}$ the process $(N_s : s \leq t)$ is a counting process that makes at most one jump. To find the intensity of this process, note that by (7.1.5) and (7.1.8) the restriction $Q^{(t,s)}_{\theta,\varphi}$ of $Q^{(t)}_{\theta,\varphi}$ to $\mathcal{F}_s$ $(s \leq t)$ is given by

$$\frac{dQ^{(t,s)}_{\theta,\varphi}}{dP^s_0} = \exp\left(\theta\int_0^{s\wedge V\wedge U} h(u)du + \varphi N_s - \Psi_t(\theta,\varphi) + C^{(t)}_s(\theta,\varphi)\right) \tag{7.7.5}$$

with

$$C^{(t)}_s(\theta,\varphi) = \begin{cases} 0 & \text{if } M_s = 1 \text{ or } N_s = 1 \\ \log\left\{\gamma^{(t)}_s(\theta) + \frac{e^{\varphi}(1-\gamma^{(t)}_s(\theta))}{1-\theta}\right\} & \text{if } M_s = N_s = 0 \end{cases} \tag{7.7.6}$$

and

$$\begin{aligned} \gamma^{(t)}_s(\theta) &= P_\theta(N_t - N_s = 0 | U > s, V > s) \\ &= \frac{\int_s^\infty (1 - F_\theta(v\wedge t))dG(v)}{(1-F_\theta(s))(1-G(s))}. \end{aligned} \tag{7.7.7}$$

From (7.7.5), (7.7.6), (7.7.7), and Lemma B.2.1 it follows that for $u < s \le t$

$$\begin{aligned}
&\lim_{s \downarrow u}(s-u)^{-1}Q^{(t,s)}_{\theta,\varphi}(N_s = 1 | M_u = N_u = 0) \\
&\quad = \lim_{s \downarrow u}(s-u)^{-1}E_0\Bigg[1_{\{1\}}(N_s) \\
&\qquad\qquad \times \exp\left\{\theta \int_{u\wedge V\wedge U}^{s\wedge V\wedge U} h(v)dv + \varphi - C^{(t)}_u(\theta,\varphi)\right\}\Bigg| M_u = N_u = 0\Bigg] \\
&\quad = \{\gamma^{(t)}_u(\theta)(e^{-\varphi} - (1-\theta)^{-1}) + (1-\theta)^{-1}\}^{-1} h(u) 1_{\{u \le V\}} 1_{\{N_{u-}=0\}}.
\end{aligned}$$

This is the intensity of $(N_s : s \le t)$ with respect to the filtration $\{\mathcal{F}_s\}$ generated by N and M under $Q^{(t)}_{\theta,\varphi}$, so under this measure, observation of N and M in $[0,t]$ is equivalent to censored observation of a random variable with hazard function

$$\{\gamma^{(t)}_s(\theta)(e^{-\varphi} - (1-\theta)^{-1}) + (1-\theta)^{-1}\}^{-1} h(s), \quad s \le t. \tag{7.7.8}$$

Consider the situation where the censoring distribution G is concentrated on $[t,\infty)$ (type 1 censoring). Then $\gamma^{(t)}_s(\theta)$ is given by

$$\gamma^{(t)}_s(\theta) = \frac{1 - F_\theta(t)}{1 - F_\theta(s)},$$

which is an increasing function of s. The factor modifying h in (7.7.8) is in this situation increasing or decreasing depending on whether $\varphi > \log(1-\theta)$ or $\varphi < \log(1-\theta)$.

As another example, suppose $h \equiv 1$ and $G(x) = 1 - e^{-\mu x}$. Then

$$\gamma^{(t)}_s(\theta) = \frac{\mu}{1-\theta+\mu} + \frac{1-\theta}{1-\theta+\mu} e^{-(1-\theta+\mu)s}.$$

Here $\gamma^{(t)}_s(\theta)$ is a decreasing function of s, and the hazard function given by (7.7.8) is monotonically increasing or decreasing depending on whether $\varphi < \log(1-\theta)$ or $\varphi > \log(1-\theta)$. □

7.8 Exercises

7.1 Show by direct calculations that the model given by (7.4.14) has the likelihood function (7.4.11).

7.2 Derive (7.6.4) and (7.6.5) by means of Proposition 7.3.2. Hints: First, show that $E(\exp[wZ^2]) = (1 - 2\sigma^2 w)^{-\frac{1}{2}} \exp(w\mu^2/(1-2\sigma^2 w))$ when $Z \sim N(\mu, \sigma^2)$. Then consider $\Theta^* = (0,\infty)$, and obtain the whole envelope family by analytic continuation. It is helpful to notice that $\sqrt{-x}\coth(\sqrt{-x}) = \sqrt{x}\cot(\sqrt{x})$ for $0 < x < \pi$.

7.3 Show that the model given by (7.6.9) has the likelihood function (7.6.6) by means of the general expression (A.10.14) for the likelihood function of diffusion models and Ito's formula.

7.4 Consider the class of linear birth-and-death processes introduced in Section 3.4. The likelihood function is given by (3.4.7). Puri (1966) has shown that

$$E_{\lambda,\mu}(e^{uS_t}) =$$
$$\left(r_+(u) + \frac{r_-(u) - r_+(u)}{1-(1-r_+(u))^{-1}}(1-r_-(u))\exp[-\lambda(r_+(u)-r_-(u))t]\right)^{x_0},$$

where x_0 is the initial population size,

$$r_\pm(u) = (2\lambda)^{-1}\{\lambda+\mu-u \pm \sqrt{(\lambda+\mu-u)^2 - 4\lambda\mu}\},$$

and $u \leq \lambda + \mu - 2\sqrt{\lambda\mu}$.

Use this result to verify that for the linear birth-and-death process, $\gamma(\theta) \in \operatorname{int}\tilde{\Gamma}_t$ for all $t > 0$ provided $\mu \neq \lambda$.

Let, as usual, $\{Q^t_{\theta,\varphi} : (\theta,\varphi) \in \tilde{\Gamma}_t\}$ denote the envelope family on $\mathcal{F}_t$ and $Q^{(t,s)}_{\theta,\varphi} = Q^t_{\theta,\varphi}|_{\mathcal{F}_s}$, $s \leq t$. Show that $\{Q^{(t,s)}_{\theta,\varphi} : (\theta,\varphi) \in \tilde{\Gamma}_t\}$ is an exponential family for all $s \leq t$. Hint: use (7.2.2).

Show that under $Q^t_{\theta,\varphi}$ the process $\{X_s : s \leq t\}$ is a birth-and-death process, and find the birth intensity and the death intensity, which are neither time-homogeneous nor linear.

7.9 Bibliographic notes

The envelope families were introduced by Küchler and Sørensen (1996a), where most of the material of this chapter was first presented. The envelope family for the class of multivariate Gaussian diffusions was studied by Stockmarr (1996). The idea of exploiting the expression for the conditional Laplace transform of A_t given S_t, which is used in the proof of Proposition 7.3.1, was first used by Jensen (1987) to obtain conditional expansions. The cumulant transform in (7.4.3) was first studied by White (1958). The simultaneous cumulant transform of $X_t - x_0$ and $\int_0^t X_s ds$ appearing in (7.5.4) was first given by Puri (1966); see also Keiding (1974). The test for the distribution of censored data investigated in Example 7.7.1 was first proposed in the particular case $h \equiv 1$ (a censored exponential distribution) by Væth (1980), who later gave an interpretation of the alternative hypothesis by means of a biased sampling scheme (Væth, 1982).

8

Likelihood Theory

In this chapter we shall consider likelihood theory for general exponential families of stochastic processes. We have already in Chapter 2 given results for natural exponential families of Lévy processes, for which the situation is not essentially different from that of classical repeated sampling from an exponential family. In Chapter 5 we considered likelihood theory for the rather general type of exponential families obtained by random time transformations from natural exponential families of Lévy processes. We shall return to the subject of likelihood theory again in Chapter 11, where we will exploit the special structure of semimartingale models to go into more depth than is possible in this chapter.

8.1 Likelihood martingales

Likelihood martingales are martingales obtained from the likelihood function process by the usual statistical operations, e.g., taking logarithms or differentiating with respect to θ. We consider an exponential family on a filtered space with the general exponential representation given by

$$L_t(\theta) = \frac{dP_\theta^t}{dP_0^t} = \exp[\gamma_t(\theta)^T B_t - \phi_t(\theta)], \quad \theta \in \Theta \subseteq \mathbb{R}^k, \tag{8.1.1}$$

where $\gamma_t(0) = \phi_t(0) = 0$ for all $t \geq 0$. As usual, we assume that γ_t, B_t, and ϕ_t, as functions of t, are right-continuous with limits from the left. We shall also need the envelope exponential families (7.1.2) in this chapter, in particular their full canonical parameter sets $\tilde{\Gamma}_t$ defined by (7.1.1). It is not always possible to find explicit expressions for the cumulant transform $\Psi_t(\gamma)$ given by (7.1.3), and hence not for the exponential representation (7.1.2) of the envelope family. Therefore, we shall also give results that only involve γ_t and ϕ_t, which are more widely applicable. In the following a dot always denotes differentiation with respect to θ. In particular (provided that the derivatives exist), $\dot{\gamma}_t(\theta)_i$ and $\ddot{\gamma}_t(\theta)_{ij}$, $i, j = 1, \cdots, k$, are the vector of the partial derivatives of the coordinates of $\gamma_t(\theta)$ with respect to θ_i and the vector of second partial derivatives with respect to θ_i and θ_j, respectively. By $\dot{\gamma}_t(\theta)$ we denote the matrix where the ith column is the vector $\dot{\gamma}_t(\theta)_i$, while $\dot{\phi}_t(\theta)$ will denote the vector of partial derivatives of $\phi_t(\theta)$. In this section we shall need cumulants of multivariate distributions and their relation to the moments. For a discussion of this subject see Barndorff-Nielsen and Cox (1989, Chapter 5.5).

Lemma 8.1.1 *Suppose $\gamma_t(\theta) \in \operatorname{int} \tilde{\Gamma}_t$. Then B_t has moments of all orders under P_θ. The function $\Psi_t(\gamma)$ given by (7.1.3) is infinitely often differentiable with respect to γ at $\gamma_t(\theta)$, and the cumulant of B_t of order $(i_1, \cdots, i_m)$ under P_θ is for all vectors $(i_1, \cdots, i_m)$ with $i_j \in \{0, 1, 2, \cdots\}$ given by*

$$\kappa^{i_1, \cdots, i_m}(\theta) = \frac{\partial^n}{\partial \gamma_1^{i_1} \cdots \partial \gamma_m^{i_m}} \Psi_t(\gamma)|_{\gamma=\gamma_t(\theta)}, \tag{8.1.2}$$

where $n = i_1 + \cdots + i_m$. In particular,

$$E_\theta(B_t) = \dot{\Psi}_t(\gamma_t(\theta)) \tag{8.1.3}$$

and

$$V_\theta(B_t) = \ddot{\Psi}_t(\gamma_t(\theta)). \tag{8.1.4}$$

If, moreover, $\gamma_t(\theta)$ is twice differentiable with respect to θ, we have

$$\dot{\gamma}_t(\theta)^T E_\theta(B_t) = \dot{\phi}_t(\theta), \tag{8.1.5}$$

$$\dot{\gamma}_t(\theta)_i^T V_\theta(B_t) \dot{\gamma}_t(\theta)_j + \ddot{\gamma}_t(\theta)_{ij}^T E_\theta(B_t) = \ddot{\phi}_t(\theta)_{ij}, \tag{8.1.6}$$

for $i, j = 1, \cdots, k$.

Proof: The results about the existence of moments and (8.1.2)–(8.1.4) follow from classical exponential family theory (Barndorff-Nielsen, 1978, p. 114). From (8.1.3)–(8.1.4) the equations (8.1.5)–(8.1.6) follow by noting that $\phi_t(\theta) = \Psi_t(\gamma_t(\theta))$. □

The likelihood function corresponding to observation of events in $\mathcal{F}_t$ is $L_t(\theta)$ given by (8.1.1). Let $l_t(\theta) = \log(L_t(\theta))$ be the log-likelihood function, and let $\dot{l}_t(\theta)$ denote the *score function*, i.e., the vector of partial derivatives of $l_t(\theta)$. We shall denote by $L_t(\theta)_{/i_1, \cdots, i_n}$ the nth partial derivative of $L_t(\theta)$ with respect to $\theta_{i_1}, \cdots, \theta_{i_n}$, i.e.,

$$L_t(\theta)_{/i_1, \cdots, i_n} = \frac{\partial^n}{\partial \theta_{i_1} \cdots \partial \theta_{i_n}} L_t(\theta).$$

Similarly for other functions of θ. Some of the indices $i_1, \cdots, i_n$ might coincide.

Theorem 8.1.2 *Suppose that $\gamma_t(\theta)$ for all $t \geq 0$ is n times continuously differentiable with respect to θ and that $\gamma_t(\theta) \in \operatorname{int} \tilde{\Gamma}_t$ for all $t \geq 0$. Then $L_t(\theta)$ is n times continuously differentiable with respect to θ, and for all $i_1, \cdots, i_n \in \{1, 2, \cdots, k\}$ the process*

$$L_t(\theta)_{/i_1,\cdots,i_n} \tag{8.1.7}$$

is a martingale with respect to P_0.

Proof: Under the conditions imposed, it follows by Lemma 8.1.1 that $\phi_t(\theta)$ is n times continuously differentiable at θ. Let $A \in \mathcal{F}_s$. Then for all $t \geq s$

$$\int_A L_t(\theta) dP_0 = \int_A L_s(\theta) dP_0. \tag{8.1.8}$$

If differentiation with respect to θ and integration can be interchanged, it follows from (8.1.8) that

$$\int_A L_t(\theta)_{/i_1,\cdots,i_n} dP_0 = \int_A L_s(\theta)_{/i_1,\cdots,i_n} dP_0, \tag{8.1.9}$$

which again implies that (8.1.7) is a martingale. The interchange of differentiation and integration is allowed because (8.1.7) is locally dominated integrable. This follows by arguments that are well-known from classical exponential family theory, because for all t the measure P_θ^t belongs to the interior of the envelope exponential family. For details, see Barndorff-Nielsen (1978, pp. 105–106) or Brown (1986, pp. 32–34). □

The following corollary follows immediately from Theorem 8.1.2 and Lemma 8.1.1.

Corollary 8.1.3 *Under the conditions of Theorem 8.1.2 the process*

$$\frac{L_t(\theta)_{/i_1,\cdots,i_n}}{L_t(\theta)} \tag{8.1.10}$$

is a square integrable martingale with mean zero under P_θ. In particular ($n = 1, 2, 3$), the processes

$$\dot{l}_t(\theta)_i = L_t(\theta)_{/i}/L_t(\theta) = \dot{\gamma}_t(\theta)_i^T B_t - \dot{\phi}_t(\theta)_i, \tag{8.1.11}$$

$$L_t(\theta)_{/ij}/L_t(\theta) = \dot{l}_t(\theta)_i \dot{l}_t(\theta)_j + \ddot{\gamma}_t(\theta)_{ij}^T B_t - \ddot{\phi}_t(\theta)_{ij}, \tag{8.1.12}$$

and

$$\begin{aligned} L_t(\theta)_{/ijk}/L_t(\theta) \quad &= \quad \dot{l}_t(\theta)_i \dot{l}_t(\theta)_j \dot{l}_t(\theta)_m \\ &\quad + \dot{l}_t(\theta)_i (\ddot{\gamma}_t(\theta)_{jm}^T B_t - \ddot{\phi}_t(\theta)_{jm})[3] \end{aligned} \tag{8.1.13}$$

for $i, j, m = 1, \cdots, k$, are square integrable zero-mean P_θ-martingales. Here a number in a square bracket after a term means the sum over that number of similar terms obtained by permuting the indices. □

Note that Corollary 8.1.3 gives conditions ensuring that the score function $\dot{l}_t(\theta)$ is a P_θ-martingale. The following proposition gives a different kind of conditions that ensure the same conclusion.

Proposition 8.1.4 *Suppose for every $t > 0$ that $L_t(\theta)$ is differentiable at the point $\theta^* \in \operatorname{int}\Theta$ and that the function $\theta_i \mapsto L_t(\theta)$, where $\theta_j = \theta_j^*$ for $j \neq i$ (θ_j is the jth coordinate of θ, $j = 1, \cdots, k$), is either convex or concave for $\theta_i \in (\theta_i^l, \theta_i^*]$ and either convex or concave for $\theta_i \in [\theta_i^*, \theta_i^r)$, where $\theta_i^l < \theta_i^* < \theta_i^r$. Assume, moreover, that the first moments of the coordinates of B_t are finite under P_θ for all $t > 0$ and for all $\theta \in \Theta$. Then $L_t(\theta^*)_{/i}$ is a P_0-martingale, and $\dot{l}_t(\theta^*)_i$ is a P_{θ^*}-martingale.*

Proof: Suppose first that $\theta_i \mapsto L_t(\theta)$ is concave in (θ_i^*, θ_i^r). Let θ_{in} be a sequence in (θ_i^*, θ_i^r) such that $\theta_{in} \downarrow \theta_i^*$. Define

$$D_n = (\theta_{in} - \theta_i^*)^{-1}[L_t(\theta_n) - L_(\theta^*)],$$

where the ith coordinate of θ_n is θ_{in}, while the other coordinates equal those of θ^*. Then

$$L_t(\theta_1)_{/i} \leq D_n \leq L_t(\theta^*),$$

where by the assumption on B, the random variables $L_t(\theta^*)_{/i}$ and $L_t(\theta_1)_{/i}$ are P_0-integrable. We can therefore apply the dominated convergence theorem, and by treating an increasing sequence similarly we find that

$$E_0(1_A L_t(\theta^*)_{/i}) = E_0(1_A L_t(\theta^*))_{/i},$$

for $A \in \mathcal{F}_t$. From this the conclusion of the proposition follows as in the proof of Theorem 8.1.2. The argument is similar when $\theta_i \mapsto L_t(\theta)$ is convex. □

The conditions of Proposition 8.1.4 are usually easy to check and are satisfied for many of the models used in applications. It could, for instance, be applied to the following example. We use instead Corollary 8.1.3 in order to get results about higher-order derivatives of the log-likelihood function too.

Example 8.1.5 For the class of *pure birth processes* considered in Section 7.5 the conditions of Theorem 8.1.2 are satisfied. The three likelihood martingales (8.1.11)–(8.1.13) are

$$m_t^{(1)}(\theta) = X_t - x_0 - (1-\theta)\int_0^t X_s ds \tag{8.1.14}$$

$$m_t^{(2)}(\theta) = m_t^{(1)}(\theta)^2 - (X_t - x_0) \tag{8.1.15}$$

and

$$m_t^{(3)}(\theta) = m_t^{(2)}(\theta)m_t^{(1)}(\theta) - 2(X_t - x_0)(m_t^{(1)}(\theta) - 1). \tag{8.1.16}$$

That (8.1.14) and (8.1.15) are P_θ-martingales is well-known and follows from properties of the intensity of a counting process. □

Let us end this section by giving a more explicit expression for the square integrable zero-mean P_θ-martingales (8.1.10). To do this we use the index notation of Barndorff-Nielsen and Cox (1989, pp. 136–140). Specifically, let $I = i_1, \cdots, i_\nu$ ($i_j \in \{0, 1, 2, \cdots\}$) be an arbitrary index set, and set $|I| = \nu$. Some of the elements $i_1, \cdots, i_\nu$ may have the same index value. For a function $f : \mathbb{R}^m \mapsto \mathbb{R}$ we denote as above the partial derivative of f with respect to $x_{i_1}, \cdots, x_{i_\nu}$ by $f(x)_{/I}$. By

$$\sum_{I/n} f(x)_{/I_1} \cdots f(x)_{/I_n}$$

we denote the sum over all partitions of the index set I into n blocks $I_1, \cdots, I_n$. For instance, if $I = 1\,2\,1$, then all partitions into two blocks are $I_1 = 1$, $I_2 = 2\,1$; $I_1 = 1\,2$, $I_2 = 1$; and $I_1 = 1\,1$, $I_2 = 2$, and

$$\sum_{I/2} f(x)_{/I_1} f(x)_{/I_2} = 2\frac{\partial}{\partial x_1} f(x) \frac{\partial^2}{\partial x_1 \partial x_2} f(x) + \frac{\partial^2}{\partial x_1^2} f(x) \frac{\partial}{\partial x_2} f(x).$$

With this notation we have

Corollary 8.1.6 *Let I be an index set. Suppose that for all $t \geq 0$ the function $\gamma_t(\theta)$ is $|I|$ times continuously differentiable with respect to θ and that $\gamma_t(\theta) \in \operatorname{int} \tilde{\Gamma}_t$ for all $t \geq 0$. Then the process*

$$m_t^I(\theta) = \sum_{i=1}^{|I|} \sum_{I/n} l_t(\theta)_{/I_1} \cdots l_t(\theta)_{/I_n} \tag{8.1.17}$$

is a square integrable zero-mean martingale under P_θ.

Proof: By formula (5.20) in Barndorff-Nielsen and Cox (1989) with $g(y) = \exp(y)$ and $f(\theta) = l_t(\theta)$,

$$\begin{aligned} L_t(\theta)_{/I} &= \sum_{i=1}^{|I|} \exp(l_t(\theta)) \sum_{I/n} l_t(\theta)_{/I_1} \cdots l_t(\theta)_{/I_n} \\ &= m_t^I(\theta) L_t(\theta). \end{aligned}$$

Thus $m_t^I(\theta)$ with $I = i_1, \cdots, i_n$ equals the process given by (8.1.10). □

In proving that $m_t^I(\theta)$ is a martingale we have used the exponential family structure of the statistical model only to ensure that we can interchange differentiation and integration in the proof of Theorem 8.1.2. Therefore, $m_t^I(\theta)$ is a martingale for any statistical model for a class of stochastic processes provided only that this interchange is allowed.

Example 8.1.7 Consider an exponential family of *diffusion-type processes* given by the class of stochastic differential equations (3.5.1) considered in Section 3.5. Here we assume that the parameter θ is one-dimensional. Under the conditions of Section 3.5, the likelihood function is

$$L_t(\theta) = \exp(\theta A_t - \tfrac{1}{2}\theta^2 B_t), \tag{8.1.18}$$

where A_t and B_t are given by (3.5.5) and (3.5.2), respectively. Since θ is one-dimensional, we can, for simplicity, denote $m_t^I(\theta)$ by $m_t^n(\theta)$ for $|I| = n$. Suppose $\theta \in \tilde{\Gamma}_t$ for all θ. Then by (8.1.17) we find

$$m_t^n(\theta) = c_n H_n(A_t - \theta B_t, B_t), \tag{8.1.19}$$

where $H_n(x, y)$ is the Hermite polynomial with generating function

$$\exp(tx - \tfrac{1}{2}t^2 y) = \sum_{n=0}^{\infty} \frac{t^n}{n!} H_n(x, y), \tag{8.1.20}$$

and c_n is a non-random constant. The polynomial $H_n(x, y)$ can also be expressed as

$$H_n(x, y) = \frac{(-y)^n}{n!} \exp\left(\frac{x^2}{2y}\right) \frac{\partial^n}{\partial x^n} \exp\left(-\frac{x^2}{2y}\right). \tag{8.1.21}$$

The first four likelihood martingales are

$$\begin{aligned}
m_t^1(\theta) &= A_t - \theta B_t, \\
m_t^2(\theta) &= (A_t - \theta B_t)^2 - B_t, \\
m_t^3(\theta) &= (A_t - \theta B_t)^3 - 3(A_t - \theta B_t) B_t, \\
m_t^4(\theta) &= (A_t - \theta B_t)^4 - 6(A_t - \theta B_t)^2 B_t + 3B_t^2.
\end{aligned}$$

□

8.2 Existence and uniqueness of the maximum likelihood estimator

As most exponential families of stochastic processes are curved exponential families, the question of existence and uniqueness of the maximum likelihood estimator does not follow immediately from the classical results for full exponential families. We shall first consider an exponential family of processes with the general representation (8.1.1). For such a model we let

$\mathcal{B}_t$ denote the closure of the convex hull of the support of the canonical statistic B_t under P_0 and assume that γ_t and ϕ_t are continuously differentiable with respect to θ. If γ_t is continuously differentiable with respect to θ, this often implies that ϕ_t is so too, because $\phi_t(\theta) = \Psi_t(\gamma_t(\theta))$, where $\Psi_t(\gamma)$ is the cumulant transform given by (7.1.3). We begin with the simplest case. The concept of a steep convex function was discussed in Section 2.2.

Theorem 8.2.1 *Assume, for fixed t, that the representation (8.1.1) is minimal, that γ_t is injective, and that $\{\gamma_t(\theta) : \theta \in \Theta\} = \tilde{\Gamma}_t$, where $\tilde{\Gamma}_t$ is given by (7.1.1). Suppose, moreover, that the cumulant transform of B_t under P_0, i.e., $x \mapsto \phi_t(\gamma_t^{-1}(x))$, is steep. Then the maximum likelihood estimator exists and is unique if and only if $B_t \in \operatorname{int} \mathcal{B}_t$. The maximum likelihood estimator is the solution of the equation*

$$(\dot{\gamma}_t(\theta)^T)^{-1}\dot{\phi}_t(\theta) = B_t. \tag{8.2.1}$$

If $B_t \in \operatorname{bd} \mathcal{B}_t$, the likelihood funtion has no maximum.

Proof: The theorem follows directly from the classical exponential family results; see Barndorff-Nielsen (1978, Chapter 9). That the likelihood equation has the form (8.2.1) follows from (8.1.5) or by using that the cumulant transform of B_t under P_0 is $x \mapsto \phi_t(\gamma_t^{-1}(x))$. □

Theorem 8.2.1 covers full exponential families of Lévy processes, which are non-curved, and envelope families. For the envelope family $\{Q_{\theta,\varphi}^{(t)} : (\theta, \varphi) \in \tilde{\Gamma}_t\}$ at time t (fixed) for the pure birth process (Section 7.5), the conditions of the theorem are satisfied at all times $s \in \mathbb{R}$. In particular, the parameter function $\gamma_s(\theta, \varphi)$ occurring in the representation of the measures $\{Q_{\theta,\varphi}^{(t,s)} : (\theta, \varphi) \in \tilde{\Gamma}_t\}$ satisfies $\{\gamma_s(\theta, \varphi) : (\theta, \varphi) \in \tilde{\Gamma}_t\} = \tilde{\Gamma}_s$. For the envelope families considered in Section 7.4 and in Section 7.6, the conditions of Theorem 8.2.1 are satisfied only at time $s = t$. For the autoregressive process, $\gamma_s(\theta, \varphi)$ is three-dimensional for $s \neq t$, whereas $\tilde{\Gamma}_s$ is two-dimensional. For the Ornstein-Uhlenbeck process, $\{\gamma_s(\theta, \varphi) : (\theta, \varphi) \in \tilde{\Gamma}_t\}$ is a proper subset of $\tilde{\Gamma}_s$ for $s \neq t$.

For many models, $P_\theta(B_t \in \operatorname{bd} \mathcal{B}_t) = 0$, so that under the conditions of Theorem 8.2.1 the maximum likelihood estimator exists and is unique with probability one. This is, of course, the case if the distribution of B_t is continuous. In several other cases, $P_\theta(B_t \in \operatorname{bd} \mathcal{B}_t)$ will go to zero rapidly as $t \to \infty$. This is, for instance, the case for the envelope families belonging to the pure birth process (Section 7.5). Here $\{B_t \in \operatorname{bd} \mathcal{B}_t\}$ is the event that no birth has happened before time t.

Next let us consider general curved exponential families of processes.

Theorem 8.2.2 *Suppose that for t fixed, the cumulant transform (7.1.3) of B_t under P_0 is steep and that the representation (7.1.2) of the envelope* *family is minimal. Assume further that* $\{\gamma_t(\theta) : \theta \in \Theta\}$ *is a relatively closed subset of* $\tilde{\Gamma}_t$. *Then if* $B_t \in \operatorname{int} \mathcal{B}_t$, *the likelihood function has at least one maximum in* Θ. *If a maximum is in* $\operatorname{int}\Theta$, *it solves the equation*

$$\dot{\gamma}_t(\theta)^T B_t = \dot{\phi}_t(\theta). \tag{8.2.2}$$

Remarks: By (8.1.5) we can rewrite (8.2.2) as

$$\dot{\gamma}_t(\theta)^T (B_t - E_\theta(B_t)) = 0. \tag{8.2.3}$$

All maxima solve this equation if Θ is open. Quite often, the likelihood function for θ is strictly convex (see Corollary 8.2.6), so that the maximum likelihood estimator is unique. If $B_t \in \operatorname{bd} \mathcal{B}_t$, the likelihood function might also have a maximum in Θ; see, e.g., Example 8.2.5.

Proof: The theorem follows directly from Theorem 5.7 in Brown (1986). Equation (8.2.2) follows simply by differentiating the log-likelihood function. □

In the rest of this section we consider curved exponential families of processes with a natural representation of the form (7.1.7), for which we can say more. Almost all curved exponential families encountered in statistical practice have this form. We assume that $\alpha_t(\theta)$ and $\phi_t(\theta)$ are twice continuously differentiable for all $t > 0$. We need the following notation.

Define for $x \in \mathbb{R}^{m-k}$ and $t > 0$ the function $\kappa_t(\cdot\,; x) : \mathbb{R}^k \mapsto \mathbb{R}$ by

$$\kappa_t(\theta; x) = \begin{cases} \alpha_t(\theta)^T x + \phi_t(\theta) & \text{if } \theta \in \Theta \\ \\ \infty & \text{otherwise,} \end{cases} \tag{8.2.4}$$

and define a subset $\mathcal{K}_t$ of $\mathbb{R}^{m-k}$ by requiring that $x \in \mathcal{K}_t$ if and only if $\kappa_t(\cdot\,; x)$ is a lower semi-continuous convex function that is strictly convex with positive definite matrix of second partial derivatives on $\operatorname{int}\Theta$. For $x \in \mathcal{K}_t$ the function

$$\xi_t(\theta; x) = \nabla_\theta \kappa_t(\theta; x) = \dot{\alpha}_t(\theta)^T x + \dot{\phi}_t(\theta) \tag{8.2.5}$$

is an injection, and

$$\mathcal{R}_{t,x} = \xi_t(\operatorname{int}\Theta, x) \tag{8.2.6}$$

is open.

Theorem 8.2.3 *Suppose* $S_t \in \mathcal{K}_t$. *Then the maximum likelihood estimator exists and is unique if and only if* $A_t \in \mathcal{R}_{t,S_t}$. *The maximum likelihood estimator solves the equation*

$$A_t = \dot{\alpha}_t(\theta)^T S_t + \dot{\phi}_t(\theta). \tag{8.2.7}$$

If $A_t \notin \mathcal{R}_{t,S_t}$, *the likelihood function has no maximum.*

Remarks: Equation (8.2.7) is identical to (8.2.2) and (8.2.3). Of course the set $\mathcal{K}_t$ might be empty.

Proof: The theorem follows directly from Theorem 5.5 in Brown (1986). □

Example 8.2.4 Consider a one-dimensional process that under P_θ can be written in the form

$$X_t = \theta t + N_t + W_t, \tag{8.2.8}$$

where W is a standard Wiener process independent of the counting process N. The intensity of N at time t under P_θ is $e^\theta H_t > 0$ for some predictable process H. We assume that $\theta \in \mathbb{R}$. The likelihood function is

$$L_t(\theta) = \exp\left(\theta X_t - (e^\theta - 1)\int_0^t H_s ds - \tfrac{1}{2}\theta^2 t\right). \tag{8.2.9}$$

For every $x > 0$ and $t > 0$ the function $(e^\theta - 1)x + \frac{1}{2}\theta^2 t$ is a steep, strictly convex cumulant transform, so $\mathcal{K}_t = \mathbb{R}$ for all $t > 0$. Since $\xi_t(\theta; x) = xe^\theta + \theta t$, we see that $\mathcal{R}_{t,x} = \mathbb{R}$ for all $t > 0$ and $x > 0$. □

Example 8.2.5 Consider a *linear birth-and-death process* X with birth rate $\lambda > 0$ and death rate $\mu > 0$. We assume that $X_0 = x_0$. Let $A_t^{(1)}$ and $A_t^{(2)}$ denote the number of births and deaths in $[0, t]$, respectively, and define

$$S_t = \int_0^t X_s ds.$$

Then the likelihood function corresponding to observing X in $[0, t]$ is (cf. (3.4.7))

$$L_t(\theta_1, \theta_2) = \exp\left[\theta_1 A_t^{(1)} + \theta_2 A_t^{(2)} - (e^{\theta_1} + e^{\theta_2} - 2)S_t\right], \tag{8.2.10}$$

where $\theta_1 = \log\,\lambda$ and $\theta_2 = \log\,\mu$. The closed convex support of $(A_t^{(1)}, A_t^{(2)}, S_t)$ is $\mathcal{B}_t = \{(x, y, z) : x \geq 0,\ 0 \leq y \leq x + x_0,\ z \geq 0\}$. For every $x > 0$ the function $(\theta_1, \theta_2) \mapsto (e^{\theta_1} + e^{\theta_2} - 2)x$, where $(\theta_1, \theta_2) \in \mathbb{R}^2$, is a cumulant transform (of the product of two independent Poisson distributions), and $\xi_t(\theta; x) = (e^{\theta_1} + e^{\theta_2})x$, so $\mathcal{R}_{t,x} = \mathbb{R}_+^2$. Hence the maximum likelihood estimator exists if $A_t^{(i)} > 0$, $i = 1, 2$, and $S_t > 0$. Obviously, $P_\theta(S_t > 0) = 1$ and $P_\theta(A_t^{(i)} = 0) \to 0$ for $t \to \infty$, $i = 1, 2$. The maximum likelihood estimator is given by $\hat{\lambda}_t = A_t^{(1)}/S_t$ and $\hat{\mu}_t = A_t^{(2)}/S_t$. Note that the maximum likelihood estimator exists on part of bd $\mathcal{B}_t$, namely where $A_t^{(2)} = A_t^{(1)} + x_0$. The event that the canonical process hits this boundary (where it will then stay) is the probability of extinction, i.e., 1 for $\mu \geq \lambda$ and $(\mu/\lambda)^{x_0}$ for $\mu < \lambda$. □

Let $\mathcal{A}_x$ be the closure of the convex hull of the support of A_t when $S_t = x$. We assume that $\mathcal{A}_x$ does not depend on t. Define a subset $\mathcal{M}$ of $\mathbb{R}^{k-m}$ by requiring that $x \in \mathcal{M}$ if and only if there exists a stopping time τ_x such that $S_{\tau_x} = x$ and $P_\theta(\tau_x < \infty) = 1$ for all $\theta \in \Theta$. Of course, $\mathcal{M}$ might be empty.

Corollary 8.2.6 *Consider a time-homogeneous exponential family (i.e., α_t does not depend on t) with $\phi_t(\theta) = 0$. Suppose* $\operatorname{int}\Theta \neq \emptyset$ *and that the components of A_t are affinely independent for all $t > 0$. Then the function $\alpha(\theta)^T x$ is a cumulant transform for all $x \in \mathcal{M}$. Suppose, moreover, that $S_t \in \mathcal{M}$ and that $\alpha(\theta)^T S_t$ is steep with domain Θ. Then the maximum likelihood estimator exists if and only if $A_t \in \operatorname{int}\mathcal{A}_{S_t}$. When the maximum likelihood estimator exists, it solves the equation*

$$A_t = \dot{\alpha}_t(\theta) S_t. \tag{8.2.11}$$

Proof: By the fundamental identity of sequential analysis (Corollary B.1.2) we have for every $x \in \mathcal{M}$ that

$$\frac{dP_\theta^{\tau_x}}{dP_0^{\tau_x}} = \exp(\theta^T A_{\tau_x} - \alpha(\theta)^T x), \tag{8.2.12}$$

where $P_\theta^{\tau_x}$ denotes the restriction of P_θ to $\mathcal{F}_{\tau_x}$. Under the first conditions of the corollary, $\{P_\theta^{\tau_x} : \theta \in \Theta\}$ is an exponential family of measures on $\mathcal{F}_{\tau_x}$ with a minimal representation (8.2.12). In particular, $\alpha(\theta)^T x$ is the cumulant transform of A_{τ_x} under P_0. If, moreover, $\alpha(\theta)^T x$ is steep with domain Θ, the family $\{P_\theta^{\tau_x} : \theta \in \Theta\}$ is steep and full, so $\mathcal{R}_{t,x} = \operatorname{int}\mathcal{A}_x$. □

Corollary 8.2.6 can be applied to the models studied in Sections 5.1 and 5.2 and implies Theorem 5.2.1. The next example shows that it can also be applied to more general models.

Example 8.2.7 Consider a Markov process X with discrete time and with state space $\{1, 2\}$. Suppose $X_0 = 1$ and that the transition probabilities (p_{ij}) are all strictly positive. We parametrize the model by $\theta_1 = \log(p_{21}p_{12})$ and $\theta_2 = \log(p_{22})$, and assume that $(\theta_1, \theta_2) \in \mathbb{R}_- \times \mathbb{R}_-$. The likelihood function based on the observations of $X_1, \cdots, X_t$ is (cf. (3.2.5))

$$\begin{aligned} & L_t(\theta_1,\theta_2) && (8.2.13) \\ = \ & \exp\{\theta_1 N_{12}(t) + \theta_2 N_{22}(t) + \log(1-e^{\theta_2})[N_{21}(t) - N_{12}(t)] \\ & + \log(1-e^{\theta_1}/(1-e^{\theta_2}))N_{11}(t)\}, \end{aligned}$$

where $N_{ij}(t)$ is the number of one-step transitions from state i to state j before time t. The possible values of $S_t = (N_{21}(t) - N_{12}(t),\ N_{11}(t))$ are $\{0,1\} \times (\mathbb{N} \cup \{0\})$. Define for $N \in \mathbb{N}$ the stopping time

$$\tau = \inf(t : N_{11}(t) = N).$$

Then $P_\theta(\tau < \infty) = 1$ for all $\theta \in \mathbb{R}_- \times \mathbb{R}_-$, and $S_\tau = (0, N)$, so $(0, N) \in \mathcal{M}$ for all $N \in \mathbb{N}$. The conditions of Corollary 8.2.6 are satisfied, so when $N_{12}(t) - N_{21}(t) = 0$ and $N_{11}(t) \geq 1$, the maximum likelihood estimator exists and is unique if and only if $N_{21}(t) > 0$ and $N_{22}(t) > 0$. The case $N_{12}(t) - N_{21}(t) = 1$ can be treated similarly after a reparametrization. □

8.3 Consistency and asymptotic normality of the maximum likelihood estimator

In this section we give conditions ensuring consistency and asymptotic normality of the maximum likelihood estimator when it exists. Here we assume that γ_t and ϕ_t are twice continuously differentiable with respect to θ. If $\gamma_t(\theta) \in \operatorname{int} \tilde{\Gamma}_t$, it is enough to assume that γ_t is twice continuously differentiable, as it then follows from Lemma 8.1.1 that ϕ_t is twice continuously differentiable. In the first theorem, which does not use the exponential family structure of the likelihood function, the conditions are on the normalized score function and the second derivative of the log-likelihood function

$$j_t(\theta) = -\ddot{l}_t(\theta). \tag{8.3.1}$$

This matrix is the *observed information matrix*. We also need the matrix $j_t(\theta^{(1)}, \cdots, \theta^{(k)})$, where $\theta^{(i)} \in \Theta$, $i = 1, \cdots, k$. This matrix is obtained from $j_t(\theta)$ by replacing θ by $\theta^{(i)}$ in the ith row ($i = 1, \cdots, k$). For a $k \times k$-matrix $A = \{a_{ij}\}$ we define $\|A\|^2 = \sum_{i=1}^k \sum_{j=1}^k a_{ij}^2$, and for a vector a the usual Euclidean norm is denoted by $\|a\|$. If A is positive semi-definite, $A^{1/2}$ denotes the positive semi-definite square root of A.

A class $\mathcal{R}$ of random vectors is said to be *stochastically bounded* if given $\epsilon > 0$, there exists a $K \in (0, \infty)$ such that $\sup_{X \in \mathcal{R}} P(|X| > K) < \epsilon$. It is not difficult to see that if $\{X_n : n \in \mathbb{N}\}$ converges in distribution as $n \to \infty$, then $\{X_n : n \in \mathbb{N}\}$ is stochastically bounded.

Theorem 8.3.1 *Suppose there exists a family of non-random invertible matrices $D_t(\theta)$, $t > 0$, such that*

$$D_t(\theta) \to 0$$

as $t \to \infty$, and a (possibly) random symmetric positive semi-definite matrix $W(\theta)$ such that under P_θ

$$\sup_{\theta^{(i)} \in M_t^{(\alpha)}} \|D_t(\theta)^T j_t(\theta^{(1)}, \cdots, \theta^{(k)}) D_t(\theta) - W(\theta)\| \to 0 \qquad (8.3.2)$$

in probability on C_θ as $t \to \infty$ for all $\alpha > 0$. Here $M_t^{(\alpha)}$ is defined by

$$M_t^{(\alpha)} = \{\bar{\theta} : \|D_t(\theta)^{-1}(\bar{\theta} - \theta)\| \leq \alpha\} \qquad (8.3.3)$$

and

$$C_\theta = \{\det W(\theta) > 0\}. \qquad (8.3.4)$$

Assume further that there exists $t_1 > 0$ such that the family of random variables

$$W(\theta)^{-\frac{1}{2}} D_t(\theta)^T \dot{l}_t(\theta), \;\; t > t_1,$$

is stochastically bounded under P_θ. Then, under P_θ, an estimator $\hat{\theta}_t$ of θ exists on C_θ that solves the likelihood equation (8.2.2) with a probability tending to $P_\theta(C_\theta)$ as $t \to \infty$. Moreover, $\hat{\theta}_t \to \theta$ in probability on C_θ, and

$$D_t(\theta)^T j_t(a_t^{(1)}, \cdots, a_t^{(k)}) D_t(\theta) \to W(\theta) \qquad (8.3.5)$$

in probability on C_θ, where for each t, $a_t^{(i)}$ is a convex combination of θ and $\hat{\theta}_t$, $i = 1, \ldots, k$.

Proof: First note that (8.3.2) implies that

$$Y_t^{(\alpha,\theta)} = \sup_{\theta^{(i)} \in S_t^{(\alpha)}} \|D_t(\theta)^T j_t(\theta^{(1)}, \cdots, \theta^{(k)}) D_t(\theta) - W(\theta)\| \to 0 \qquad (8.3.6)$$

in P_θ-probability on C_θ as $t \to \infty$, where

$$S_t^{(\alpha)} = \{\bar{\theta} : \|W(\theta)^{\frac{1}{2}} D_t(\theta)^{-1}(\bar{\theta} - \theta)\| \leq \alpha\}. \qquad (8.3.7)$$

This is because $\bar{\theta} \in S_t^{(\alpha)}$ implies that $\|D_t(\theta)^{-1}(\bar{\theta} - \theta)\| \leq \alpha / \lambda_W^-$, where λ_W^- is the smallest eigenvalue of $W(\theta)$. Therefore, $Y_t^{(\alpha,\theta)}$ is dominated by the supremum over $M_t^{(\alpha/\epsilon)}$ on $\{\lambda_W^- \geq \epsilon\}$, $\epsilon > 0$, and we can make the probability of the event $\{\lambda_W^- \geq \epsilon\}$ as small as we like by choosing ϵ sufficiently small.

If no solution of (8.2.2) exists, we set $\hat{\theta}_t = \infty$. Otherwise, choose the solution closest to θ in the sense that for some $n \geq 1$, $\hat{\theta}_t \in S_t^{(n)}$, while there is no solution in $S_t^{(m)}$, $m = 1, \cdots, n-1$. Denote by $E_{t,c}$ the set $\{\hat{\theta}_t \in S_t^{(c)}\}$, and consider the Taylor expansion

$$\dot{l}_t(\bar{\theta}) = \dot{l}_t(\theta) - j_t(\theta^{(1)}, \cdots, \theta^{(k)})(\bar{\theta} - \theta), \qquad (8.3.8)$$

where $\bar{\theta}$ is chosen such that $\|W(\theta)^{\frac{1}{2}} D_t(\theta)^{-1}(\bar{\theta}-\theta)\| = c$. Each vector $\theta^{(i)}$ is thus an element of $S_t^{(c)}$.

In the following we work under P_θ. Fix $\epsilon > 0$. Since $\{W(\theta)^{-\frac{1}{2}} D_t(\theta)^T \dot{l}_t(\theta): t > t_1\}$ is stochastically bounded, we can find $K > 0$ such that the event

$$A_t = \{\|W(\theta)^{-\frac{1}{2}} D_t(\theta)^T \dot{l}_t(\theta)\| \leq K\} \cap C_\theta$$

has probability larger than $P_\theta(C_\theta) - \epsilon$ for $t > t_1$. Now fix $\delta > 0$ and choose c large enough that $cK + \delta - c^2 < 0$. Note that (8.3.8) implies that for $\bar{\theta} \in \operatorname{bd} S_t^{(c)}$ we have

$$\begin{aligned}(\bar{\theta}-\theta)^T \dot{l}_t(\bar{\theta}) &= (\bar{\theta}-\theta)^T (D_t(\theta)^{-1})^T W(\theta)^{\frac{1}{2}} W(\theta)^{-\frac{1}{2}} D_t(\theta)^T \dot{l}(\theta) \quad (8.3.9)\\ &\quad - (\bar{\theta}-\theta)^T (D_t(\theta)^{-1})^T D_t(\theta)^T j_t(\theta^{(1)}, \cdots, \theta^{(k)}) D_t(\theta)\\ &\quad \times D_t(\theta)^{-1}(\bar{\theta}-\theta)\end{aligned}$$

for some $\theta^{(i)} \in S_t^{(c)}$, $i = 1, \cdots, k$. Let U_t denote the maximum over $\bar{\theta} \in \operatorname{bd} S_t^{(c)}$ of the absolute value of the difference between $-c^2$ and the last term in (8.3.9). It follows from (8.3.6) that we can find $t_2 > 0$ such that $P_\theta(B_t) \geq P_\theta(C_\theta) - \epsilon$ for $t > t_2$, where $B_t = \{U_t < \delta\} \cap C_\theta$.

On $A_t \cap B_t$ we find that $(\bar{\theta}-\theta)^T \dot{l}_t(\bar{\theta}) \leq 0$ for all $\bar{\theta} \in \operatorname{bd} S_t^{(c)}$. Since $\dot{l}_t(\bar{\theta})$ is a continuous function of $\bar{\theta}$, the likelihood function has at least one solution in $S_t^{(c)}$. This follows from Brouwer's fixed point theorem; see Aitchison and Silvey (1958, Lemma 2). In conclusion, $A_t \cap B_t \subset E_{t,c}$, and since $P_\theta(A_t \cap B_t) \geq P_\theta(C_\theta) - 2\epsilon$ for $t > \max(t_1, t_2)$, we have proved that $P_\theta(\hat{\theta}_t < \infty) \to P_\theta(C_\theta)$ as $t \to \infty$.

Moreover, $\|\hat{\theta}_t - \theta\| \leq c/\lambda^-(t)$ on $E_{t,c}$, where $\lambda^-(t)$ is the smallest eigenvalue of $W(\theta)^{\frac{1}{2}} D_t(\theta)^{-1}$. Since $c/\lambda^-(t)$ tends to zero in probability on C_θ, it follows that $\hat{\theta}_t \to \theta$ in probability on C_θ. Similarly, (8.3.5) follows from (8.3.6). □

The maximum likelihood estimator found in Theorem 8.3.1 need not be unique. However, since we consider exponential families, we know in many situations from results in Section 8.2 that the maximum likelihood estimator is unique, and in such cases Theorem 8.3.1 tells us that it is consistent and satisfies (8.3.5). Usually, $P_\theta(\{\hat{\theta}_t \text{ exists uniquely}\} \cap C_\theta) \to P_\theta(C_\theta)$ as $t \to \infty$. In many models $P_\theta(C_\theta) = 1$, but the following example shows that it is necessary to include the possibility $P_\theta(C_\theta) < 1$ in Theorem 8.3.1.

Example 8.2.5 (continued). For a linear birth-and-death process X with birth rate λ and death rate μ starting at x_0, the likelihood function for observation of X in $[0, t]$ is given by (8.2.10).Thus the observed information matrix is

$$j_t(\theta) = \begin{pmatrix} e^{\theta_1} & 0 \\ 0 & e^{\theta_2} \end{pmatrix} S_t,$$

where $\theta_1 = \log(\lambda)$, $\theta_2 = \log(\mu)$, and

$$S_t = \int_0^t X_s ds.$$

We consider the case $\lambda > \mu$. Then it is known (see Exercise 8.4) that under P_θ,

$$(\lambda - \mu)e^{-(\lambda-\mu)t} S_t \to x_0 U,$$

almost surely as $t \to \infty$, where U is a non-negative random variable satisfying $\{U = 0\} = \{X_t \to 0\}$. The probability of this event, the probability of extinction, is $(\mu/\lambda)^{x_0} > 0$. We see that the condition (8.3.2) is satisfied with

$$D_t(\theta) = \sqrt{\lambda - \mu}\, e^{-\frac{1}{2}(\lambda-\mu)t} \operatorname{diag}\left(e^{-\frac{1}{2}\theta_1}, e^{-\frac{1}{2}\theta_2}\right)$$

and

$$W(\theta) = x_0 U I_2,$$

where I_2 is the 2×2 identity matrix. The uniformity of the convergence in $M_t^{(\alpha)}$ follows since $j_t(\theta)$ is a product of a non-random continuous function of θ and a stochastic process independent of θ. It follows that $\{\det W(\theta) = 0\} = \{U = 0\} = \{X_t \to 0\}$. If $\lambda \leq \mu$, $P_\theta(C_\theta) = 0$. □

In Theorem 8.3.1 the exponential family structure of the likelihood function was not used at all. However, for exponential families it is much easier to check the conditions than it is for more general models. For an exponential family with a general representation (8.1.1), the observed information matrix is given by

$$j_t(\theta)_{ij} = \ddot{\phi}_t(\theta)_{ij} - \ddot{\gamma}_t(\theta)_{ij}^T B_t. \tag{8.3.10}$$

The fact that the terms in the components of $j_t(\theta)$ are a product of a non-random function of θ and a stochastic process independent of θ makes it relatively easy to verify the uniformity of the convergence in the condition (8.3.2); see Example 8.2.5 and Exercise 8.5.

In the following discussion of asymptotic normality of the score function and the maximum likelihood estimator, we need the quadratic variation of the canonical process B. This concept is well-defined when B is a semimartingale. For a more comprehensive discussion of these concepts see Appendix A, Section 4. It is always possible to find an exponential representation (8.1.1) such that B is a semimartingale. The quadratic variation $[B]$ of B is defined by

$$\sum_{j=1}^{2^n} (B_{jt2^{-n}} - B_{(j-1)t2^{-n}})(B_{jt2^{-n}} - B_{(j-1)t2^{-n}})^T \to [B]_t \tag{8.3.11}$$

in P_θ-probability as $n \to \infty$. The sum converges in probability for every semimartingale B.

We shall also need the *expected information matrix* defined by

$$i_t(\theta) = E_\theta(j_t(\theta)). \tag{8.3.12}$$

This information matrix is also sometimes called the Fisher information. When $\gamma_t(\theta) \in \operatorname{int} \tilde{\Gamma}_t$, it follows from (8.3.10), (8.1.6), and (8.1.4) that

$$i_t(\theta) = \dot{\gamma}_t(\theta)^T V_\theta(B_t)\dot{\gamma}_t(\theta) = \dot{\gamma}_t(\theta)^T \ddot{\Psi}_t(\gamma_t(\theta))\dot{\gamma}_t(\theta). \tag{8.3.13}$$

In the rest of this section we restrict attention to time-homogeneous exponential families of processes. This is done because in this case, which is by far the most common in practice, the theory is particularly simple. However, rather similar results hold for families that are not time-homogeneous. The results will be given under the following condition, in which $\Delta B_t = B_t - B_{t-}$, where $B_{t-} = \lim_{s\uparrow t} B_s$ is well-defined when B is a semimartingale because a semimartingale has trajectories that are right-continuous with limits from the left. The condition also involves the invertible matrices $\{D_t(\theta), t > 0\}$, which appeared in Theorem 8.3.1, and the matrices

$$H_t(\theta) = \dot{\gamma}(\theta) D_t(\theta). \tag{8.3.14}$$

Condition 8.3.2 *The following holds under P_θ as $t \to \infty$:*

$$D_t(\theta) \to 0; \tag{8.3.15}$$

$$D_t^*(\theta) E_\theta \left(\sup_{s \le t} |\Delta B_s^{(i)}| \right) \to 0, \quad i = 1, \cdots, k, \tag{8.3.16}$$

where $D_t^*(\theta) = \max_{i,j} |D_t(\theta)_{i,j}|$;

$$H_t(\theta)^T [B]_t H_t(\theta) \to W(\theta) \tag{8.3.17}$$

in probability, where $W(\theta)$ is the (possibly) random positive semi-definite matrix appearing in (8.3.2); and

$$D_t(\theta)^T i_t(\theta) D_t(\theta) \to \Sigma, \tag{8.3.18}$$

where Σ is a positive definite matrix. We assume that $P_\theta(\det(W(\theta)) > 0) > 0$.

In the next section conditions will be given under which the limit of the random matrices in (8.3.17) equals the random matrix in (8.3.2). In the following we assume that this is the case.

The next result serves two purposes. It can be used to prove asymptotic normality of the maximum likelihood estimator and to check the condition in Theorem 8.3.1 that $W(\theta)^{-\frac{1}{2}} D_t(\theta)^T \dot{l}_t(\theta)$ is stochastically bounded.

Theorem 8.3.3 *Suppose $\gamma(\theta) \in \operatorname{int}\tilde{\Gamma}_t$ for all $t > 0$, that the funtion $t \mapsto \dot{\phi}_t(\theta)$ is continuous with bounded variation in compact intervals, and that Condition 8.3.2 holds. Then*

$$(D_t(\theta)^T \dot{l}_t(\theta), H_t(\theta)^T [B]_t H_t(\theta)) \to (W(\theta)^{\frac{1}{2}} Z, W(\theta)) \tag{8.3.19}$$

and, conditionally on $\{\det(W(\theta)) > 0\}$,

$$W(\theta)^{-\frac{1}{2}} D_t(\theta)^T \dot{l}_t(\theta) \to Z \tag{8.3.20}$$

in distribution under P_θ as $t \to \infty$. Here Z is a k-dimensional standard normal distributed random vector independent of $W(\theta)$. The convergence result (8.3.19) also holds conditionally on $\{\det(W(\theta)) > 0\}$.

Remark: The condition $\gamma(\theta) \in \operatorname{int}\tilde{\Gamma}_t$ can be replaced by the condition that the score vector $\dot{l}_t(\theta)$ is a square integrable martingale. This is also the case for Theorem 8.3.5.

Proof: The result follows directly from the central limit theorem for martingales (Theorem A.7.7) applied to the square integrable martingale $\dot{l}_t(\theta)$ with $K_t = D_t(\theta)^T$. First note that (A.7.1) follows from (8.3.16) because $|\Delta \dot{l}_s(\theta)_i| \leq \sum_j |\dot{\gamma}(\theta)_{ij}||\Delta B_s^{(j)}|$. Since $\dot{\phi}_t(\theta)$ is a continuous function of t with bounded variation on compact intervals, it follows from (8.1.11) that the quadratic variation of $\dot{l}(\theta)$ is $\dot{\gamma}(\theta)^T [B] \dot{\gamma}(\theta)$; cf. (A.4.6). Therefore, (A.7.2) follows from (8.3.17). Finally, (A.7.3) is the same condition as (8.3.18) because by (8.3.13), $i_t(\theta)$ is the covariance matrix of $\dot{l}_t(\theta)$. □

The distribution of $W(\theta)^{\frac{1}{2}} Z$ is the normal variance-mixture with characteristic function $u \mapsto E_\theta(\exp[-\frac{1}{2} u^T W(\theta) u])$.

The asymptotic normality of the maximum likelihood estimator and the asymptotic distribution of the likelihood ratio test statistic for a point hypothesis about θ can now be deduced by standard arguments.

Theorem 8.3.4 *Suppose the conditions of Theorem 8.3.3 and (8.3.5) hold and that the maximum likelihood estimator $\hat{\theta}_t$ exists for t large enough. Then, conditionally on* $\{\det(W(\theta)) > 0\}$,

$$\{H_t(\theta)^T [B]_t H_t(\theta)\}^{-\frac{1}{2}} D_t(\theta)^T j_t(\theta)(\hat{\theta}_t - \theta) \to N(0, I_k), \tag{8.3.21}$$

$$\{D_t(\theta)^T j_t(\hat{\theta}_t) D_t(\theta)\}^{-\frac{1}{2}} D_t(\theta)^T j_t(\hat{\theta}_t)(\hat{\theta}_t - \theta) \to N(0, I_k), \tag{8.3.22}$$

and

$$2\{l_t(\hat{\theta}_t) - l_t(\theta)\} \to \chi^2(k) \tag{8.3.23}$$

in distribution under P_θ as $t \to \infty$. Here I_k denotes the $k \times k$ identity matrix.

Proof: On series expansion we see that

$$\dot{l}_t(\theta) = j_t(\theta_t^{(1)}, \cdots, \theta_t^{(k)})(\hat{\theta}_t - \theta),$$

where $\theta_t^{(i)}$ is a convex combination of θ and $\hat{\theta}_t$ for each $i = 1, \cdots, k$. Therefore,

$$\begin{aligned}&\{H_t(\theta)^T[B]_t H_t(\theta)\}^{-\frac{1}{2}} D_t(\theta)^T j_t(\theta_t^{(1)}, \cdots, \theta_t^{(k)})(\hat{\theta}_t - \theta) \\ &\quad = \{H_t(\theta)^T[B]_t H_t(\theta)\}^{-\frac{1}{2}} D_t(\theta)^T \dot{l}_t(\theta)\end{aligned} \tag{8.3.24}$$

converges in distribution to Z conditionally on $\{\det(W(\theta)) > 0\}$. The difference between (8.3.21) and (8.3.24) is

$$\{H_t(\theta)^T[B]_t H_t(\theta)\}^{-\frac{1}{2}} D_t(\theta)^T\{j_t(\theta) - j_t(\theta_t^{(1)}, \cdots, \theta_t^{(k)})\} D_t(\theta) D_t(\theta)^{-1}(\hat{\theta}_t - \theta). \tag{8.3.25}$$

By (8.3.5), $D_t(\theta)^T\{j_t(\theta) - j_t(\theta_t^{(1)}, \cdots, \theta_t^{(k)})\} D_t(\theta)$ tends to zero in probability under P_θ on the set $\{\det(W(\theta)) > 0\}$. Moreover, for $\det(W(\theta)) > 0$ we have for t large enough that

$$D_t(\theta)^{-1}(\hat{\theta}_t - \theta) = D_t(\theta)^{-1} j_t(\theta_t^{(1)}, \cdots, \theta_t^{(k)})^{-1} [D_t(\theta)^T]^{-1} D_t(\theta)^T \dot{l}_t(\theta), \tag{8.3.26}$$

which by (8.3.5) and (8.3.19) is stochastically bounded under P_θ. Hence (8.3.25) tends to zero in probability under P_θ on $\{\det(W(\theta)) > 0\}$. In a similar way it follows that the difference between (8.3.21) and (8.3.22) tends to zero in probability. Finally, (8.3.23) follows, since

$$2\{l_t(\hat{\theta}_t) - l_t(\theta)\} = (\hat{\theta}_t - \theta)^T j_t(\tilde{\theta}_t^{(1)}, \cdots, \tilde{\theta}_t^{(k)})(\hat{\theta}_t - \theta), \tag{8.3.27}$$

where again $\tilde{\theta}_t^{(i)}$ is a convex combination of θ and $\hat{\theta}_t$ for each $i = 1, \cdots, k$. The difference between (8.3.27) and the squared norm of (8.3.22) is

$$(\hat{\theta}_t - \theta)^T\{j_t(\hat{\theta}_t) - j_t(\tilde{\theta}_t^{(1)}, \cdots, \tilde{\theta}_t^{(k)})\}(\hat{\theta}_t - \theta), \tag{8.3.28}$$

and by (8.3.5) and the fact that (8.3.26) is stochastically bounded on $\{\det(W(\theta)) > 0\}$ it follows that (8.3.28) tends to zero in probability on $\{\det(W(\theta)) > 0\}$. □

Note that in the proof we used that the squared norm of (8.3.22) is simply the Wald test statistic $(\hat{\theta}_t - \theta)^T j_t(\hat{\theta}_t)(\hat{\theta}_t - \theta)$ based on the observed information, and that

$$(\hat{\theta}_t - \theta)^T j_t(\hat{\theta}_t)(\hat{\theta}_t - \theta) \to \chi^2(k) \tag{8.3.29}$$

in distribution under P_θ as $t \to \infty$.

For discrete time models the assumption in Theorem 8.3.3 that $\dot{\phi}_t(\theta)$ is a continuous function of t is satisfied only if $\dot{\phi}_t(\theta)$ is zero. This is often the case, but if it is not, the following result is useful.

Theorem 8.3.5 *Suppose* $\gamma(\theta) \in \operatorname{int} \tilde{\Gamma}_t$ *for all* $t > 0$; *that (8.3.15) holds; that as* $t \to \infty$

$$\bar{D}_t(\theta)_i E_\theta\left(\sup_{s \leq t} |\Delta \dot{l}_t(\theta)|\right) \to 0, \quad i = 1, \cdots, k, \tag{8.3.30}$$

where $\bar{D}_t(\theta)_i = \sum_{j=1}^k |D_t(\theta)_{ji}|$, *and*

$$D_t(\theta)^T[\dot{l}(\theta)]_t D_t(\theta) \to W(\theta) \tag{8.3.31}$$

in P_θ*-probability; and that (8.3.18) holds. Here* $W(\theta)$ *is as in Condition 8.3.2. Then the conclusions of Theorem 8.3.3 hold with* $H_t(\theta)$ *replaced by* $D_t(\theta)$ *and* $[B]_t$ *replaced by* $[\dot{l}(\theta)]_t$.

Suppose, furthermore, that the maximum likelihood estimator $\hat{\theta}_t$ *exists for t large enough and that (8.3.5) holds. Then the conclusions of Theorem 8.3.4 hold, again with* $H_t(\theta)$ *replaced by* $D_t(\theta)$ *and* $[B]_t$ *replaced by* $[\dot{l}(\theta)]_t$.

Proof: The first part of the theorem follows immediately from the central limit theorem for martingales (Theorem A.7.7). The second part is proved in the same way as Theorem 8.3.4. □

In Theorem 8.3.5, $[\dot{l}(\theta)]$ denotes the quadratic variation of the process $\dot{l}(\theta)$. This process is defined like $[B]$, cf. (8.3.11), and is discussed further in the next section, see (8.4.4), and in Appendix A.

Example 8.3.6 *Multi-dimensional Gaussian diffusions.* In this example we consider the class of k-dimensional Gaussian diffusions given as solutions to the stochastic differential equation

$$dX_t = BX_t dt + dW_t, \quad X_0 = 0, \tag{8.3.32}$$

where $X_t = (X_{1t}, \cdots, X_{kt})^T$, B is a $k \times k$-matrix, and W is a k-dimensional standard Wiener process.

If $\theta = (b_{11}, \cdots, b_{1k}, \cdots, b_{k1}, \cdots, b_{kk})^T$, then the likelihood function based on observation of X_s for $s \in [0, t]$ is

$$L_t(\theta) = \exp(N_t\theta - \frac{1}{2}\theta^T I_t \theta), \tag{8.3.33}$$

where N_t is the k^2-dimensional vector

$$\left(\int_0^t X_{1s}dX_{1s}, \cdots, \int_0^t X_{ks}dX_{1s}, \cdots, \int_0^t X_{1s}dX_{ks}, \cdots, \int_0^t X_{ks}dX_{ks}\right)^T$$

and I_t is the $k^2 \times k^2$-matrix

$$I_t = \begin{pmatrix} K_t & 0 & \cdots & 0 \\ 0 & K_t & & \vdots \\ \vdots & & \ddots & 0 \\ 0 & \cdots & 0 & K_t \end{pmatrix},$$

where K_t is the $k \times k$-matrix $\int_0^t X_s X_s^T ds$, which is almost surely invertible. The maximum likelihood estimator is $\hat{\theta}_t = I_t^{-1} N_t$. Note that

$\hat{\theta}_t = I_t^{-1}(I_t\theta + M_t) = \theta + I_t^{-1}M_t$, where M_t is the k^2-dimensional square integrable martingale

$$M_t = \left(\int_0^t X_{1s}dW_{1s}, \cdots, \int_0^t X_{ks}dW_{1s}, \cdots, \int_0^t X_{1s}dW_{ks}, \cdots, \int_0^t X_{ks}dW_{ks}\right).$$

Since $j_t(\theta) = I_t$, the asymptotic normality of $(D_t(\theta)^T I_t D_t(\theta))^{\frac{1}{2}} D_t(\theta)^{-1}(\hat{\theta}_t - \theta)$ follows from Theorem 8.3.4 if we can find a family of matrices $\{D_t(\theta) : t > 0\}$ such that as $t \to \infty$ we have $D_t(\theta)^T I_t D_t(\theta) \to \eta^2(\theta)$ and $D_t(\theta)^T i_t(\theta) D_t(\theta) \to \Sigma(\theta)$ in probability, where $\eta^2(\theta)$ and $\Sigma(\theta)$ are positive definite matrices, of which $\eta^2(\theta)$ might be random. This would imply Condition 8.3.2, and (8.3.5) would then be automatically satisfied because j_t does not depend on θ. In the following we find such matrices $\{D_t(\theta)\}$ in the cases where B can be diagonalized and has real eigenvalues different from zero. Also the conditions of the law of large numbers Corollary A.7.8 will then be satisfied, and by applying this result to the martingale M, the consistency of $\hat{\theta}$ follows since $[M]_t = I_t$.

The solution to (8.3.32) is

$$X_t = \int_0^t e^{B(t-s)} dW_s. \tag{8.3.34}$$

Suppose B can be diagonalized, i.e., that there exists an invertible matrix C and a diagonal matrix D such that $B = C^{-1}DC$. Since $B^n = C^{-1}D^nC$, we find that

$$X_t = \int_0^t C^{-1} e^{D(t-s)} C dW_s$$

or

$$CX_t = \int_0^t e^{D(t-s)} d\tilde{W}_s, \tag{8.3.35}$$

where $\tilde{W}_s = CW_s$ is a k-dimensional Wiener process with coordinate processes that are typically not independent. In fact, $\langle \tilde{W} \rangle_t = tCC^T$. Hence (i,j)th entry of

$$CK_tC^T = \int_0^t CX_sX_s^TC^T ds \tag{8.3.36}$$

is

$$a_{ij}(t) = \int_0^t \int_0^s e^{d_i(s-u)} d\tilde{W}_{iu} \int_0^s e^{d_j(s-u)} d\tilde{W}_{ju} ds, \tag{8.3.37}$$

where $d_i, i = 1, \cdots, k$, are the diagonal elements of D, i.e., the eigenvalues of B. If $d_i > 0$ and $d_j > 0$, the Toeplitz lemma implies that

$$e^{-t(d_i+d_j)} a_{ij}(t) \to (d_i + d_j)^{-1} \int_0^\infty e^{-d_iu} d\tilde{W}_{iu} \int_0^\infty e^{-d_ju} d\tilde{W}_{ju}$$

almost surely as $t \to \infty$. If $d_i < 0$ and $d_j < 0$,

$$t^{-1} a_{ij}(t) \to -\zeta_{ij}/(d_i + d_j)$$

almost surely as $t \to \infty$, where ζ_{ij} is the (i,j)th entry of CC^T. This is because the two processes under the Lebesgue integral in (8.3.37) are in this case ergodic. Finally, suppose $d_i > 0$ and $d_j < 0$. Then

$$t^{-\frac{1}{2}} e^{-d_i t} a_{ij}(t) \to 0$$

in probability as $t \to \infty$ by a generalization of the Toeplitz lemma because $\int_0^t e^{d_j(t-s)} d\tilde{W}_{js}$ converges in distribution and hence is stochastically bounded.

Thus, if all eigenvalues of B are real and different from zero, we can define $D_t(\theta)$ by

$$D_t(\theta)^T = \begin{pmatrix} A_t C & 0 & \cdots & 0 \\ 0 & A_t C & & \vdots \\ \vdots & & \ddots & \vdots \\ 0 & \cdots & 0 & A_t C \end{pmatrix},$$

where A_t is the diagonal matrix $A_t = \text{diag}(\varphi_1(t), \cdots, \varphi_k(t))$, with

$$\varphi_i(t) = \begin{cases} e^{-d_i t} & \text{if } d_i > 0 \\ t^{-\frac{1}{2}} & \text{if } d_i < 0. \end{cases}$$

We have shown that $D_t(\theta)^T I_t D_t(\theta) \to \eta^2(\theta)$ in probability as $t \to \infty$, where $\eta^2(\theta)$ is random if one or more of the eigenvalues of B are positive. It is not a restriction of the generality to assume that $d_i < 0$ for $i = 1, \cdots, m$ and $d_i > 0$ for $i = m+1, \cdots, k$, where $0 \leq m \leq k$. Then η^2 has the form

$$\eta^2 = \begin{pmatrix} \eta_1^2 & O \\ O^T & \eta_2^2 \end{pmatrix},$$

where O is an $m \times (k-m)$-matrix with all entries equal to zero. To prove that η^2 is positive definite it is enough to prove that each of the two matrices η_1^2 and η_2^2 is positive definite. Obviously, they are positive semi-definite, so it is enough to prove that they are invertible. To impose necessary conditions for this to hold, we need the concept of controllability. Let R and V be $d \times d$-matrices, where V is positive semi-definite; then (R, V) are called *controllable* if the rank of the $d \times d^2$-matrix $[V, \ RV, \cdots, R^{d-1}V]$ is d. We need the following result, which is well-known in control theory (for a proof see Davis (1977, Chapter 4) or Chaleyat-Maurel and Elie (1981, Lemma 1.10)): The matrix $\int_0^t \exp(sR) V \exp(sR^T) ds$ is invertible (i.e., positive definite) for $t > 0$ if and only if (R, V) are controllable.

Now, $\eta_1^2 = \int_0^\infty \exp(sR_1)V_1 \exp(sR_1)ds$, where $R = \text{diag}(d_1, \cdots, d_m)$ and V_1 is the upper left $m \times m$-submatrix of CC^T, so η_1^2 is positive definite if (R_1, V_1) are controllable. This is, for instance, the case if V_1 is invertible. The random matrix η_2^2 is positive definite if the matrix δ with entries $\delta_{ij} = (d_i + d_j)^{-1}$ is positive definite. Since $\delta = \int_0^\infty \exp(sR_2)V_2 \exp(sR_2)ds$, where $R_2 = \text{diag}(-d_{m+1}, \cdots, -d_k)$ and V_2 is the $(k-m) \times (k-m)$-matrix with all entries equal to one, we see that η_2^2 is positive definite if (R_2, V_2) are controllable. This is the case if and only if the eigenvalues $d_{m+1}, \cdots, d_k$ are all different.

We have now given conditions ensuring that (8.3.17) is satisfied. Let us turn to condition (8.3.18). Since

$$E(a_{ij}(t)) = \zeta_{ij}(d_i + d_j)^{-2} \left\{ e^{(d_i+d_j)t} - 1 - t(d_i + d_j) \right\},$$

we see that

$$A_t C E_\theta(K_t) C^T A_t \to \Sigma(\theta),$$

where

$$\Sigma_{ij}(\theta) = \begin{cases} \zeta_{ij}(d_i + d_j)^{-2} & \text{if } d_i > 0 \text{ and } d_j > 0 \\ -\zeta_{ij}(d_i + d_j)^{-1} & \text{if } d_i < 0 \text{ and } d_j < 0 \\ 0 & \text{if } d_i > 0 \text{ and } d_j < 0. \end{cases}$$

As $i_t(\theta) = E_\theta(I_t)$, we see that we just have to check that Σ is positive definite. Under the assumption made earlier, that $d_i < 0$ for $i = 1, \cdots, m$ and $d_i > 0$ for $i = m+1, \cdots, k$, where $0 \le m \le k$, the matrix Σ is of the form

$$\Sigma = \begin{pmatrix} \Sigma_{11} & O \\ O^T & \Sigma_{22} \end{pmatrix}.$$

We have already seen that $\Sigma_{11} = \eta_1^2$ is positive definite if (R_1, V_1) is controllable. Note that $\Sigma_{22} = \int_0^\infty \exp(sR_2)V_3 \exp(sR_2)ds$, where the (i,j)th entry of the $(k-m) \times (k-m)$-matrix V_3 is $\zeta_{ij}/(d_i + d_j)$, $m+1 \le i, j \le k$. If V_3 is invertible, (R_2, V_3) are controllable, so that by the above result from control theory the matrix Σ_{22} is positive definite. That V_3 is invertible follows in the same way as the invertibility of η_1^2, if we assume that (R_2, V_4) are controllable, where V_4 is the lower right $(k-m) \times (k-m)$-submatrix of CC^T. In particular, we can assume that V_4 is invertible.

We have thus proved consistency and, after normalization as described above, asymptotic normality of the maximum likelihood estimator provided that all eigenvalues of B are real and different from zero, that all positive eigenvalues are different, and that CC^T is positive definite.

It is important that $D_t(\theta)$ is not restricted to be a diagonal matrix. Consider, for instance, the 2-dimensional diffusion with

$$B = \begin{pmatrix} 1 & 0 \\ 1 & -1 \end{pmatrix}.$$

Then B^2 equals the identity matrix, so $e^{Bu} = I\cosh u + B\sinh u$. Hence

$$X_t = \begin{pmatrix} \int_0^t e^{(t-s)} dW_{1s} \\ \int_0^t \sinh(t-s) dW_{1s} + \int_0^t e^{-(t-s)} dW_{2s} \end{pmatrix}.$$

It is not difficult to see that all entries of $E(K_t)$ are of order e^{2t} and that

$$e^{-2t}E(K_t) \to \begin{pmatrix} \frac{1}{4} & \frac{1}{8} \\ \frac{1}{8} & \frac{1}{16} \end{pmatrix},$$

which is singular, so there is no way of normalizing K_t and $E(K_t)$ with a diagonal matrix to obtain a non-singular limit. However, B can be diagonalized:

$$\begin{pmatrix} 1 & 0 \\ 1 & -1 \end{pmatrix} = \begin{pmatrix} 2 & 0 \\ 1 & 1 \end{pmatrix} \begin{pmatrix} 1 & 0 \\ 0 & -1 \end{pmatrix} \begin{pmatrix} \frac{1}{2} & 0 \\ -\frac{1}{2} & 1 \end{pmatrix},$$

so the result proved above holds.

8.4 Information matrices

A symmetric $k \times k$-matrix $\mathcal{I}_t(\theta)$ of $\{\mathcal{F}_t\}$-adapted (possibly non-random) processes is called an *information matrix* if the following holds for all $\theta \in \Theta$:

(i) $P_\theta(\mathcal{I}_t(\theta)$ is positive semi-definite$) \to 1$ as $t \to \infty$;

(ii) either

$$(\hat{\theta}_t - \theta)^T \mathcal{I}_t(\theta)(\hat{\theta}_t - \theta) \to F, \tag{8.4.1}$$

or, provided that $\mathcal{I}_t(\theta)$ is invertible,

$$\dot{l}_t(\theta)^T \mathcal{I}_t(\theta)^{-1} \dot{l}_t(\theta) \to G \tag{8.4.2}$$

in distribution under P_θ as $t \to \infty$, where F and G are non-degenerate distribution functions on $\mathbb{R}_+$.

We also call $\mathcal{I}_t(\theta)$ an information matrix in cases where the conditions hold only conditionally on an event C_θ with $P_\theta(C_\theta) > 0$. Condition (8.4.1) of course only makes sense in cases where the maximum likelihood estimator exists for t large enough. In many cases, F and G both equal the χ^2-distribution with k degrees of freedom.

We have already encountered the two classical information matrices, the observed information matrix $j_t(\theta)$ given by (8.3.1) and the expected information matrix (Fisher information) $i_t(\theta)$ given by (8.3.12). That $j_t(\theta)$ is an information matrix in the above sense follows from (8.3.5) and (8.3.29). That $i_t(\theta)$ is an information matrix follows from (8.3.13), (8.3.18), and (8.3.19).

We shall consider two other important information matrices too. The first is the *incremental observed information* $J_t(\theta)$ defined as the quadratic variation of the score martingale $\dot{l}_t(\theta)$ under P_θ, i.e.,

$$J_t(\theta) = [\dot{l}(\theta)]_t, \tag{8.4.3}$$

or, using (A.4.3),

$$\sum_{j=1}^{2^m} [\Delta_j^m \dot{l}_t(\theta)][\Delta_j^m \dot{l}_t(\theta)]^T \to J_t(\theta) \tag{8.4.4}$$

in P_θ-probability as $m \to \infty$, where

$$\Delta_j^m \dot{l}_t(\theta) = \dot{l}_{jt2^{-m}}(\theta) - \dot{l}_{(j-1)t2^{-m}}(\theta). \tag{8.4.5}$$

The concept of the quadratic variation of a semimartingale is discussed in more detail in Appendix A. Here we just state the following facts. If the process $\dot{l}(\theta)$ has no continuous martingale part, the incremental observed information is simply given by

$$J_t(\theta) = \sum_{s \le t} [\Delta \dot{l}_s(\theta)][\Delta \dot{l}_s(\theta)]^T; \tag{8.4.6}$$

cf. (A.4.8). For discrete time models in particular we have that

$$J_t(\theta) = \sum_{j=1}^{t} [\dot{l}_j(\theta) - \dot{l}_{j-1}(\theta)][\dot{l}_j(\theta) - \dot{l}_{j-1}(\theta)]^T. \tag{8.4.7}$$

If $\dot{\phi}_t(\theta)$ is a continuous function of t with bounded variation in compact intervals, it follows from (8.1.11) that

$$J_t(\theta) = \dot{\gamma}(\theta)^T [B]_t \dot{\gamma}(\theta); \tag{8.4.8}$$

cf. (A.4.8).

Now suppose $\gamma_t(\theta) \in \operatorname{int} \tilde{\Gamma}_t$ for all $t > 0$. Then the score process $\dot{l}_t(\theta)$ is a square integrable martingale under P_θ (Corollary 8.1.3). Therefore, it has a quadratic characteristic $\langle \dot{l}(\theta) \rangle$, see Appendix A, Section 3. The *incremental expected information* $I_t(\theta)$ is defined as

$$I_t(\theta) = \langle \dot{l}(\theta) \rangle_t. \tag{8.4.9}$$

This is a predictable process such that $\dot{l}_t(\theta)\dot{l}_t(\theta)^T - I_t(\theta)$ is a P_θ-martingale. In most models used in statistical practice $I_t(\theta)$ can be calculated by the formula

$$\sum_{j=1}^{2^m} E_\theta([\Delta_j^m \dot{l}_t(\theta)][\Delta_j^m \dot{l}_t(\theta)]^T | \mathcal{F}_{(j-1)t2^{-m}}) \to I_t(\theta) \tag{8.4.10}$$

in P_θ-probability as $m \to \infty$, where $\Delta_j^m \dot{l}_t(\theta)$ is given by (8.4.5). This holds, for instance, if $\dot{l}(\theta)$ is a quasi-left-continuous process. For a definition of this concept see Appendix A, Section 3. It essentially means that the jumps of the process cannot be predicted. Generally, the convergence in (8.4.10) is weak convergence in $L_1(P_\theta)$. For discrete time processes,

$$I_t(\theta) = \sum_{j=1}^{t} E_\theta([\dot{l}_j(\theta) - \dot{l}_{j-1}(\theta)][\dot{l}_j(\theta) - \dot{l}_{j-1}(\theta)]^T | \mathcal{F}_{j-1}). \tag{8.4.11}$$

The following are consequences of general results in Appendix A about the quadratic variation and the quadratic characteristic. We suppose that $\gamma_t(\theta) \in \operatorname{int} \tilde{\Gamma}_t$ for all $t > 0$. The matrix-valued processes $J_t(\theta)$ and $I_t(\theta)$ are positive semi-definite for all $t > 0$, and they are increasing in the partial ordering of the positive semi-definite matrices, i.e., $I_t(\theta) - I_s(\theta)$ and $J_t(\theta) - J_s(\theta)$ are positive semi-definite for $t > s$. The predictable compensator of the process $J_t(\theta)$ is $I_t(\theta)$; in fact, each component of $J_t(\theta) - I_t(\theta)$ is a P_θ-martingale. Since $\dot{l}_t(\theta)\dot{l}_t(\theta)^T - I_t(\theta)$ is a P_θ-martingale, (8.1.12) and (8.3.10) imply that $j_t(\theta) - I_t(\theta)$ is a P_θ-martingale too. This shows that $j_t(\theta)$ is a matrix submartingale in the sense that $E_\theta(j_t(\theta)|\mathcal{F}_s) - j_s(\theta) = E_\theta(I_t(\theta) - I_s(\theta)|\mathcal{F}_s)$ is positive semi-definite for $t > s$. In particular, each diagonal element of $j_t(\theta)$ is a submartingale, and the unique increasing predictable process in the Doob-Meyer decomposition of $j_t(\theta)_{ii}$ is $I_t(\theta)_{ii}$. If $\ddot{\gamma}_t(\theta)_{ij}^T B_t$ is predictable and $j_t(\theta)$ has finite variation, then $j_t(\theta) = I_t(\theta)$.

It is an interesting question when the same limiting distribution is obtained for two different information matrices $\mathcal{I}_t^{(1)}(\theta)$ and $\mathcal{I}_t^{(2)}(\theta)$ in (8.4.1) and (8.4.2). This is the case if $\mathcal{I}_t^{(1)}(\theta)$ and $\mathcal{I}_t^{(2)}(\theta)$ are *asymptotically equivalent* in the sense that for all $\theta \in \Theta$ there exists a family of non-random invertible $k \times k$-matrices $D_t(\theta)$ such that

$$D_t(\theta)^T \{\mathcal{I}_t^{(1)}(\theta) - \mathcal{I}_t^{(2)}(\theta)\} D_t(\theta) \to 0 \tag{8.4.12}$$

and

$$D_t(\theta)^T \mathcal{I}_t^{(1)}(\theta) D_t(\theta) \to W(\theta) \tag{8.4.13}$$

in P_θ-probability as $t \to \infty$. Here $W(\theta)$ is of the type considered earlier. Note that (8.4.12) and (8.4.13) imply that

$$D_t(\theta)^{-1} \mathcal{I}_t^{(1)}(\theta)^{-1} \mathcal{I}_t^{(2)}(\theta) D_t(\theta) \to I_k \tag{8.4.14}$$

in P_θ-probability on the event $\{\det(W(\theta)) > 0\}$. On this event, $\mathcal{I}_t^{(1)}(\theta)$ is invertible for t large enough. Note also that (8.4.1) and (8.4.13) imply that the process $D_t(\theta)^{-1}(\hat{\theta}_t - \theta)$ is stochastically bounded under P_θ on $\{\det(W(\theta)) > 0\}$ for t large enough, and that (8.4.2) and (8.4.13) imply a similar result for $D_t(\theta)^T \dot{l}_t(\theta)$.

Suppose $\mathcal{I}_t^{(1)}(\theta) = J_t(\theta)$ and that $\mathcal{I}_t^{(1)}(\theta)$ and $\mathcal{I}_t^{(2)}(\theta)$ satisfy (8.4.14). Under the conditions of Theorem 8.3.3 (or in discrete time Theorem 8.3.5),

$J_t(\theta)$ satisfies (8.4.2) conditionally on $\det(W(\theta)) > 0$, and (8.4.14) implies that (under the same conditions) $\mathcal{I}_t^{(2)}(\theta)$ satisfies (8.4.2) too. Thus (8.4.14) ensures that $\mathcal{I}_t^{(2)}(\theta)$ is an information matrix. It is not difficult to prove that this is also the case if I_k is replaced by another non-random matrix in (8.4.14). In fact, I_k can even be replaced by a random matrix. Analogous remarks can be made concerning (8.4.1).

We have seen that the differences between any two of $j_t(\theta)$, $J_t(\theta)$ and $I_t(\theta)$ are P_θ-martingales. If such a martingale is square integrable, we can apply Liptser's law of large numbers for martingales (Theorem A.7.3) to prove asymptotic equivalence of the two information matrices, provided of course that they are asymptotically equivalent. Suppose, specifically, that $\mathcal{I}_t^{(1)}(\theta) - \mathcal{I}_t^{(2)}(\theta)$ is a square integrable P_θ-martingale, and let $Q_t^{(i,j)}(\theta)$ denote the quadratic characteristic under P_θ of the (i,j)th component of this martingale. For simplicity, assume that $D_t(\theta)$ is a diagonal matrix. Then

$$\int_0^\infty D_t(\theta)_{ii}^2 D_t(\theta)_{jj}^2 dQ_t^{(i,j)}(\theta) < \infty \tag{8.4.15}$$

for $i, j = 1, \cdots, k$ and $D_t(\theta) \to 0$ as $t \to \infty$ imply (8.4.12).

Models where the observed and the expected information matrices are not asymptotically equivalent are sometimes called *non-ergodic models*, while models for which these two information matrices are equivalent are called ergodic models. For non-ergodic models, $W(\theta)$ in (8.4.13) with $\mathcal{I}_t^{(1)}(\theta) = j_t(\theta)$ is a random matrix. For non-ergodic models, the three random information matrices are often asymptotically equivalent too. It is, however, not difficult to give examples where this is not the case; see Exercise 8.6.

In the rest of this section we assume that the conditions of Theorem 8.3.4 (or for discrete time Theorem 8.3.5) hold and that the matrices $j_t(\theta)$, $J_t(\theta)$, and $I_t(\theta)$ are asymptotically equivalent. Then for $\mathcal{I}_t(\theta)$ equal to one of $j_t(\theta)$, $J_t(\theta)$, or $I_t(\theta)$ we have under P_θ that

$$\dot{l}_t(\theta)^T \mathcal{I}_t(\theta)^{-1} \dot{l}_t(\theta) \to \chi^2(k) \tag{8.4.16}$$

and

$$(\hat{\theta}_t - \theta)^T \mathcal{I}_t(\theta)(\hat{\theta}_t - \theta) \to \chi^2(k) \tag{8.4.17}$$

in distribution as $t \to \infty$ conditionally on $\det(W(\theta)) > 0$. If, moreover, the conditions of Theorem 8.3.1 and an equivalent of (8.3.2) for $I_t(\theta)$ and $J_t(\theta)$ hold with $D_t(\theta)$ and $W(\theta)$ as in Condition 8.3.2, then we have the more easily applicable results

$$\dot{l}_t(\theta)^T \mathcal{I}_t(\hat{\theta}_t)^{-1} \dot{l}_t(\theta) \to \chi^2(k) \tag{8.4.18}$$

and

$$(\hat{\theta}_t - \theta)^T \mathcal{I}_t(\hat{\theta}_t)(\hat{\theta}_t - \theta) \to \chi^2(k) \tag{8.4.19}$$

in distribution under P_θ as $t \to \infty$.

Under the conditions given, normalization with an information matrix that is not asymptotically equivalent to $j_t(\theta)$ and $J_t(\theta)$ will result in a limit distribution that is different from the $\chi^2(k)$-distribution. Typically, it will be the square of a normal variance mixture. This is, for instance, the case when the maximum likelihood estimator is normalized by the expected information in a non-ergodic model. Therefore (and for other reasons, see Barndorff-Nielsen and Sørensen, 1994), it is important to use a random information matrix in non-ergodic models.

Suppose all eigenvalues of $J_t(\theta)$ tend to infinity at the same rate. Then $D_t(\theta)$ can be defined by

$$D_t(\theta) = i_t(\theta)_{11}^{-\frac{1}{2}} I_k, \tag{8.4.20}$$

and (8.3.21) becomes

$$J_t(\theta)^{-\frac{1}{2}} j_t(\theta)(\hat{\theta}_t - \theta) \to N(0, I_k). \tag{8.4.21}$$

Hence under the conditions imposed above,

$$\mathcal{I}_t(\theta)^{-\frac{1}{2}} \dot{l}_t(\theta) \to N(0, I_k) \tag{8.4.22}$$

and

$$\mathcal{I}_t(\hat{\theta}_t)^{\frac{1}{2}} (\hat{\theta}_t - \theta) \to N(0, I_k) \tag{8.4.23}$$

in distribution under P_θ as $t \to \infty$. As above, $\mathcal{I}_t(\theta)$ can equal $j_t(\theta)$, $J_t(\theta)$, or $I_t(\theta)$.

8.5 Local asymptotic mixed normality

In this section we shall briefly discuss conditions ensuring that exponential families of stochastic processes are locally asymptotically mixed normal.

Consider a filtered statistical space $(\Omega, \mathcal{F}, \{\mathcal{F}_t\}, \mathcal{P})$, $\mathcal{P} = \{P_\theta : \theta \in \Theta\}$, $\Theta \subseteq \mathbb{R}^k$, with $\operatorname{int}\Theta \neq \emptyset$. We assume that there exists a probability measure P on $\mathcal{F}$ such that $P_\theta^t \ll P^t$ for all $\theta \in \Theta$ and all $t > 0$. This assumption can be weakened, but for exponential families there is no reason to do so. As usual, the log-likelihood function on $\mathcal{F}_t$ is given by $l_t(\theta) = \log(dP_\theta^t / dP^t)$.

Definition 8.5.1 *The filtered statistical space* $(\Omega, \mathcal{F}, \{\mathcal{F}_t\}, \mathcal{P})$ *is called* locally asymptotically mixed normal *at* $\theta_0 \in \operatorname{int}\Theta$ *if*

(i) There exists a family $\{Z_t(\theta_0)\}$ *of* k*-dimensional random vectors and a family* $\{T_t(\theta_0)\}$ *of symmetric positive semidefinite random* $k \times k$*-matrices such that* $Z_t(\theta_0)$ *and* $T_t(\theta_0)$ *are* $\mathcal{F}_t$*-measurable for all* $t > 0$*,* $T_t(\theta_0)$ *is almost surely positive definite for* t *large enough under* P_{θ_0}*, and*

$$l_t(\theta_0 + \delta_t h) - l_t(\theta_0) - h^T T_t(\theta_0)^{\frac{1}{2}} Z_t(\theta_0) - \tfrac{1}{2} h^T T_t(\theta_0) h$$

converges to zero in P_{θ_0}-probability for every $h \in \mathbb{R}^k$, where $\{\delta_t\}$ is a family of positive definite matrices satisfying $\delta_t \to 0$ as $t \to \infty$.

(ii) There exists a P_{θ_0}-almost surely positive definite random matrix $T(\theta_0)$ such that

$$(Z_t(\theta_0), T_t(\theta_0)) \to (Z(\theta_0), T(\theta_0))$$

in distribution under P_{θ_0} as $t \to \infty$. Here $Z(\theta_0)$ is a k-dimensional standard normal distributed random vector independent of $T(\theta_0)$. □

Remarks: The vector $\theta_0 + \delta_t h$ need not belong to Θ for all $t > 0$. However, since $\theta_0 \in \operatorname{int}\Theta$, it will belong to Θ when t is large enough for every $h \in \mathbb{R}^d$. Some authors do not require that the matrices $\{\delta_t\}$ be positive definite. The statistical space is called *locally asymptotically normal* if $T(\theta_0)$ is non-random. □

For filtered statistical spaces that are locally asymptotically mixed normal we have the following result about the asymptotic distribution of estimators. This is a version of the so-called *convolution theorem.* Here $\mathcal{L}(X_t|P_\theta^t)$ denotes the distribution of the random vector X_t under P_θ^t and $\mathcal{L}(X_t|P_\theta^t) \to \mathcal{L}(X)$ denotes convergence in distribution.

Theorem 8.5.2 *Suppose the filtered statistical space $(\Omega, \mathcal{F}, \{\mathcal{F}_t\}, \mathcal{P})$ is locally asymptotically mixed normal at $\theta_0 \in \operatorname{int}\Theta$, and let $\{V_t\}$ be a family of estimators of θ such that V_t is $\mathcal{F}_t$-measurable and such that for every $h \in \mathbb{R}^k$,*

$$\mathcal{L}(T_t(\theta_0), \delta_t^{-1}(V_t - \theta_0 - \delta_t h)|P_{\theta_0+\delta_t h}^t) \to \mathcal{L}(T(\theta_0), V(\theta_0)) \tag{8.5.1}$$

as $t \to \infty$ for some random k-dimensional vector $V(\theta_0)$. Let $\mathcal{L}_{T(\theta_0)}$ be a regular conditional distribution of $V(\theta_0)$ given $T(\theta_0)$. Then there exists a stochastic kernel $K_{T(\theta_0)}$ such that almost surely

$$\mathcal{L}_{T(\theta_0)} = K_{T(\theta_0)} * N(0, T(\theta_0)^{-1}). \tag{8.5.2}$$

For general models we have that if the log-likelihood function is twice continuously differentiable at θ_0, then

$$l_t(\theta_0 + \delta_t h) - l_t(\theta_0) = h^T \delta_t \dot{l}_t(\theta_0) - \tfrac{1}{2} h^T \delta_t j_t(\tilde{\theta}_t) \delta_t h, \tag{8.5.3}$$

where $j_t(\theta)$ is the observed information matrix and $\tilde{\theta}_t$ is a convex combination of θ_0 and $\theta_0 + \delta_t h$. Hence

$$\begin{aligned} l_t(\theta_0 + \delta_t h) - l_t(\theta_0) - h^T T_t(\theta_0)^{\frac{1}{2}} Z_t(\theta_0) - \tfrac{1}{2} h^T T_t(\theta_0) h \quad (8.5.4)\\ = \tfrac{1}{2} h^T \delta_t [J_t(\theta_0) - j_t(\tilde{\theta}_t)] \delta_t h, \end{aligned}$$

where $J_t(\theta)$ is the incremental observed information matrix,

$$Z_t(\theta_0) = -(\delta_t J_t(\theta_0) \delta_t)^{-\frac{1}{2}} \delta_t \dot{l}_t(\theta_0), \tag{8.5.5}$$

and

$$T_t(\theta_0) = -\delta_t J_t(\theta_0)\delta_t. \tag{8.5.6}$$

Now consider a general exponential family on a filtered space with exponential representation (8.1.1). Under the conditions of Theorem 8.3.3 with $D_t(\theta_0) = \delta_t$ (or in the case of discrete time possibly Theorem 8.3.5), we have conditionally on $\{\det(W(\theta)) > 0\}$ that

$$(Z_t(\theta_0), T_t(\theta_0)) \to (Z(\theta_0), W(\theta_0)) \tag{8.5.7}$$

in distribution under P_{θ_0} as $t \to \infty$, where $Z(\theta_0)$ is a k-dimensional standard normally distributed random vector independent of $W(\theta_0)$. Of course, $Z_t(\theta_0)$ is only well-defined if $T_t(\theta_0)$ is strictly positive definite. On the event $\{\det(W(\theta_0)) > 0\}$ the probability that $T_t(\theta_0)$ is not strictly positive definite tends to zero as $t \to \infty$. We will assume that for t large enough $J_t(\theta_0)$ (and hence $T_t(\theta_0)$) is almost surely strictly positive definite. In order to ensure that the model is locally asymptotically mixed normal we only need that the right-hand side of (8.5.4) converges to zero in P_{θ_0}-probability on the event $\{\det(W(\theta_0)) > 0\}$. This is the case provided that $J_t(\theta_0)$ and $j_t(\theta_0)$ are asymptotically equivalent in the sense of (8.4.12) and that $j_t(\theta_0)$ satisfies (8.3.5). We see that under the conditions imposed in Section 8.3 the exponential family is locally asymptotically mixed normal conditional on $\{\det(W(\theta)) > 0\}$. Ergodic models are locally asymptotically normal under the same conditions. In Chapter 9 we shall see that statistical inference for stochastic process models can be much more complicated than it is for the rather well-behaved locally asymptotically mixed normal models.

8.6 Exercises

8.1 Write, in the case of a one-dimensional parameter θ, a detailed expression for the fourth likelihood martingale ($|I| = 4$) by means of (8.1.17).

8.2 Let Z and $W(\theta)$ be as in Theorem 8.3.3. Prove that the characteristic function of the distribution of $W(\theta)^{1/2}Z$ under P_θ is $u \to E_\theta(\exp[-\frac{1}{2}u^T W(\theta)u])$, where $u = (u_1, \cdots, u_k)^T$.

8.3 Prove Theorem 8.3.5 in detail.

8.4 Consider the linear birth-and-death process introduced in Section 3.4. Let the notation be as in that section and suppose that $\lambda \neq \mu$. Show that

$$X_t - x_0 - (\lambda - \mu)\int_0^t X_s ds$$

is a zero-mean martingale. Hint: Corollary 8.1.3 and Exercise 7.3.

Deduce from this that $E_{\lambda,\mu}(X_t|\mathcal{F}_s) = e^{(\lambda-\mu)(t-s)}X_s$, and show that $e^{-(\lambda-\mu)t}X_t$ converges almost surely to a non-negative random variable U as $t \to \infty$. Hint: Theorem A.7.1. Finally, conclude that $(\lambda - \mu)e^{-(\lambda-\mu)t}S_t \to U$ almost surely for $\lambda > \mu$. Hint: the Toeplitz lemma (Lemma B.3.1).

8.5 Consider an exponential family of processes with the exponential representation (8.1.1), and let $D_t(\theta)$ be defined as in Theorem 8.3.1. Show that (8.3.2) is satisfied if the following conditions hold:

$$D_t(\theta)^T \ddot{\phi}_t(\eta_1, \cdots, \eta_k) D_t(\theta) \to G(\eta_1, \cdots, \eta_k; \theta)$$

and

$$D_t(\theta)^T \left[\sum_{l=1}^{m} \ddot{\gamma}_t^{(l)}(\eta_1, \cdots, \eta_k) B_t^{(l)} \right] D_T(\theta)$$
$$\to \left\{ \sum_{l=1}^{m} h_{pq}^{(l)}(\eta_1, \cdots, \eta_k; \theta) V_{pq}^{(l)}(\theta) \right\}_{p,q=1,\cdots,k}$$

as $t \to \infty$ uniformly in $(\eta_1, \cdots, \eta_k)$ when $(\eta_1, \cdots, \eta_k)$ belongs to a compact set containing $(\theta, \cdots, \theta)$ as an interior point. The latter convergence is P_θ-almost sure. Here $V^{(l)}(\theta)$ $(l = 1, \cdots, m)$ are random matrices such that the matrix

$$W(\theta) - G(\theta, \cdots, \theta; \theta) - \sum_{l=1}^{m} V^{(l)}(\theta)$$

is positive semi-definite, while the functions $G(\eta_1, \cdots, \eta_k; \theta)_{pq}$ and $h_{pq}^{(l)}(\eta_1, \cdots, \eta_k; \theta)$ $(p, q = 1, \cdots, k,\ l = 1, \cdots, m)$ are continuous in $(\eta_1, \cdots, \eta_k)$ when $(\eta_1, \cdots, \eta_k)$ belongs to a neighbourhood of $(\theta, \cdots, \theta)$, and $h_{pq}^{(l)}(\theta, \cdots, \theta; \theta) = 1$ $(p, q = 1, \cdots, k,\ l = 1, \cdots, m)$. Note how the conditions simplify when the exponential family is time-homogeneous and $D_t(\theta) = d_t(\theta)I_k$, where $d_t(\theta)$ is real-valued and I_k is the $k \times k$ identity matrix.

8.6 Consider the Gaussian autoregression defined by

$$X_t = \theta X_{t-1} + Z_t, \quad t = 1, 2, \cdots,$$

where $\theta \in \mathbb{R}$, $X_0 = 0$, and the Z_t's are independent random variables with a standard normal distribution. Show that this is an exponential family of processes with incremental observed information (8.4.3) given by

$$J_t(\theta) = \sum_{k=1}^{t} X_{k-1}^2 (X_k - \theta X_{k-1})^2,$$

and that the incremental expected information (8.4.9) equals the expected information (8.3.1) and is given by

$$I_t = \sum_{k=1}^{t} X_{k-1}^2.$$

From now on consider only the case $|\theta| > 1$. Show that there exists a random variable $Y(\theta)$ such that

$$\theta^{-t} X_t \to Y(\theta)$$

P_θ-almost surely as $t \to \infty$ (Hint: the martingale convergence theorem, Theorem A.7.2), and conclude from this that

$$\theta^{-2t} I_t \to Y(\theta)^2/(\theta^2 - 1)$$

P_θ-almost surely as $t \to \infty$ (Hint: the Toeplitz lemma, Lemma B.3.1). Finally, show that

$$\theta^{-2t} J_t(\theta) \to Y(\theta)^2 U_\infty(\theta)$$

P_θ-almost surely as $t \to \infty$, where $U_\infty(\theta)$ is the P_θ-almost sure limit of

$$U_t(\theta) = \sum_{j=1}^{t} \theta^{-2(t-j+1)} Z_j^2.$$

Hint: To prove that $U_t(\theta)$ converges, apply the supermartingale convergence theorem (Theorem A.7.1) to $U_t(\theta) - E_\theta(U_t(\theta))$.

Since $U_\infty(\theta)$ is a non-degenerate random variable (with mean value $(\theta^2 - 1)^{-1}$), it has now been proved that $J_t(\theta)$ and I_t are not asymptotically equivalent. Specifically,

$$J_t(\theta)/I_t \to U_\infty(\theta)/(\theta^2 - 1)$$

P_θ-almost surely.

8.7 Let X be the inhomogeneous Poisson process with intensity $\exp(\theta_1 + \theta_2 t)$, $(\theta_1, \theta_2) \in \mathbb{R}^2$, and let $\tau_1, \tau_2, \cdots$ denote the jump times of X. Show that this is an exponential family of processes and that the score vector, when X has been observed continuously in $[0, t]$, is given by

$$\dot{l}_t(\theta_1, \theta_2) = \begin{pmatrix} X_t - \int_0^t \exp(\theta_1 + \theta_2 s) ds \\ \sum_{j=1}^{X_t} \tau_j - \int_0^t s \exp(\theta_1 + \theta_2 s) ds \end{pmatrix}$$

and is square integrable. Next show that the incremental observed information is

$$J_t(\theta_1, \theta_2) = \begin{bmatrix} X_t & \sum_{j=1}^{X_t} \tau_j \\ \sum_{j=1}^{X_t} \tau_j & \sum_{j=1}^{X_t} \tau_j^2 \end{bmatrix}$$

and that the incremental expected information, the observed information, and the expected information coincide and are given by

$$j_t(\theta_1, \theta_2) = \begin{bmatrix} \int_0^t \exp(\theta_1 + \theta_2 s) ds & \int_0^t s \exp(\theta_1 + \theta_2 s) ds \\ \int_0^t s \exp(\theta_1 + \theta_2 s) ds & \int_0^t s^2 \exp(\theta_1 + \theta_2 s) ds \end{bmatrix}.$$

Finally, show that $J_t(\theta_1, \theta_2)_{kk} / j_t(\theta_1, \theta_2)_{kk} \to \infty$ P_{θ_1, θ_2}-almost surely as $t \to \infty$ for $k = 1, 2$. Hint: the law of large numbers for martingales (Theorem A.7.3).

8.7 Bibliographic notes

The material in the Sections 8.1 and 8.2 has not been published before, except Proposition 8.1.4, which is from Franz and Winkler (1987). The results in Section 8.2 are mainly adaptations of results for classical non-curved exponential families to the curved exponential families of stochastic processes. The classical results can be found in Barndorff-Nielsen (1978) and Brown (1986). Example 8.2.7 has been studied by Stefanov (1995). Theorem 8.3.1 is an improved version of a result in Barndorff-Nielsen and Sørensen (1984), which was again based on work by Sweeting (1980) and Aitchison and Silvey (1958). The results on asymptotic normality are applications of the central limit theorem for multivariate martingales in Küchler and Sørensen (1996b). For a general discussion of asymptotic likelihood theory for stochastic processes, see Barndorff-Nielsen and Sørensen (1994). The material on Gaussian diffusions is taken from Küchler and Sørensen (1996b). Likelihood theory for multivariate Gaussian diffusions has been investigated by several other authors including Le Breton (1977, 1984), Le Breton and Musiella (1982, 1985), and Stockmarr (1996). The discussion of information matrices in Section 8.4 is to a large extent based on Barndorff-Nielsen and Sørensen (1994), where also some further information matrices are discussed, in particular model robust information matrices. Definition 8.5.1 and Theorem 8.5.2 are taken from Jeganathan (1982); see also Basawa and Scott (1983), Le Cam and Yang (1990), and Genon-Catalot and Picard (1993).

The result (8.3.19) and similar results in Sections 5.2 and 5.5 indicate that inference on θ should be drawn conditionally on the incremental observed information or the observed information in cases where $W(\theta)$ is random. This question has not been studied in depth, but some results on conditional inference for stochastic processes can be found in Feigin and Reiser (1979), Sweeting (1980, 1986, 1992), Basawa and Brockwell (1984), and Jensen (1987).

9

Linear Stochastic Differential Equations with Time Delay

In this chapter we consider examples of exponential families of processes generated by linear stochastic differential equations with time delay. These processes are diffusion-type processes as introduced in Section 3.5. The examples illustrate that the asymptotic behaviour of the maximum likelihood estimator can depend strongly on the value of the statistical parameter. In contrast to the case of independent observations, where the picture is quite nice, it turns out that even in relatively simply defined exponential families of stochastic processes complicated asymptotic properties may occur. The results in this chapter are mainly taken from Gushchin and Küchler (1996).

9.1 The differential equations and the maximum likelihood estimator

Let $W = (W(t), t \geq 0)$ be a standard Wiener process, let $r_0, r_1, \ldots, r_N$ be fixed nonnegative real numbers with $0 = r_0 < r_1 < \ldots < r_N = r$, and let

$$\vartheta^T = (\vartheta_0, \vartheta_1, \ldots, \vartheta_N)$$

be a parameter with $\vartheta \in \Theta = \mathbb{R}^{N+1}$. Consider the linear stochastic differential equation with time delay in the drift:

$$dX(t) = \sum_{j=0}^{N} \vartheta_j X(t - r_j)dt + dW(t), \qquad t \geq 0, \tag{9.1.1}$$

$$X(s) = X_0(s), \quad s \in [-r, 0], \tag{9.1.2}$$

where the given initial process $(X_0(s), s \in [-r,0])$ is supposed to be continuous and independent of $(W(t), t \geq 0)$, and to satisfy $E_\vartheta(X_0^2(s)) < \infty$, $s \in [-r, 0]$.

As a special case we find for $N = 0$ the Langevin equation with the Ornstein-Uhlenbeck process as solution (see Section 5.2). For $N \geq 1$ and $\sum_{j=1}^N |\theta_j| \neq 0$ the solutions of (9.1.1) are not Markov processes because of the influence of the past in the drift coefficient.

The equation (9.1.1) is an example of more general linear stochastic functional differential equations of the form

$$\begin{aligned} dX(t) &= \int_{-r}^{0} X(t+s)a(ds)dt + dM(t), & (9.1.3) \\ X(s) &= X_0(s), \quad s \in [-r, 0], & (9.1.4) \end{aligned}$$

where a is a function of bounded variation on $[-r, 0]$ and M is any process with appropriate trajectories, e.g., a semimartingale. (See, e.g., Mohammed and Scheutzow (1990)). From a statistical point of view, (9.1.3) is a nonparametric model. Some first results concerning estimation of $a(\cdot)$ can be found in Rothkirch (1993).

Here we restrict ourselves to differential equations of the form (9.1.1)–(9.1.2). They provide exponential families of processes in the sense considered in this book, because the parameter space $\Theta = \mathbb{R}^{N+1}$ is finite-dimensional. The more general case (9.1.3)–(9.1.4) yields solutions forming what could be called an infinite-dimensional exponential family in a rather reasonable sense. By infinite-dimensional is meant that the parameter of the exponential family is infinite. This will not be treated in this book. Infinite-dimensional exponential families have been only rarely studied in the literature; see, e.g., Soler (1977), who treated families of distributions, and, more recently and in a process context, Grigelionis (1996). The restriction to the case where the driving term is a Wiener process allows us to identify all effects arising in the asymptotic properties as effects caused by the time delay. The use of more general semimartingales may well generate additional complications.

Proposition 9.1.1 *The equation (9.1.1) with the initial condition (9.1.2) has a uniquely determined solution $(X(t), t \geq -r)$ given by*

$$\left.\begin{aligned} X(t) = \;& \sum_{j=1}^{N} \vartheta_j \int_{-r_j}^{0} x_0(t-s-r_j)X_0(s)ds \\ & + x_0(t) \cdot X_0(0) + \int_0^t x_0(t-s)dW(s), \quad t > 0 \\ X(t) = \;& X_0(t), \qquad\qquad\qquad\qquad t \in [-r, 0]. \end{aligned}\right\} \tag{9.1.5}$$

The function $x_0(.)$ in (9.1.5) denotes the continuous solution of the deterministic differential equation corresponding to (9.1.1), i.e.,

$$\dot{x}(t) = \sum_{j=0}^{N} \vartheta_j x(t - r_j), \qquad t \geq 0, t \neq r_j, j = 0, 1, \ldots, N, \tag{9.1.6}$$

satisfying the initial condition

$$x_0(s) = 0, \quad s \in [-r, 0), \quad x_0(0) = 1. \tag{9.1.7}$$

Remark: This solution is often called the fundamental solution of (9.1.6). For existence and uniqueness see, e.g., Hale and Lunel (1993).
Proof: Insert (9.1.5) into the integrated version of (9.1.1)–(9.1.2)

$$X(t) = X(0) + \int_0^t \sum_{j=0}^{N} \vartheta_j X(s - r_j) ds + W(t)$$

to affirm that (9.1.5) is a solution of (9.1.1)–(9.1.2). Assume $Y(t)$ is another solution of (9.1.1)–(9.1.2). Then $Z(t) = X(t) - Y(t)$ satisfies

$$dZ(t) = \sum_{j=0}^{N} \vartheta_j Z(t - r_j) dt, \; t \geq 0, \quad Z(t) = 0, \; t \in [-r, 0],$$

which implies $Z \equiv 0$. In order to see this, use the inequality

$$|Z(t)| \leq \sum_{j=0}^{N} |\vartheta_j| \int_0^t |Z(s - r_j)| ds \leq \sum_{j=0}^{N} |\vartheta_j| \int_0^t |Z(s)| ds,$$

from which $Z \equiv 0$ follows by Gronwall's Lemma (see Appendix B). □

Obviously, $(X(t),\, t \geq 0)$ is continuous. Thus, for every parameter value $\vartheta \in \mathbb{R}^{N+1}$ the process $(X(t),\, t \geq 0)$ generates a probability measure P_ϑ on $C([0, \infty))$, the set of real continuous functions on $[0, \infty)$. Observe that $(X(t),\, t \geq 0)$ is an Ito-process in the sense of Liptser and Shiryaev (1977). Indeed, it satisfies a stochastic differential equation of the type

$$dX(t) = \sum_{j=0}^{N} b_j(t, \omega) dt + dW(t), \quad t \geq 0,$$

where $b_j(t, \cdot) = \vartheta_j \cdot X(t - r_j)$ is $\mathcal{F}_t$-measurable with $\mathcal{F}_t = \mathcal{F}_t^X \vee \mathcal{F}^{X_0}, t \geq 0$.

Therefore, P_ϑ is locally absolutely continuous with respect to P_0, the probability measure generated by the Wiener process W. Moreover, we have for $L_t = \frac{dP_\vartheta^t}{dP_0^t}$ the representation (see Section 3.5)

$$\ell_t = \log L_t = \vartheta^T V_t^0 - \tfrac{1}{2} \vartheta^T I_t^0 \vartheta, \qquad t > 0, \tag{9.1.8}$$

where

$$V_t^0 = \left(\int_0^t X(s-r_j)dX(s) \right)_{j=0,\dots,N} \tag{9.1.9}$$

is an $(N+1)$-dimensional column vector, and where

$$I_t^0 = \left(\int_0^t X(s-r_k)X(s-r_\ell)ds \right)_{k,\ell=0,\dots,N} \tag{9.1.10}$$

denotes the expected information matrix (Fisher information). In particular, $(P_\vartheta, \vartheta \in \mathbb{R}^{N+1})$ forms an exponential family of continuous Ito-processes.

Lemma 9.1.2 *For arbitrary, but fixed, $\vartheta \in \Theta$ the matrix I_t^0 is positive definite P_ϑ-a.s. for every $t > r$.*

Proof: Obviously, I_t^0 is nonnegative definite. Assume that for some $t > r$ there exists a non-zero vector $\lambda^T = (\lambda_0, \dots, \lambda_N)$ with $\lambda^T I_t^0 \lambda = 0$ on a set A of positive probability: $P_\vartheta(A) > 0$. Then we have that on A

$$0 = \lambda^T I_t^0 \lambda = \sum_{k,\ell=0}^N \lambda_k \lambda_\ell \int_0^t X(s-r_k)X(s-r_\ell)ds = \int_0^t \left(\sum_{k=0}^N \lambda_k X(s-r_k) \right)^2 ds.$$

Thus it follows that on A the equation $\sum_{k=0}^N \lambda_k X(s-r_k) = 0$ holds Lebesgue a.e. on $[0,t]$. Define $k_0 = \min\{k \geq 0 \mid \lambda_k \neq 0\}$. For $s \in [r_{k_0}, r_{k_0+1}]$, in particular, we get that on A

$$\begin{aligned} 0 &= \sum_{k=k_0}^N \lambda_k X(s-r_k) \qquad (9.1.11) \\ &= \lambda_{k_0} X(s-r_{k_0}) + \sum_{i=k_0+1}^N \lambda_k X(s-r_k) \\ &= \lambda_{k_0} \int_0^{s-r_{k_0}} x_0(s-r_{k_0}-u)dW(u) + \lambda_{k_0} x_0(s-r_{k_0})X_0(0) \\ &\quad + \lambda_{k_0} \sum_{j=0}^N \int_{-r_j}^0 \vartheta_j x_0(s-r_{k_0}-u-r_j)X_0(u)du \\ &\quad + \sum_{k=k_0+1}^N \lambda_k X(s-r_k). \end{aligned}$$

Note that for $k > k_0$ we have $s - r_k < 0$, and hence $x_0(s - r_k) = 0$ on $[r_{k_0}, r_{k_0+1}]$. This implies that the last sum in (9.1.11) contains no contribution from $\{W(u) : u \geq 0\}$. The term in (9.1.11) involving W and the other three terms, which only involve X_0, are independent by assumption, and thus the whole sum has a density. This contradicts that $P(A) > 0$. □

From (9.1.8) it follows that the maximum likelihood estimator $\hat{\vartheta}_T$, based on continuous observation of $X(s), s \in [0, t]$, is given by

$$\hat{\vartheta}_t = (I_t^0)^{-1} V_t^0 = \vartheta + (I_t^0)^{-1} V_t,$$

where V_t denotes the column vector

$$V_t = \left(\int_0^t X(s - r_j) dW(s) \right)_{j=0,\ldots,N}$$

The problems treated in the next sections are the consistency of $\hat{\vartheta}_t$ and the asymptotic distribution of the appropriately normalized vector $\hat{\vartheta}_t - \vartheta$ as $t \to \infty$.

Remark: We have fixed the time delays $r_0, r_1, \ldots, r_N$. If we allow them or a part of them to be parameters too, then the corresponding family of probability measure loses the character of an exponential family. Moreover, the maximum likelihood estimator cannot be calculated explicitly, and the likelihood function seems not to be differentiable with respect to the parameters r_j. Nevertheless, one can prove asymptotic properties of the maximum likelihood estimator, see Dietz (1992) and Küchler and Kutoyants (1997).

9.2 The fundamental solution $x_0(\cdot)$

Before studying the asymptotic properties of I_t^0 and V_t occurring in the maximum likelihood estimator above, we summarize some properties of the fundamental solution $x_0(\cdot)$ of (9.1.6)–(9.1.7). For a detailed study of deterministic differential equations with time delay see, e.g., Hale and Lunel (1993).

Integrating (9.1.6)–(9.1.7) we get

$$x_0(t) = 1 + \int_0^t \sum_{j=0}^N \vartheta_j x_0(s - r_j) ds, \qquad t \geq 0,$$

and thus we obtain

$$|x_0(t)| \leq 1 + \int_0^t \Big(\sum_{j=0}^N |\vartheta_j| \Big) |x_0(s)| ds, \qquad t \geq 0.$$

From Gronwall's Lemma (see Appendix B) it follows that

$$|x_0(t)| \leq C \exp(at), \qquad t \geq 0, \tag{9.2.1}$$

for some $C > 0$ and for $a = \sum_{j=0}^{N} |\vartheta_j|$. Therefore, for $\alpha > a$ the Laplace transform of $x_0(\cdot)$ at α exists:

$$\hat{x}_0(\alpha) = \int_0^\infty e^{-\alpha t} x_0(t) dt.$$

Obviously, $\hat{x}_0(\lambda)$ is defined for every complex number λ with $\operatorname{Re}\lambda > a$. Using (9.1.6)–(9.1.7) it follows that

$$\hat{x}_0(\lambda) = [\lambda - Q_\vartheta(\lambda)]^{-1}, \qquad \operatorname{Re}\lambda > a, \tag{9.2.2}$$

where $Q_\vartheta(\lambda) = \sum_{j=0}^{N} \vartheta_j \exp[-\lambda r_j], \quad \lambda \in \mathbb{C}$.

Except when $N = 0$ or $\sum_{i=1}^{N} |\vartheta_i| = 0$, the function $Q_\vartheta(.)$ is an entire transcendental function. Thus the set $\Lambda = \{\lambda \in \mathbb{C} : \lambda = Q_\vartheta(\lambda)\}$ is non-void, countably infinite, and does not have a finite accumulation point. Using

$$|\lambda| \leq \sum_{j=0}^{N} |\vartheta_j| \exp(-r_j \cdot \operatorname{Re}\lambda), \qquad \lambda \in \Lambda, \tag{9.2.3}$$

it follows that for every $c \in \mathbb{R}$ the set

$$\Lambda_c = \{\lambda \in \Lambda : \operatorname{Re}\lambda > c\}$$

is finite. Therefore, the value $v_0 = \max\{\operatorname{Re}\lambda : \lambda \in \Lambda\}$ is always well-defined and finite. Moreover, if $N \geq 1$ and $\sum_{i=1}^{N} |\vartheta_i| > 0$, we have $-\infty < v_1 = \max\{\operatorname{Re}\lambda : \lambda \in \Lambda, \operatorname{Re}\lambda < v_0\} < v_0$.

The inverse Laplace transform yields

$$x_0(t) = \frac{1}{2\pi i} \lim_{c \to \infty} \int_{\alpha - ic}^{\alpha + ic} \frac{\exp[\mu t]}{\mu - Q_\vartheta(\mu)} d\mu, \qquad t \geq 0, \tag{9.2.4}$$

where $\alpha > a$ as above. Choose a real number β with $\beta < v_0$ and $\beta \neq \operatorname{Re}\lambda$ for all $\lambda \in \Lambda$. Then the integral occuring in (9.2.4) can, by Cauchy's residue theorem, be expressed in terms of the sum

$$\int_{\alpha - ic}^{\beta - ic} F(t,\mu) d\mu + \int_{\beta - ic}^{\beta + ic} F(t,\mu) d\mu + \int_{\beta + ic}^{\alpha + ic} F(t,\mu) d\mu + \sum_{\substack{\lambda \in \Lambda \\ \operatorname{Re}\lambda > \beta}} \operatorname*{Res}_{\mu = \lambda} F(t,\mu) \tag{9.2.5}$$

with

$$F(t,\mu) = \exp(\mu t)/(\mu - Q_\vartheta(\mu)).$$

It can be shown, see Hale and Lunel (1993), that as $c \to \infty$ the first and the third integrals in (9.2.5) tend to zero, uniformly for t in compact intervals. Moreover, choose a $v \in (\beta, v_0)$ such that there is no $\lambda \in \Lambda$ with $\operatorname{Re}\lambda \in (\beta, v]$. Then the second integral in (9.2.5) and its limit as $c \to \infty$ can be estimated by $o(\exp[vt])$. Finally, if $\lambda \in \Lambda$ and λ has multiplicity m as a solution of $\lambda = Q_\vartheta(\lambda)$, then

$$\frac{1}{2\pi i} \operatorname*{Res}_{\mu=\lambda} \frac{\exp[\mu t]}{\mu - Q_\vartheta(\mu)} = \frac{d^{m-1}}{d\mu^{m-1}} \left[\frac{(\mu-\lambda)^m}{\mu - Q_\vartheta(\mu)} \exp[\mu t] \right]_{\mu=\lambda} = P_\lambda(t)\exp[\lambda t], \tag{9.2.6}$$

where $P_\lambda(t)$ is a polynomial of degree $m-1$, the coefficients of which depend on λ.

Summarizing these facts, we get that for every $v < v_0$,

$$x_0(t) = \sum_{\lambda \in \Lambda_v} P_\lambda(t)\exp[\lambda t] + o(\exp[vt]), \quad t \geq 0. \tag{9.2.7}$$

Observing that $\lambda \in \Lambda$ if and only if $\bar{\lambda} \in \Lambda$ ($\bar{\lambda}$ denotes the complex conjugate of λ), we obtain further that for every $v < v_0$,

$$\begin{aligned} x_0(t) &= \sum_{\substack{\lambda \in \Lambda_v \\ \operatorname{Im}\lambda > 0}} \left(Q_\lambda^{(1)}(t)\cos[(\operatorname{Im}\lambda)t] + Q_\lambda^{(2)}(t)\sin[(\operatorname{Im}\lambda)t] \right) \exp[(\operatorname{Re}\lambda)t] \\ &\quad + \sum_{\substack{\lambda \in \Lambda_v \\ \lambda \text{ real}}} P_\lambda(t)\exp[\lambda t] + o(\exp[vt]), \quad t > 0, \end{aligned} \tag{9.2.8}$$

where $Q_\lambda^{(1)}$ and $Q_\lambda^{(2)}$ are real polynomials in t of degree at most $m(\lambda) - 1$. Here $m(\lambda)$ denotes the multiplicity of the root λ. Indeed, we have $Q_\lambda^{(1)}(t) = 2\operatorname{Re} P_\lambda(t)$, $Q_\lambda^{(2)}(t) = -2\operatorname{Im} P_\lambda(t)$, and $P_\lambda(t)$ is calculated from (9.2.6).

Note that for $N = 0$ or $\sum_{j=1}^N |\vartheta_j| = 0$ we have $v_0 = \vartheta_0$ and $x_0(t) = \exp[\vartheta_0 t]$, $t \geq 0$.

9.3 Asymptotic likelihood theory

Now we shall use the results of the preceding section to sketch how to study the asymptotic properties of the maximum likelihood estimator $\hat{\vartheta}_t$ as $t \to \infty$. Throughout the rest of this chaper we suppose $N \geq 1$ and $\sum_{j=1}^N |\vartheta_j| > 0$; i.e., we exclude the Langevin equation, which was treated in detail in Example 5.2.5.

The following proposition is not surprising.

Proposition 9.3.1 *If ϑ is chosen such that $v_0 < 0$, then $\hat{\vartheta}_t$ is strongly consistent, i.e.,*

$$\lim_{t\to\infty} \hat{\vartheta}_t = \vartheta \qquad P_\vartheta\text{-a.s.}$$

Moreover, $\hat{\vartheta}_t$ is asymptotically normal distributed under P_ϑ:

$$t^{\frac{1}{2}}(\hat{\vartheta}_t - \vartheta) \xrightarrow{\mathcal{D}} N(0, \Sigma_\vartheta^{-1})$$

as $t \to \infty$, where Σ_ϑ denotes the invertible symmetric $(N+1) \times (N+1)$-matrix with the elements

$$\sigma_{k\ell}^2 = \int_0^\infty x_0(s)x_0(s + |r_k - r_\ell|)ds, \ k, \ell = 0, \ldots, N.$$

Proof: If $v_0 < 0$, then $x_0(\cdot)$ converges exponentially to zero (see (9.2.7)), and hence the solution $(X(t), t \geq -r)$ of (9.1.1)–(9.1.2) given by (9.1.5) tends almost surely to the stationary solution $(Y(t), t \geq 0)$ given by

$$Y(t) = \int_{-\infty}^{t} x_0(t-s)dW(s), \qquad t \geq 0;$$

i.e., it holds that $X(t) - Y(t) \to 0$ P_ϑ-a.s. as $t \to \infty$. Here $(W(s), s \leq 0)$ denotes a standard Wiener process independent of $(W(s), s \geq 0)$. The existence of this stationary solution follows from $\int_0^\infty x_0^2(s)ds < \infty$, which is a consequence of $v_0 < 0$ and (9.2.7). The stationary process $(Y(t), t \geq 0)$ is centered, Gaussian, and ergodic, due to the fact that its covariance function

$$K_Y(h) = E(Y(t+h)Y(t)) = \int_0^\infty x_0(s+h)x_0(s)ds$$

decreases exponentially fast to zero for $h \to \infty$. Using $X(t) - Y(t) \to 0$ (P_ϑ-a.s.) as $t \to \infty$, we obtain from the strong law of large numbers for $(Y(t), t > 0)$ and for stochastic integrals that

$$\lim_{t\to\infty} t^{-1} \int_0^t X(s - r_k)dW(s) = 0 \qquad P_\vartheta\text{-a.s.}, \ k = 0, 1, \ldots, N,$$

and

$$\lim_{t\to\infty} t^{-1} \int_0^t X(s - r_k)X(s - r_\ell)ds = \sigma_{k\ell}^2 \quad P_\vartheta\text{-a.s.}, \ k, \ell = 0, 1, \ldots, N. \tag{9.3.1}$$

This implies the consistency of $\hat{\vartheta}_t$ stated in Proposition 9.3.1. The proof of the positive definiteness of Σ_ϑ is left as an exercise.

Now consider

$$t^{\frac{1}{2}}(\hat{\vartheta}_t - \vartheta) = t \cdot (I_t^0)^{-1} \cdot t^{-\frac{1}{2}} \left(\int_0^t X(s - r_k) dW(s) \right)_{k=0,1,\dots,N}$$

Let U be an orthogonal matrix such that

$$U^T \Sigma_\vartheta U = D = \operatorname{diag}(d_{ii}, i = 0, 1, \dots, N),$$

where D is a diagonal matrix with $d_{ii} > 0, i = 0, 1, \dots, N$, and introduce $Z(t)$ given by

$$Z(t) = U^T X(t), \quad t \geq 0,$$

where $X(t)^T = (X(t-r_0), \dots, X(t-r_N))$. Then we have from (9.3.1) that

$$\frac{1}{t} \int_0^t Z(s) Z^T(s) ds \underset{t \to \infty}{\longrightarrow} D \qquad P_\vartheta\text{-a.s.}$$

Thus a central limit theorem for the vector martingale (see Appendix A)

$$\int_0^t Z(s) dW(s), \quad t \geq 0,$$

yields

$$t^{-\frac{1}{2}} \int_0^t Z(s) dW(s) \xrightarrow{\mathcal{D}} N(0, D).$$

Now, inserting $UZ(s) = X(s)$ and noting that $t^{-1} I_t^0 \to \Sigma_\vartheta$, P_ϑ-a.s., we obtain

$$t^{-\frac{1}{2}} \int_0^t X(s) dW(s) \xrightarrow{\mathcal{D}} N(0, UDU^T) = N(0, \Sigma_\vartheta)$$

and

$$t^{\frac{1}{2}}(\hat{\vartheta}_t - \vartheta) \xrightarrow{\mathcal{D}} N(0, \Sigma_\vartheta^{-1}).$$

□

A necessary and sufficient condition on the parameter $\vartheta^T = (\vartheta_0, \dots, \vartheta_N)$ ensuring that $v_0 < 0$ is given in the Appendix of Hale and Lunel (1993).

The following special case illustrates what may happen when $v_0 < 0$ does not hold.

Proposition 9.3.2 *Assume that $v_0 \notin \Lambda$, $v_0 > 0$, and that the multiplicity of v_0 equals one. Suppose further that $Re\lambda < 0$ for every other $\lambda \in \Lambda$. Introduce the matrix $\varphi(t) = \varphi^{(1)}(t)\varphi^{(2)}(t)$ where*

$$\varphi^{(1)}(t) = \begin{pmatrix} 1 & 1 & 1 & \cdots & 1 \\ 0 & -e^{v_0 r_1} & 0 & \cdots & 0 \\ 0 & 0 & -e^{v_0 r_2} & & \\ \vdots & \vdots & & \ddots & \\ 0 & 0 & & & -e^{v_0 r_N} \end{pmatrix},$$

$$\varphi^{(2)}(t) = \begin{pmatrix} e^{-v_0 t} & 0 & \cdots & 0 \\ 0 & t^{-\frac{1}{2}} & & \\ \vdots & & \ddots & \\ 0 & & & t^{-\frac{1}{2}} \end{pmatrix}.$$

Then $\varphi(t) \to 0$ as $t \to \infty$, and

$$\varphi_t^T I_t^0 \varphi_t \xrightarrow{P} \begin{pmatrix} \frac{U_0^2}{2v_0} & 0 & \cdots & 0 \\ 0 & \gamma_{11} & \cdots & \gamma_{1N} \\ \vdots & \vdots & & \vdots \\ 0 & \gamma_{N1} & \cdots & \gamma_{NN} \end{pmatrix} = I_\infty, \tag{9.3.2}$$

where

$$U_0 = X_0(0) + \sum_{j=1}^{N} \int_{-r_j}^{0} \vartheta_j e^{-(s+r_j)v_0} X_0(s)ds + \int_0^\infty e^{-v_0 s} dW(s)$$

and

$$\gamma_{k\ell} = \int_0^\infty [x_0(s) - e^{v_0 r_k} x_0(s-r_k)][x_0(s) - e^{v_0 r_\ell} x_0(s-r_\ell)]ds,$$

for $k, \ell = 1, \dots, N$. The matrix I_∞ is positive definite P_ϑ-a.s. Moreover, there exists a normalized $(N+1)$-dimensional Gaussian random vector Z, independent of I_∞, such that

$$(\varphi_t^T V_t^0, \varphi_t^T I_t^0 \varphi_t) \xrightarrow{\mathcal{D}} (V_\infty, I_\infty), \tag{9.3.3}$$

where the processes V_t^0 and I_t^0 are given by (9.1.9) and (9.1.10), and where $V_\infty = I_\infty^{\frac{1}{2}} Z$.

Proof: It is easy to conclude from (9.2.8) and the assumptions made in the proposition that

$$x_0(t) = C\exp[v_0 t] + y(t), \qquad t \geq 0, \tag{9.3.4}$$

where C is a constant equal to $(1-\sum_{j=1}^{N}\vartheta_j r_j e^{-v_0 r_j})^{-1}$ and $y(t)=0(e^{vt})$ for some $v<0$. Thus, using (9.3.4) and (9.1.5) we obtain

$$C^{-1}e^{-v_0 t}X(t) = X_0(0)+\sum_{j=1}^{N}\int_{-r_j}^{0}\vartheta_j e^{-(s+r_j)v_0}X_0(s)ds + \int_0^t e^{-v_0 s}dW(s)+0(e^{vt}) \tag{9.3.5}$$

for some $v<0$. Therefore, it follows that

$$\lim_{t\to\infty} e^{-v_0 t}X(t)=C\cdot U_0. \tag{9.3.6}$$

Inserting (9.3.5) into

$$(I_t^0)_{k\ell}=\int_0^t X(s-r_k)X(s-r_\ell)ds$$

and using (9.3.6) and $x_0(s)-e^{v_0 r_k}x_0(s-r_k)-y(s)-e^{v_0 r_k}y(s-r_k)=o(e^{vs})$ $(k\geq 1)$ for some $v<0$, we finally get (9.3.2). The conclusion (9.3.3) follows from the central limit theorem for martingales in Appendix A. □

The complete explicit description of the asymptotic behaviour of the distribution of the maximum likelihood estimator $\hat\vartheta_t$ in its dependence on the parameter ϑ involves a detailed analysis of the equation $\lambda=Q_\vartheta(\lambda)$. This seems a rather formidable task in the general case. Nevertheless, at least in the case $N=1$ a complete analysis is possible, as we shall see in the next section.

9.4 The case $N=1$

In this section we consider the case $N=1$, where

$$\left.\begin{aligned} dX(t)&=\vartheta_0 X(t)dt+\vartheta_1 X(t-r_1)dt+dW(t), && t>0\\ X(t)&=X_0(t), && t\in[-r_1,0]\end{aligned}\right\}. \tag{9.4.1}$$

In the following we present a complete description of the limit distribution of the maximum likelihood estimator $\hat\vartheta_t$ as $t\to\infty$ in its dependence on the underlying parameter $\vartheta\in\mathbb{R}^2$. This is possible because the characteristic equation

$$\lambda=Q_\vartheta(\lambda)=\vartheta_0+\vartheta_1 e^{-\lambda r_1}$$

can be treated explicitly. We can assume $r_1 = 1$ for simplicity. This can always be obtained by a change of the time scale. The presentation below is based on the paper Gushchin and Küchler (1996), where the local asymptotic properties of the likelihood process in the sense of LeCam are studied in detail. The results given here are easy consequences of the cited paper. Therefore, we omit most of the proofs. Throughout this section we will suppose that $\vartheta_1 \neq 0$ to exclude the Langevin case.

The results in this section show that the behaviour of the maximum likelihood estimator within an exponential family of stochastic processes can have very different asymptotic properties depending on the value of the parameter of the family. Moreover, the extent to which the addition of a single term $\vartheta_1 X(t-1)$ to the Langevin equation changes the picture illustrates the statistical complexity of stochastic process models.

We start with some preliminary definitions and properties that are easy to check. Obviously, for every fixed $\vartheta = (\vartheta_0, \vartheta_1)^T \in \mathbb{R}^2$ the characteristic equation

$$\lambda = Q_\vartheta(\lambda) = \vartheta_0 + \vartheta_1 \exp(-\lambda), \qquad \lambda \in \mathbb{C}, \tag{9.4.2}$$

has at most two real solutions. Recall the notation $\Lambda = \{\lambda \in \mathbb{C} : \lambda$ satisfies (9.4.2)$\}$ and $v_0 = \max\{\operatorname{Re}\lambda : \lambda \in \Lambda\} < \infty$, and introduce $v_1 = \max\{\operatorname{Re}\lambda : \lambda \in \Lambda, \operatorname{Re}\lambda < v_0\}$. If there exists a real solution v of (9.4.2), then every non-real solution has a real part that is strictly less than v. Consequently, the only possible real solutions are v_0 (if there is exactly one) or v_0 and v_1 (if there are two); see Exercise 9.1.

With the definition $v(x) = -\exp[x-1]$, $x \in \mathbb{R}$, we have

$$v_0 \in \Lambda \Longleftrightarrow v(\vartheta_0) \leq \vartheta_1 \tag{9.4.3}$$

and

$$\{v_0, v_1\} \subseteq \Lambda \Longleftrightarrow v(\vartheta_0) < \vartheta_1 < 0. \tag{9.4.4}$$

If $v_0 \notin \Lambda$, i.e., if $\vartheta_1 < v(\vartheta_0)$, then there exists a unique $\lambda_0 \in \Lambda$ with $\operatorname{Re}\lambda_0 = v_0$ and $\xi_0 = \operatorname{Im}\lambda_0 > 0$. In this case we have $\xi_0 \in (0, \pi)$ (exercise). For λ from Λ, let $m(\lambda)$ be the multiplicity of λ. We have $m(\lambda) = 1$ for all $\lambda \in \Lambda$ except $\lambda = \vartheta_0 - 1$. The number $\lambda = \vartheta_0 - 1$ belongs to Λ if and only if $\vartheta_1 = v(\vartheta_0)$, and in this case $m(\vartheta_0 - 1) = 2$.

For all $v < v_0$ the fundamental solution $x_0(\cdot)$ of the deterministic equation corresponding to (9.4.1) can be represented in the form

$$x_0(t) = \psi_0(t)\exp[v_0 t] + \sum_{\lambda_k \in \Lambda_v} c_k \exp[\lambda_k t] + y_0(t)$$

for certain constants c_k, and where $y_0(t) = o(\exp(\gamma t))$ for some $\gamma < v$. The function $\psi_0(t)$ is given by

$$\psi_0(t) = \begin{cases} \dfrac{1}{v_0 - \vartheta_0 + 1} & \text{if } \vartheta_1 > v(\vartheta_0), \\[2ex] 2t + \frac{2}{3} & \text{if } \vartheta_1 = v(\vartheta_0), \\[2ex] \dfrac{2(v_0 - \vartheta_0 + 1)}{(v_0 - \vartheta_0 + 1)^2 + \xi_0^2} \cos(\xi_0 t) & \\ \quad + \dfrac{2\xi_0}{(v_0 - \vartheta_0 + 1)^2 + \xi_0^2} \sin(\xi_0 t) & \text{if } \vartheta_1 < v(\vartheta_0). \end{cases}$$

Note that the three cases for ψ_0 correspond to $v_0 \in \Lambda$ with multiplicity one, $v_0 \in \Lambda$ with multiplicity two, and $v_0 \notin \Lambda$, respectivley.

If $\vartheta_1 > v(\vartheta_0)$ and $\vartheta_1 \neq 0$, then for some purposes it is necessary to separate a further term from the sum (9.2.7). We get

$$x_0(t) = \frac{1}{v_0 - \vartheta_0 + 1} \exp[v_0 t] + \psi_1(t) \exp[v_1 t] + y_1(t),$$

where

$$\psi_1(t) = \begin{cases} \dfrac{1}{v_1 - \vartheta_1 + 1} & \text{if } v(\vartheta_0) < \vartheta_1 < 0, \\[2ex] \dfrac{2(v_1 - \vartheta_1 + 1)}{(v_1 - \vartheta_1 + 1)^2 + \xi_1^2} \cos(\xi_1 t) & \\ \quad + \dfrac{2\xi_1}{(v_1 - \vartheta_1)^2 + \xi_1^2} \sin(\xi_1 t) & \text{if } \vartheta_1 > 0, \end{cases}$$

and where $y_1(t) = o(\exp(\gamma t))$ for some $\gamma < v_1$. Here ξ_1 denotes the uniquely determined positive number with $\lambda_1 = v_1 + i\xi_1 \in \Lambda$, when $\vartheta_1 > 0$. It is noted that in this case $\xi_1 \in (\pi, 2\pi)$.

We are now ready to formulate the results.

Proposition 9.4.1 (Weak consistency) *For every $\vartheta \in \mathbb{R}^2$ we have that*

$$\hat{\vartheta}_t \underset{t\to\infty}{\longrightarrow} \vartheta \tag{9.4.5}$$

in probability under P_ϑ.

Proof: See Gushchin and Küchler (1996). □

To deal with the limit distribution of $\hat{\vartheta}_t$ we have to differentiate among several cases. First, we deal with the case $v_0 < 0$, for which we already know the result, Proposition 9.3.1. In the present case ($N = 1$) one can go one step further in characterizing the set of parameters $\vartheta \in \mathbb{R}^2$ for which $v_0 < 0$. Indeed, we have

$$v_0 < 0 \iff \vartheta \in S_0 = \{(\vartheta_0, \vartheta_1) \in \mathbb{R}^2 \mid \vartheta_0 < 1, u(\vartheta_0) < \vartheta_1 < -\vartheta_0\},$$

where $s \to u(s)$, $s < 1$, is defined as follows. For $s \in \mathbb{R}$ define $\xi(s)$ as the solution of

$$\begin{aligned} &\xi = s \tan \xi, \ 0 < \xi < \pi, && \text{if } s \neq 0, \ s < 0, \\ &\xi = \tfrac{\pi}{2} && \text{if } s = 0, \end{aligned}$$

and put $u(s) = -\xi(s) \sin[\xi(s)] - s \cos[\xi(s)]$, $s \in \mathbb{R}$.

Proposition 9.4.2 *If $\vartheta \in S_0$, i.e., if $v_0 < 0$ (case 1), then the maximum likelihood estimator $\hat{\vartheta}_t$ is asymptotically normal distributed with*

$$t^{\frac{1}{2}}(\hat{\vartheta}_t - \vartheta) \xrightarrow{\mathcal{D}} \Sigma_\vartheta^{-\frac{1}{2}} Z,$$

where Z is a standardized normal distributed two-dimensional vector, and where

$$\Sigma_\vartheta = \begin{pmatrix} \int_0^\infty x_0^2(s)ds & \int_0^\infty x_0(s)x_0(s+1)ds \\ \int_0^\infty x_0(s)x_0(s+1)ds & \int_0^\infty x_0^2(s+1)ds \end{pmatrix}$$

is nonsingular if $\vartheta_1 \neq 0$. There are no other cases with asymptotic normality.

If $\vartheta \in S_0$, then (9.4.5) can be strengthened to

$$\hat{\vartheta}_t \underset{t \to \infty}{\longrightarrow} \vartheta \qquad P_\vartheta\text{-a.s.}.$$

As the next cases, we summarize all parameters $\vartheta \in \mathbb{R}^2$ for which an asymptotic distribution of $(\hat{\vartheta}_t - \vartheta)$ after some suitable normalization exists and is mixed normal.

The case $v_0 < 0$ was treated in Proposition 9.4.2 and led to asymptotic normality. We will refer to this case as Case 1. Now assume $v_0 > 0$. In this case the limit behaviour depends on the sign of v_1 and, if $v_1 \geq 0$, also on whether or not $v_1 \in \Lambda$. The following Propositions 9.4.3–9.4.5 treat all the different cases with $v_0 > 0$ in detail. In order to describe these cases in terms of the sets of values of the parameter $\vartheta = (\vartheta_0, \vartheta_1)^T$ for which they occur, let us introduce a function $w(s)$, $s \in \mathbb{R}$, as follows. First, define $\zeta(s)$, $s \in \mathbb{R}$, implicitly by

$$\begin{aligned} &\zeta = s \tan \zeta, \ \pi < \zeta < 2\pi, && \text{if } s \neq 0, \\ &\zeta = 3\pi/2 && \text{if } s = 0. \end{aligned}$$

Now, w is given by $w(s) = -\zeta(s) \sin[\zeta(s)] - s \cos[\zeta(s)]$, $s \in \mathbb{R}$. With this definition we have

$$v_0 > 0, v_0 \in \Lambda, v_1 < 0 \quad \Leftrightarrow \quad -\vartheta_0 < \vartheta_1 < w(\vartheta_0),$$

$$v_0 > 0, v_0 \in \Lambda, v_1 = 0, v_1 \in \Lambda \;\Leftrightarrow\; \vartheta_0 > 1, \vartheta_1 = -\vartheta_0,$$

$$v_0 > 0, v_0 \in \Lambda, v_1 = 0, v_1 \notin \Lambda \;\Leftrightarrow\; \vartheta_1 = w(\vartheta_0),$$

$$v_0 > 0, v_0 \in \Lambda, v_1 > 0, v_1 \in \Lambda \;\Leftrightarrow\; \vartheta_0 > 1, v(\vartheta_0) < \vartheta_1 < -\vartheta_0,$$

$$v_0 > 0, v_0 \in \Lambda, v_1 > 0, v_1 \notin \Lambda \;\Leftrightarrow\; \vartheta_1 > w(\vartheta_0).$$

The following proposition treats the cases in which local asymptotic mixed normality occurs and describes the corresponding limit distributions. To simplify the formulation, we suppose from now on that $X_0 \equiv 0$. If this assumption is omitted, only the form of the random variables U, U_0, U_1, and U_2 changes; see Gushchin and Küchler (1996).

Proposition 9.4.3 *Assume that $\vartheta \in \mathbb{R}^2$ satisfies one of the following conditions:*
Case 2: *$-\vartheta_0 < \vartheta_1 < w(\vartheta_0)$ (i.e., $0 < v_0 \in \Lambda$ and $v_1 < 0$).*

Case 3: *$\vartheta_0 > 1$ and $v(\vartheta_0) < \vartheta_1 < -\vartheta_0$ (i.e., $0 < v_0 \in \Lambda$ and $0 < v_1 \in \Lambda$).*

Case 4: *$\vartheta_0 > 1$ and $v(\vartheta_0) = \vartheta_1$ (i.e., $0 < v_0 \in \Lambda,\ m(v_0) = 2$).*

Define a family of matrices φ_t $(t > 0)$ by $\varphi_t = \varphi_t^{(1)}\varphi_t^{(2)}$, where

$$\varphi_t^{(1)} = \begin{pmatrix} 1 & 1 \\ 0 & -e^{v_0} \end{pmatrix}$$

in Case 2 and Case 3,

$$\varphi_t^{(1)} = \begin{pmatrix} 1 & 1 \\ 0 & -(1+t^{-1})e^{v_0} \end{pmatrix}$$

in Case 4, and

$$\varphi_t^{(2)} = \begin{pmatrix} \varphi_{11}(t) & 0 \\ 0 & \varphi_{22}(t) \end{pmatrix}$$

with

$$\varphi_{11}(t) = \begin{cases} \exp[-v_0 t] & \text{in Cases 2 and 3,} \\ t^{-1}\exp[-v_0 t] & \text{in Case 4,} \end{cases}$$

$$\varphi_{22}(t) = \begin{cases} t^{-1/2} & \text{in Case 2,} \\ \exp[-v_1 t] & \text{in Case 3,} \\ T\exp[-v_0 t] & \text{in Case 4.} \end{cases}$$

With this choice of the normalizing matrix φ_t we have that

$$\varphi_t^{-1}(\hat{\vartheta}_t - \vartheta) \xrightarrow{\mathcal{D}} U^{-\frac{1}{2}} Z$$

under P_ϑ as $t \to \infty$, where Z is a standardized normal distributed two-dimensional vector, and where U is a positive definite random matrix, independent of Z, which is given by

$$\begin{pmatrix} \dfrac{U_0^2}{2v_0(v_0 - \vartheta_0 + 1)^2} & 0 \\ 0 & \displaystyle\int_0^\infty (x_0(t) - e^{v_0} x_0(t-1))^2 dt \end{pmatrix}$$

in Case 2,

$$\begin{pmatrix} \dfrac{U_0^2}{2v_0(v_0 - \vartheta_0 + 1)^2} & \dfrac{U_0 U_1 (e^{v_0 - v_1} - 1)}{(v_0 + v_1)(v_0 - \vartheta_0 + 1)(\vartheta_0 - v_1 - 1)} \\ \dfrac{U_0 U_1 (e^{v_0 - v_1} - 1)}{(v_0 + v_1)(v_0 - \vartheta_0 + 1)(\vartheta_0 - v_1 - 1)} & \dfrac{U_1^2 (e^{v_0 - v_1} - 1)^2}{2v_1(\vartheta_0 - v_1 - 1)^2} \end{pmatrix}$$

in Case 3, and

$$\begin{pmatrix} \dfrac{2U_0^2}{v_0} & \dfrac{U_0^2}{v_0^2} + \dfrac{U_0(\frac{4}{3}U_0 + 2U_2)}{v_0} \\ \dfrac{U_0^2}{v_0^2} + \dfrac{U_0(\frac{4}{3}U_0 + 2U_2)}{v_0} & \dfrac{(\frac{4}{3}U_0 + 2U_2)^2}{2v_0} + \dfrac{U_0(\frac{4}{3}U_0 + 2U_2)}{v_0^2} + \dfrac{U_0^2}{v_0^3} \end{pmatrix}$$

in Case 4, where

$$\begin{aligned} U_i &= \int_0^\infty e^{-v_i s} dW(s), \quad i = 0, 1, \\ U_2 &= \int_0^\infty s e^{-v_0 s} dW(s). \end{aligned}$$

The next type of asymptotic behaviour to be studied consists in what could be called "periodic asymptotical mixed normality." This means that there is no deterministic normalization leading to a certain limit distribution, but if one runs through time points on a grid with a particular, uniquely determined step length, then one does gets a limit distribution, which, however, depends on the starting point.

Proposition 9.4.4 *Assume that ϑ satisfies one of the conditions*

Case 5: $\vartheta_1 < u(\vartheta_0)$ *if* $\vartheta_0 < 1$ *and* $\vartheta_1 < v(\vartheta_0)$ *if* $\vartheta_0 \geq 1$ *(i.e.,* $0 < v_0 \notin \Lambda$*)*

Case 6: $\vartheta_1 > w(\vartheta_0)$ *(i.e.,* $0 < v_0 \in \Lambda$ *and* $0 < v_1 \notin \Lambda$*).*

Recall that $\xi_0 = Im\,\lambda > 0$ *for* $\lambda \in \Lambda$ *with* $Re\,\lambda = v_0$*, and put* $\xi_1 = Im\,\lambda > 0$ *for* $\lambda \in \Lambda$ *with* $Re\,\lambda = v_1$ *in Case 6. Define the normalizing matrix* φ_t *by* $\varphi_t = \varphi_t^{(1)}\varphi_t^{(2)}$ *with*

$$\varphi_t^{(1)} = I_2, \quad \varphi_t^{(2)} = e^{-v_0 t} I_2$$

in Case 5 and

$$\varphi_t^{(1)} = \begin{pmatrix} 1 & 1 \\ 0 & -e^{v_0} \end{pmatrix}, \quad \varphi_t^{(2)} = \begin{pmatrix} e^{-v_0 t} & 0 \\ 0 & -e^{v_1 t} \end{pmatrix}$$

in Case 6. Then for every fixed $u \in (0, \xi)$*,*

$$\varphi_t^{-1}(\hat{\vartheta}_t - \vartheta) \xrightarrow{\mathcal{D}} U_u^{-\frac{1}{2}} Z,$$

provided that t runs through the grid $u + n\xi$ $(n = 0, 1, \cdots)$*. Here* $\xi = \xi_0$ *in Case 5 and* $\xi = \xi_1$ *in Case 6, Z is a standard normal distributed vector, and* U_u *is a positive definite matrix with stochastic elements that are independent of Z. In Case 5,* U_u *is given by*

$$\begin{pmatrix} \int_0^\infty e^{-2v_0 t} U_1^2(u-t)\,dt & \int_0^\infty e^{-2v_0 t} U_1(u-t) U_2(u-t)\,dt \\ \int_0^\infty e^{-2v_0 t} U_1(u-t) U_2(u-t)\,dt & \int_0^\infty e^{-2v_0 t} U_2^2(u-t)\,dt \end{pmatrix},$$

where

$$U_i(t) = \int_0^\infty \psi_i(t-s) e^{-v_0 s}\,dW(s), \quad i = 1, 2,$$

with

$$\psi_i(t) = A_i \cos(\xi_0 t) + B_i \sin(\xi_0 t), \quad i = 1, 2,$$

$$A_1 = \frac{2(v_0 - \vartheta_0 + 1)}{(v_0 - \vartheta_0 + 1)^2 + \xi_0^2}, \quad B_1 = \frac{2\xi_0}{(v_0 - \vartheta_0 + 1)^2 + \xi_0^2},$$

and

$$\begin{pmatrix} A_2 \\ B_2 \end{pmatrix} = e^{-v_0} \begin{pmatrix} \cos\xi_0 & -\sin\xi_0 \\ \sin\xi_0 & \cos\xi_0 \end{pmatrix} \begin{pmatrix} A_1 \\ B_1 \end{pmatrix}.$$

In Case 6, the random matrix U_u *is given by*

$$\begin{pmatrix} \frac{U_0^2}{2v_0(v_0 - \vartheta_0 + 1)^2} & \frac{U_0}{v_0 - \vartheta_0 + 1} \int_0^\infty e^{-(v_0+v_1)t} U(u-t)\,dt \\ \frac{U_0}{v_0 - \vartheta_0 + 1} \int_0^\infty e^{-(v_0+v_1)t} U(u-t)\,dt & \int_0^\infty e^{-2v_1 t} U^2(u-t)\,dt \end{pmatrix},$$

where U_0 is defined in Proposition 9.4.3 and

$$U(t) = \int_0^\infty \psi(t-s)e^{-v_1 s}dW(s),$$

with

$$\psi(t) = A\cos(\xi_1 t) + B\sin(\xi_1 t).$$

Here

$$\begin{pmatrix} A \\ B \end{pmatrix} = \begin{pmatrix} A' \\ B' \end{pmatrix} - e^{v_0 - v_1} \begin{pmatrix} \cos\xi_1 & -\sin\xi_1 \\ \sin\xi_1 & \cos\xi_1 \end{pmatrix} \begin{pmatrix} A' \\ B' \end{pmatrix},$$

with

$$A' = \frac{2(v_1 - \vartheta + 1)}{(v_1 - \vartheta + 1)^2 + \xi_1^2}, \qquad B' = \frac{2\xi_1}{(v_1 - \vartheta + 1)^2 + \xi_1^2}.$$

Until now we have treated all values of the parameter ϑ for which the distribution of the maximum likelihood estimator is asymptotically normal, asymptotically mixed normal, and what was called periodically asymptotically mixed normal in Gushchin and Küchler (1996). In all other cases a limit distribution exists; but it, or at least some of its components, is not equal to a normal or a mixed normal distribution. In accordance with the usual terminology, these remaining cases are referred to as locally asymptotically quadratic (LAQ) because here the likelihood function for local parameters is asymptotically quadratic; see also the case $\theta = 0$ in Example 5.2.5. However, not all the remaining parameter values give rise to the same type of asymptotic distribution, so again we have to differentiate between a number of cases, more precisely between the following five cases:

Case 7: $\vartheta_1 = -\vartheta_0, \vartheta_0 < 1$, (i.e., $v_0 = 0 \in \Lambda, m(v_0) = 1$).

Case 8: $\vartheta_1 = -\vartheta_0, \vartheta_0 = 1$, (i.e., $v_0 = 0 \in \Lambda, m(v_0) = 2$).

Case 9: $\vartheta_1 = u(\vartheta_0), \vartheta_0 < 1$, (i.e., $v_0 = 0 \notin \Lambda$).

Case 10: $\vartheta_1 = -\vartheta_0, \vartheta_0 > 1$, (i.e., $0 < v_0 \in \Lambda, v_1 = 0 \in \Lambda$).

Case 11: $\vartheta_1 = w(\vartheta_0)$, (i.e., $0 < v_0 \in \Lambda, v_1 = 0 \notin \Lambda$).

Proposition 9.4.5 *Assume that one of the cases 7–11 holds. Then there exists a matrix normalization φ_t, which is given below for each of these cases, such that*

$$\varphi_t^{-1}(\hat{\vartheta}_t - \vartheta) \xrightarrow{\mathcal{D}} I_\infty^{-1} V_\infty,$$

where I_∞ is a random positive definite symmetric matrix and V_∞ is a random vector. Explicit expressions for I_∞ and V_∞ are given below for each of the five cases.

In Case 7, $\varphi_t = \varphi_t^{(1)}\varphi_t^{(2)}$ with

$$\varphi_t^{(1)} = \begin{pmatrix} 1 & 1 \\ 0 & -1 \end{pmatrix} \quad \text{and} \quad \varphi_t^{(2)} = \begin{pmatrix} t^{-1} & 0 \\ 0 & t^{-\frac{1}{2}} \end{pmatrix},$$

$$V_\infty = \left(\frac{1}{1-\vartheta_0} \int_0^1 \widetilde{W}(t) d\widetilde{W}(t), Z_1 \right)^T,$$

and

$$I_\infty = \begin{pmatrix} \frac{1}{(1-\vartheta_0)^2} \int_0^1 \widetilde{W}^2(t) dt & 0 \\ 0 & \sigma^2 \end{pmatrix},$$

where $(\widetilde{W}(s), s \in [0,1])$ *is a standard Wiener process,* $\sigma^2 = \int_0^\infty [x_0(t) - x_0(t-1)]^2 dt$, *and* Z_1 *an* $N(0, \sigma^2)$*-distributed random variable that is independent of* $\widetilde{W}(\cdot)$.

In Case 8, $\varphi_t = \varphi_t^{(1)} \varphi_t^{(2)}$ *with*

$$\varphi_t^{(1)} = \begin{pmatrix} 1 & 1 \\ 0 & -1 \end{pmatrix} \quad \textit{and} \quad \varphi_t^{(2)} = \begin{pmatrix} t^{-2} & 0 \\ 0 & t^{-1} \end{pmatrix},$$

$$V_\infty = \left(\int_0^1 \widetilde{X}(t) d\widetilde{W}(t), \int_0^1 \widetilde{W}(t) d\widetilde{W}(t) \right)^T$$

with $\widetilde{X}(t) = \int_0^1 \widetilde{W}(s) ds$, *and*

$$I_\infty = 2 \begin{pmatrix} \int_0^1 \widetilde{X}^2(t) dt & \int_0^1 \widetilde{X}(t) \widetilde{W}(t) dt \\ \int_0^1 \widetilde{X}(t) \widetilde{W}(t) dt & \int_0^1 \widetilde{W}^2(t) dt \end{pmatrix},$$

where $(\widetilde{W}(t))$ *is a standard Wiener process.*

In Case 9, $\varphi_t = t^{-1} I_2$, *where* I_2 *is the* 2×2 *identity matrix,*

$$V_\infty = \frac{1}{\sqrt{2}} \begin{pmatrix} A_1(\mathcal{I}_{1,1} + \mathcal{I}_{2,2}) + B_1(\mathcal{I}_{1,2} - \mathcal{I}_{2,1}) \\ A_2(\mathcal{I}_{1,1} + \mathcal{I}_{2,2}) + B_2(\mathcal{I}_{1,2} - \mathcal{I}_{2,1}) \end{pmatrix},$$

and $2I_\infty$ *equals*

$$\begin{pmatrix} (A_1^2 + B_1^2) \int_0^1 (\widetilde{W}_1^2(t) + \widetilde{W}_2^2(t)) dt & (A_1 A_2 + B_1 B_2) \int_0^1 \widetilde{W}_1(t) \widetilde{W}_2(t) dt \\ (A_1 A_2 + B_1 B_2) \int_0^1 \widetilde{W}_1(t) \widetilde{W}_2(t) dt & (A_2^2 + B_2^2) \int_0^1 (\widetilde{W}_1^2(t) + \widetilde{W}_2^2(t)) dt \end{pmatrix}.$$

Here

$$\mathcal{I}_{i,j} = \int_0^1 \widetilde{W}_i(t) d\widetilde{W}_j(t),$$

$i, j = 1, 2$, where $\widetilde{W}_i$, $i = 1, 2$, are two independent standard Wiener processes,

$$A_1 = \frac{2(1-\vartheta_0)}{(1-\vartheta_0)^2 + \xi_0^2}, \quad B_1 = \frac{2\xi_0}{(1-\vartheta_0)^2 + \xi_0^2},$$

and

$$\begin{pmatrix} A_2 \\ B_2 \end{pmatrix} = \begin{pmatrix} \cos\xi_0 & -\sin\xi_0 \\ \sin\xi_0 & \cos\xi_0 \end{pmatrix} \begin{pmatrix} A_1 \\ B_1 \end{pmatrix}.$$

In Case 10, the normalizing matrix φ_T is given by $\varphi_t = \varphi_t^{(1)} \varphi_t^{(2)}$ with

$$\varphi_t^{(1)} = \begin{pmatrix} 1 & 1 \\ 0 & -e^{v_0} \end{pmatrix} \quad \text{and} \quad \varphi_t^{(2)} = \begin{pmatrix} e^{-v_0 t} & 0 \\ 0 & t^{-1} \end{pmatrix},$$

while

$$V_\infty = \left(\frac{U_0 Z}{\sqrt{2v_0}(v_0 - \vartheta_0 + 1)}, \; \frac{e^{v_0} - 1}{\vartheta_0 - 1} \int_0^1 \widetilde{W}(t) d\widetilde{W}(t) \right)^T$$

and

$$I_\infty = \begin{pmatrix} \dfrac{U_0^2}{2v_0(v_0 - \vartheta + 1)^2} & 0 \\ 0 & \dfrac{(e^{v_0} - 1)^2}{(\vartheta - 1)^2} \displaystyle\int_0^1 \widetilde{W}^2(t) dt \end{pmatrix}.$$

Here U_0 is as in Proposition 9.4.3, Z is a standard normal distributed random variable, $\widetilde{W}$ is a standard Wiener process, and U_0, Z, and $\widetilde{W}$ are independent.

In Case 11, φ_t is as in Case 10,

$$V_\infty = \begin{pmatrix} \dfrac{U_0 Z}{\sqrt{2v_0}(v_0 - \vartheta_0 + 1)} \\ \dfrac{1}{\sqrt{2}} [A(\mathcal{I}_{1,1} + \mathcal{I}_{2,2}) + B(\mathcal{I}_{1,2} - \mathcal{I}_{2,1})] \end{pmatrix},$$

and

$$I_\infty = \begin{pmatrix} \dfrac{U_0^2}{2v_0(v_0 - \vartheta_0 + 1)^2} & 0 \\ 0 & \dfrac{1}{2}(A^2 + B^2) \displaystyle\int_0^1 [\widetilde{W}_1^2(t) + \widetilde{W}_2^2(t)] dt \end{pmatrix}.$$

Here U_0 and Z are as in Case 10; $\widetilde{W}_1$, $\widetilde{W}_2$, and $\mathcal{I}_{i,j}$ $(i,j=1,2)$ are as in Case 9; and U_0, Z, $\widetilde{W}_1$, and $\widetilde{W}_2$ are independent. Finally,

$$\begin{pmatrix} A \\ B \end{pmatrix} = \begin{pmatrix} A' \\ B' \end{pmatrix} - e^{v_0} \begin{pmatrix} \cos\xi_1 & -\sin\xi_1 \\ \sin\xi_1 & \cos\xi_1 \end{pmatrix} \begin{pmatrix} A' \\ B' \end{pmatrix}$$

with

$$A' = \frac{2(1-\vartheta_0)}{(1-\vartheta_0)^2+\xi_1^2} \quad \textit{and} \quad B' = \frac{2\xi_1}{(1-\vartheta_0)^2+\xi_1^2}.$$

9.5 Exercises

9.1 Let $x_0(\cdot)$ be the fundamental solution of (9.1.6) and (9.1.7). Prove that the matrix $\Sigma_\theta = \{\sigma_{kl}^2\}$ with σ_{kl}^2 defined in Proposition 9.3.1 is positive definite.

9.2 Show under the assumption $\vartheta_1 \neq 0$ that if there exists a real solution $\lambda = v$ of

$$\lambda = \vartheta_0 + \vartheta_1 \exp(-\lambda), \quad \lambda \in \mathbb{C},$$

then the real part of every non-real solution is less that v.

9.3 Consider the differential equation

$$\begin{aligned} dX(t) &= \vartheta X(t-1)dt + dW_t, \quad t \geq 0, \\ X(t) &= 0, \quad t < 0. \end{aligned}$$

Determine the asymptotic behaviour as $T \to \infty$ of the distribution of the maximum likelihood estimator $\hat{\vartheta}_T$ of ϑ based on observation of X in $[0,T]$, which is given by

$$\hat{\vartheta}_T = \frac{\int_0^T X(s-1)dX(s)}{\int_0^T X^2(s-1)ds}.$$

Specifically, show that there are five different cases, depending on which of the sets $(0,\infty)$, $\{0\}$, $(-\pi/2,0)$, $\{-\pi/2\}$, $(-\infty,-\pi/2)$ the parameter ϑ belongs to, and calculate the limit distribution, if it exists.

9.4 The covariance function $\Gamma(s,t) = E(X_s X_t)$ of a solution of (9.4.1) with $X_0(\cdot) \equiv 0$ may be oscillating in the sense that for every $s > 0$ there exist sequences $t_n^+ \uparrow \infty$ and $t_n^- \uparrow \infty$ such that $\Gamma(s,t_n^+) > 0$ and $\Gamma(s,t_n^-) < 0$. Show that this is the case if and only if $\vartheta_1 < -\exp(\vartheta_0 - 1)$.

9.5 Assume $v_0 < 0$, and let Y be the stationary solution

$$Y(t) = \int_{-\infty}^{t} x_0(t-s)dW_s$$

of (9.1.5)(see the proof of Proposition 9.3.1). Calculate the covariance function $K(h) = E[Y(t+h)Y(t)]$ and the spectral density $\varphi(\lambda)$, defined by

$$K(h) = \int_{-\infty}^{\infty} e^{ih\lambda}\varphi(\lambda)d\lambda.$$

9.6 Bibliographic notes

The results of this chapter have mainly been taken from Gushchin and Küchler (1996), who considered the case $N = 1$ only. They basically proved local asymptotic properties of the likelihood function and derived some further statistical properties.

Stochastic differential equations involving (partially or completely) the past of the process have been studied by, e.g., Ito and Nisio (1964), Mohammed (1984), Scheutzow (1984), Mohammed and Scheutzow (1990), Küchler and Mensch (1992), and Bell and Mohammed (1995).

Statistical properties of stochastic differential equations with memory (or time delay) have been studied in Sørensen (1983) and Dietz (1992) in the framework of likelihood theory; see also Chapters 8, 10, and 11. The statistical problem discussed in this chapter is related to studies of statistical properties of time series models of higher order; see, e.g., Jeganathan (1995).

For deterministic differential equations with time delay a large number of references can be given; see, e.g., Hale and Lunel (1993) or Bellmann and Cooke (1963). Often, such differential equations can be considered as ordinary differential equations in a Banach space, see, e.g., Curtain and Pritchard (1978) or Zabczyk (1983).

10
Sequential Methods

10.1 Preliminary results

In this chapter we shall mainly be concerned with models of the form

$$\frac{dP_\theta^t}{dP_0^t} = \exp\left(\theta^T A_t - \kappa(\theta) S_t\right), \tag{10.1.1}$$

where $\theta \in \Theta \subseteq \mathbb{R}^k$ and $\operatorname{int}\Theta \neq \emptyset$. This type of exponential family was also studied in Chapter 5. The situation is as described there, except that we will sometimes allow that S_t is not continuous, so that discrete time models can be included. In particular, $\kappa(\theta)$ and S_t are one-dimensional, A_t is k-dimensional, and S is a non-decreasing process for which $S_0 = 0$ and $S_t \to \infty$ as $t \to \infty$.

In Chapter 5 we studied the stopping times

$$\tau_u = \inf\{t : S_t > u\}. \tag{10.1.2}$$

If the underlying random phenomenon is observed in the random time interval $[0, \tau_u]$, it turned out that the corresponding model is a non-curved exponential family. More precisely, with respect to the filtration $\{\mathcal{F}_{\tau_u}\}_{u\geq 0}$ the class $\{P_\theta : \theta \in \Theta\}$ is an exponential family with a non-empty kernel.

Here we study the more general stopping times

$$\tau_u(\alpha, \beta) = \inf\left\{t : \alpha^T A_t + \beta S_t > u\right\}, \tag{10.1.3}$$

where $\alpha \in \mathbb{R}^k$, $\beta \in \mathbb{R}$, and $u > 0$ are prescribed constants that should be chosen in such a way that $P_\theta(\tau_u(\alpha, \beta) < \infty)$. In particular, $\alpha_1, \cdots, \alpha_k$ and β must not all equal zero. Usually, we will omit α and β in the notation.

Note that the observed information for the model (10.1.1) is $\ddot{\kappa}(\theta) S_t$. The stopping rule (10.1.3) with $\alpha = 0$ therefore corresponds to sampling until a prescribed level of information at a particular parameter value has been obtained. When $\alpha \neq 0$, the stopping rules correspond to sampling until a prescribed level of observed information has been obtained for a certain one-dimensional parameter function.

When observation is made in the random time interval $[0, \tau_u]$, we obtain exactly or approximately a *non-curved exponential family*. From the fundamental identity of sequential analysis (see Appendix B) it follows that

$$L_{\tau_u}(\theta) = \frac{dP_\theta^{\tau_u}}{dP_0^{\tau_u}} = \exp\left(\theta^T A_{\tau_u} - \kappa(\theta) S_{\tau_u}\right), \tag{10.1.4}$$

where $P_\theta^{\tau_u}$ denotes the restriction of P_θ to $\mathcal{F}_{\tau_u}$. If

$$P_\theta\left(\alpha^T A_{\tau_u} + \beta S_{\tau_u} = u\right) = 1, \tag{10.1.5}$$

we can find a representation with a k-dimensional canonical statistic, so the model is a non-curved exponential family. Specifically, we see that

$$S_{\tau_u} = \beta^{-1}\left(u - \alpha^T A_{\tau_u}\right), \tag{10.1.6}$$

provided that $\beta \neq 0$. Hence

$$L_{\tau_u}(\theta) = \exp\left(\left(\theta + \kappa(\theta)\beta^{-1}\alpha\right)^T A_{\tau_u} - \kappa(\theta)\beta^{-1}u\right). \tag{10.1.7}$$

This family has a non-empty kernel, so by Theorem 4.2.1 A_{τ_u} is a Lévy process. Now suppose $\alpha \neq 0$. Without loss of generality we can assume that $\alpha_1 \neq 0$. Then

$$A_{\tau_u}^{(1)} = \alpha_1^{-1}\left(u - \alpha_2 A_{\tau_u}^{(2)} - \cdots - \alpha_k A_{\tau_u}^{(k)} - \beta S_{\tau_u}\right) \tag{10.1.8}$$

and

$$\begin{aligned} L_{\tau_u}(\theta) \quad = \quad & \exp[\left(\theta_2 - \theta_1\alpha_1^{-1}\alpha_2\right) A_{\tau_u}^{(2)} + \cdots + \left(\theta_k - \theta_1\alpha_1^{-1}\alpha_k\right) A_{\tau_u}^{(k)} \\ & - \left(\kappa(\theta) + \theta_1\alpha_1^{-1}\beta\right) S_{\tau_u} + \theta_1\alpha_1^{-1}u]. \end{aligned} \tag{10.1.9}$$

Again, a family with a non-empty kernel is obtained, and again the canonical process is a Lévy process.

Since the exponential families (10.1.7) and (10.1.9) are non-curved, the maximum likelihood estimator for the mean value parameter is efficient in the sense that there is equality in the Cramér-Rao inequality. This has in much of the literature been a main motivation for studying stopping rules of the type (10.1.3). Since efficient estimators do not exist for curved exponential families (see Theorem 15.4 in Chensov, 1982, p. 219), such stopping rules are the only possibilities to obtain a so-called efficient sequential plan.

Now, usually the process A is not continuous, and in many cases it might jump past u, so that (10.1.5) does not hold. For discrete time processes, S is also not continuous. In such cases we need the random variable

$$D_{\tau_u} = \alpha^T A_{\tau_u} + \beta S_{\tau_u} - u, \tag{10.1.10}$$

which is often referred to as the overshoot. To obtain the likelihood function when (10.1.5) does not hold, u should be replaced by $u + D_{\tau_u}$ in (10.1.7) and (10.1.9). Then the exponential families are no longer non-curved, but if D_{τ_u} is small compared to A_{τ_u} and S_{τ_u}, it will approximately be so. We shall study this in detail in Section 10.3.

Let us first discuss a condition that ensures that the stopping times (10.1.3) are almost surely finite. Obviously, this is the case if $\alpha = 0$. Otherwise we can use the following results.

Proposition 10.1.1 *Assume that S is continuous. Suppose, moreover, that either S is strictly increasing or $(\theta, \kappa(\theta)) \in \operatorname{int} \tilde{\Gamma}_t$ for all $t > 0$. Then $\alpha^T \dot{\kappa}(\theta) + \beta \geq 0$ implies that*

$$P_\theta(\tau_u(\alpha, \beta) < \infty) = 1.$$

If $\alpha^T \dot{\kappa}(\theta) + \beta > 0$, then $E_\theta\left(S_{\tau_u(\alpha,\beta)}\right) < \infty$; otherwise, $E_\theta\left(S_{\tau_u(\alpha,\beta)}\right) = \infty$.

Remarks: The set $\tilde{\Gamma}_t$ is the canonical parameter set of the envelope exponential family at time t defined in Chapter 7. A condition ensuring that $(\theta, \kappa(\theta)) \in \operatorname{int} \tilde{\Gamma}_t$ is that zero be an interior point in the domain of the Laplace transform of S_t for all $t > 0$ (Corollary 7.3.3). The condition $(\theta, \kappa(\theta)) \in \operatorname{int} \tilde{\Gamma}_t$ can be replaced by the condition that the score function $A_t - \dot{\kappa}(\theta) S_t$ be a square integrable P_θ-martingale with quadratic characteristic proportional to S_t. A similar remark holds for Proposition 10.1.2 below. □

Proof: Under the conditions of the proposition, $A_t = B_{S_t}$, where B under P_θ is a Lévy process with mean $E_\theta(B_t) = \dot{\kappa}(\theta)t$; cf. Chapter 5. Therefore,

$$\alpha^T A_t + \beta S_t = Y_{S_t},$$

where $Y_t = \alpha^T B_t + \beta t$ is a Lévy process with $E_\theta(Y_t) = (\alpha^T \dot{\kappa}(\theta) + \beta)t$. Since τ_u is the first time $\alpha^T A_t + \beta S_t$ exceeds u, S_{τ_u} is the first time Y exceeds u. If $\alpha^T \dot{\kappa}(\theta) + \beta > 0$, then $Y_t \to \infty$ as $t \to \infty$, so $S_{\tau_u} < \infty$ and hence $\tau_u < \infty$. Moreover, the mean value of S_{τ_u} is finite. If $\alpha^T \dot{\kappa}(\theta) + \beta = 0$, then $\limsup Y_t = \infty$ (see, e.g., Asmussen, 1987, p. 169), so again, $S_{\tau_u} < \infty$ and $\tau_u < \infty$. In this case, however, the mean waiting time for Y to exceed u is infinite. □

The next result is also true when S is not continuous, so that also discrete time is covered. We state the result for more general models with an exponential representation of the form

$$\frac{dP_\theta^t}{dP_0^t} = \exp\left(\theta^T A_t - \varphi(\theta)^T S_t\right), \tag{10.1.11}$$

where $\theta \in \Theta \subseteq \mathrm{I\!R}^k$, and where A is as before, while $\varphi(\theta)$ and S_t need not be one-dimensional. For consistency with our general notation, we denote their dimension by $m-k$. Here we assume that each coordinate of S is a non-decreasing predictable process satisfying $S_0 = 0$ and $S_t \to \infty$ as $t \to \infty$. In the discrete time setting predictability simply means that S_n is measurable with respect to $\mathcal{F}_{n-1}$. For the more general model (10.1.11) we consider stopping time of the type

$$\tau_u(\alpha,\beta) = \inf\left\{t > 0 : \alpha^T A_t + \beta^T S_t > u\right\}, \tag{10.1.12}$$

where $\alpha \in \mathrm{I\!R}^k$, $\beta \in \mathrm{I\!R}^{m-k}$, and $u > 0$.

In accordance with earlier notation, let $\dot{\varphi}(\theta)$ denote the matrix whose ith row is the vector of partial derivatives of the ith coordinate of $\varphi(\theta)$.

Proposition 10.1.2 *Suppose there exists a one-dimensional process V_t such that under P_θ,*

$$S_t/V_t \to W(\theta)$$

in probability as $t \to \infty$, where $W(\theta)$ is a random vector. Assume, furthermore, that $(\theta, \varphi(\theta)) \in \operatorname{int}\tilde{\Gamma}_t$ for all $t > 0$ and that P_θ-almost surely,

$$(\dot{\varphi}(\theta)\alpha + \beta)^T W(\theta) > 0.$$

Then

$$P_\theta(\tau_u(\alpha,\beta) < \infty) = 1.$$

Proof:

$$\alpha^T A_t + \beta^T S_t = \alpha^T(A_t - \dot{\varphi}(\theta)^T S_t) + (\dot{\varphi}(\theta)\alpha + \beta)^T S_t,$$

where $\alpha^T(A_t - \dot{\varphi}(\theta)^T S_t)$ by Corollary 8.1.3 is a square integrable P_θ-martingale with quadratic characteristic $\left(\alpha^T \ddot{\varphi}(\theta)\alpha\right)^T S_t$. Here $\alpha^T\ddot{\varphi}(\theta)\alpha$ denotes the vector with the lth element equal to

$$\sum_{i,j} \alpha_i\alpha_j \frac{\partial^2}{\partial\theta_i\partial\theta_j}\varphi_l(\theta),$$

where $\varphi_l(\theta)$ is the lth coordinate of $\varphi(\theta)$. By the law of large numbers for martingales (Appendix A),

$$(\alpha^T A_t + \beta^T S_t)/V_t \to (\dot{\varphi}(\theta)\alpha + \beta)^T W(\theta) > 0$$

in P_θ-probability, which proves the proposition. □

10.2 Exact likelihood theory

In the following we consider models of the type (10.1.1). For models where S is continuous, we saw in Chapter 5 that κ is necessarily a cumulant transform and thus is convex. We assume that $\Theta = \operatorname{dom}\kappa$ and that κ is steep. This concept was defined in Section 2.2. In Chapter 5 we also found by classical exponential family theory that the closed convex support of the random vector A_t/S_t is contained in $C = \operatorname{cl}\dot{\kappa}(\operatorname{int}\Theta)$ for all $t > 0$. Note that we could also have parametrized the model by the compensator parameter $\mu = \dot{\kappa}(\theta)$. We need the following subsets of Θ that are defined for given $\alpha \in \mathbb{R}^k$ and $\beta \in \mathbb{R}$:

$$\begin{aligned}\Theta_{\alpha,\beta} &= \left\{\theta \in \Theta : \alpha^T \dot{\kappa}(\theta) + \beta > 0\right\}, \\ \bar{\Theta}_{\alpha,\beta} &= \left\{\theta \in \Theta : \alpha^T \dot{\kappa}(\theta) + \beta \geq 0\right\}.\end{aligned}$$

Theorem 10.2.1 *Suppose S is continuous, that κ is steep and strictly convex, and that $\theta \in \bar{\Theta}_{\alpha,\beta}$. Then*

$$P_\theta\left(A_{\tau_u(\alpha,\beta)}/S_{\tau_u(\alpha,\beta)} \in \operatorname{bd} C\right) = 0$$

implies that the maximum likelihood estimator $\hat{\theta}_{\tau_u}$ based on observation in the random time interval $[0, \tau_u(\alpha,\beta)]$ exists and is unique. It is given by

$$\hat{\theta}_{\tau_u} = \dot{\kappa}^{-1}\left(A_{\tau_u}/B_{\tau_u}\right).$$

Moreover, $\hat{\theta}_{\tau_u} \in \Theta_{\alpha,\beta}$.

Proof: Since $\theta \in \bar{\Theta}_{\alpha,\beta}$, $P_\theta(\tau_u(\alpha,\beta) < \infty) = 1$ by Proposition 10.1.1. Let $\tau_u(0,1)$ be the stopping time of the type (10.1.3) with $\alpha = 0$ and $\beta = 1$. This stopping time is finite, and we can think of $A_{\tau_u(\alpha,\beta)}/S_{\tau_u(\alpha,\beta)}$ as an observation of $A_{\tau_u(0,1)}/S_{\tau_u(0,1)}$ and proceed as in the proof of Theorem 5.2.1. If $\alpha \neq 0$, we get by (10.1.8) with u replaced by $u + D_{\tau_u}$ (D_{τ_u} given by (10.1.10)) that

$$\alpha^T \dot{\kappa}\left(\hat{\theta}_{\tau_u}\right) + \beta = \left(u + D_{\tau_u}\right)/S_{\tau_u} > 0,$$

so $\hat{\theta}_{\tau_u} \in \Theta_{\alpha,\beta}$. If $\alpha = 0$ and $\beta > 0$, then $\Theta_{\alpha,\beta} = \Theta$, so in this case the last statement is trivial. □

Next we will derive the Laplace transform of the sufficient or approximately sufficient statistic when sampling in $[0, \tau_u(\alpha,\beta)]$. If $\alpha \neq 0$, we can, as noted before, assume that $\alpha_1 \neq 0$. Then the likelihood function (10.1.9) with u replaced by $u + D_{\tau_u}$ can be written as

$$\frac{dP_\theta^{\tau_u}}{dP_0^{\tau_u}} = \exp\left(\gamma(\theta)^T B_{\tau_u} + \theta_1 \alpha_1^{-1}\left(u + D_{\tau_u}\right)\right), \tag{10.2.1}$$

where

$$B_{\tau_u} = \left(S_{\tau_u}, A_{\tau_u}^{(2)}, \cdots, A_{\tau_u}^{(k)}\right)^T \tag{10.2.2}$$

and

$$\gamma(\theta)^T = \left(-\kappa(\theta) - \theta_1\alpha_1^{-1}\beta, \theta_2 - \theta_1\alpha_1^{-1}\alpha_2, \cdots, \theta_k - \theta_1\alpha_1^{-1}\alpha_k\right). \tag{10.2.3}$$

Theorem 10.2.2 *Suppose $\alpha_1 \neq 0$ and that the conditions of Proposition 10.1.1 or Proposition 10.1.2 hold. Then the function γ given by (10.2.3) is invertible on $\Theta_{\alpha,\beta}$, and for $\theta \in \Theta_{\alpha,\beta}$ the Laplace transform of B_{τ_u} (given by (10.2.2)) under P_θ is*

$$\begin{aligned} E_\theta\left(\exp\left[s^T B_{\tau_u}\right]\right) &= \exp\left\{\alpha_1^{-1} u\left[\theta_1 - \gamma^{-1}(s+\gamma(\theta))_1\right]\right\} \\ &\quad\times \varphi^u_{\gamma^{-1}(s+\gamma(\theta))}\left[\alpha_1^{-1}\left(\theta_1 - \gamma^{-1}(s+\gamma(\theta))_1\right)\right] \end{aligned} \tag{10.2.4}$$

for $s \in \mathbb{R}^k$ such that $s + \gamma(\theta) \in \gamma(\Theta_{\alpha,\beta})$, where φ^u_θ denotes the Laplace transform of D_{τ_u} under P_θ.

If $\alpha = 0$ and $\beta > 0$, the Laplace transform of A_{τ_u} under P_θ for $\theta \in \Theta$ is

$$\begin{aligned} E_\theta\left(\exp\left[s^T A_{\tau_u}\right]\right) &= \exp\left[u\beta^{-1}\left(\kappa(\theta+s) - \kappa(\theta)\right)\right] \\ &\quad\times \varphi^u_{\theta+s}\left[\beta^{-1}\left(\kappa(\theta+s) - \kappa(\theta)\right)\right] \end{aligned} \tag{10.2.5}$$

for $s \in \mathbb{R}^k$ such that $\theta + s \in \Theta$.

Remarks: If the conditions of Proposition 10.1.1 or Proposition 10.1.2 do not hold, the conclusion of Theorem 10.2.2 still holds for those $\theta \in \Theta_{\alpha,\beta}$ for which $P_\theta(\tau_u < \infty) = 1$. Otherwise, it holds conditionally on $\{\tau_u < \infty\}$ with φ^u_θ then denoting the conditional Laplace transform.

Note also that if the likelihood function is the product of a factor like (10.1.1) and a factor depending only on a parameter φ independent of θ (and on the observations, of course), then the theorem still holds with φ suitably introduced in the notation. For examples of this situation, see Example 10.2.6 and Example 10.2.8. □

Proof: First, suppose $\alpha_1 \neq 0$. The inverse of γ is given for $y \in \gamma(\Theta_{\alpha,\beta})$ by

$$\begin{aligned} \gamma^{-1}(y)_1 &= g^{-1}_{y_2,\cdots,y_k}(y_1), \\ \gamma^{-1}(y)_2 &= y_2 + \gamma^{-1}(y)_1\alpha_1^{-1}\alpha_2, \\ &\vdots \\ \gamma^{-1}(y)_k &= y_k + \gamma^{-1}(y)_1\alpha_1^{-1}\alpha_k, \end{aligned}$$

where the function

$$g_{y_2,\cdots,y_k}(x) = -\kappa\left(x, y_2 + x\alpha_1^{-1}\alpha_2, \cdots, y_k + x\alpha_1^{-1}\alpha_k\right) - x\alpha_1^{-1}\beta$$

is defined on the set

$$M_{y_2,\cdots,y_k} = \left\{x \in \mathbb{R} : \left(x, y_2 + x\alpha_1^{-1}\alpha_2, \cdots, y_k + x\alpha_1^{-1}\alpha_k\right) \in \Theta_{\alpha,\beta}\right\},$$

which for $y \in \gamma(\Theta_{\alpha,\beta})$ is non-empty. The derivative of $g_{y_2,\cdots,y_k}$ is

$$g'_{y_2,\cdots,y_k}(x) = -\alpha_1^{-1}\left\{\alpha^T\dot{\kappa}\left(x, y_2 + x\alpha_1^{-1}\alpha_2, \cdots, y_k + x\alpha_1^{-1}\alpha_k\right) + \beta\right\},$$

which for $x \in M_{y_2,\cdots,y_k}$ is either strictly positive or strictly negative depending on the sign of α_1. Therefore, $g_{y_2,\cdots,y_k}$, and hence γ, is invertible.

Using (10.2.1) twice, we find that for $\tilde{\theta} = \gamma^{-1}(s + \gamma(\theta))$,

$$\begin{aligned} &E_\theta\left(\exp\left(s^T B_\tau\right)\right) \\ &= E_0[\exp\{(s+\gamma(\theta))^T B_\tau + \theta_1\alpha_1^{-1}(u + D_\tau)\}] \\ &= \exp(\alpha_1^{-1}u(\theta_1 - \tilde{\theta}_1))E_0[\exp(\alpha_1^{-1}(\theta_1 - \tilde{\theta}_1)D_\tau) \\ &\qquad \times \exp(\gamma(\tilde{\theta})^T B_\tau + \tilde{\theta}_1\alpha_1^{-1}(u + D_\tau))] \\ &= \exp(\alpha_1^{-1}u(\theta_1 - \tilde{\theta}_1))E_{\tilde{\theta}}[\exp(\alpha_1^{-1}(\theta_1 - \tilde{\theta}_1)D_\tau)]. \end{aligned}$$

The result for $\alpha = 0$ follows in a similar way. Here it is easiest to use the formula (10.1.4) for the likelihood function. □

Corollary 10.2.3 *Suppose (10.1.5) holds. If $\alpha_1 \neq 0$ and $\theta \in \Theta_{\alpha,\beta}$, and if the conditions of Proposition 10.1.1 or Proposition 10.1.2 are satisfied, then B_{τ_u} given by (10.2.2) is minimal sufficient and its Laplace transform under P_θ is*

$$E_\theta\left(\exp\left[s^T B_{\tau_u}\right]\right) = \exp\left[\alpha_1^{-1}u\left\{\theta_1 - \gamma^{-1}\left(s + \gamma(\theta)\right)_1\right\}\right] \tag{10.2.6}$$

for values of $s \in \mathbb{R}^k$ such that $s + \gamma(\theta) \in \gamma(\Theta_{\alpha,\beta})$.

If $\alpha = 0$ and $\beta > 0$, A_{τ_u} is minimal sufficient, and its Laplace transform under P_θ (for $\theta \in \Theta$) is

$$E_\theta\left(\exp\left[s^T A_{\tau_u}\right]\right) = \exp\left[u\beta^{-1}\left(\kappa(\theta + s) - \kappa(\theta)\right)\right] \tag{10.2.7}$$

for $s \in \mathbb{R}^k$ such that $\theta + s \in \Theta$.

Note that provided that $\theta \in \operatorname{int}\Theta_{\alpha,\beta}$, the sufficient statistic B_{τ_u} or A_{τ_u} has moments of any order. When, in particular, $S_t = t$, as it is if A_t is a Lévy process, it follows that the stopping time τ_u has moments of any order. In that somewhat special case,

$$E_\theta(\tau_u) = u\left(\alpha^T\dot{\kappa}(\theta) + \beta\right)^{-1}. \tag{10.2.8}$$

Note also that when S is continuous, (10.1.5) is automatically satisfied for $\alpha = 0$. In that case the result (10.2.7) is contained in Theorem 4.2.1.

When (10.1.5) is not satisfied, we can still say something about the moments of A_{τ_u} and B_{τ_u}.

Corollary 10.2.4 *If $\theta \in \operatorname{int}\Theta_{\alpha,\beta}$, then $A_{\tau_u(\alpha,\beta)}$ and $S_{\tau_u(\alpha,\beta)}$ have moments of any order under P_θ. In particular,*

$$E_\theta\left(A_{\tau_u(\alpha,\beta)}\right) = \dot{\kappa}(\theta)\, E_\theta\left(S_{\tau_u(\alpha,\beta)}\right) \tag{10.2.9}$$

and

$$E_\theta\left(S_{\tau_u(\alpha,\beta)}\right) = \frac{u + E_\theta\left(D_{\tau_u(\alpha,\beta)}\right)}{\alpha^T \dot{\kappa}(\theta) + \beta}. \tag{10.2.10}$$

Proof: First suppose $\alpha_1 \neq 0$. Since $\gamma(\theta) \in \operatorname{int}\gamma(\Theta_{\alpha,\beta})$ (γ^{-1} is continuous), it follows from Theorem 10.2.2 that zero is an interior point in the Laplace transform of $S_{\tau_u(\alpha,\beta)}$. Therefore, by arguments similar to the proofs of Proposition 7.3.1 and Corollary 7.3.3, $(\theta, \kappa(\theta))$ belongs to the interior of the full exponential family generated by $A_{\tau_u(\alpha,\beta)}$ and $S_{\tau_u(\alpha,\beta)}$. Hence the first claim follows, and (10.2.9) is proved by differentiation under the expectation in $E_0\left(L_{\tau_u(\alpha,\beta)}(\theta)\right) = 1$. The last equation is simply a consequence of $\alpha^T A_\tau + \beta S_\tau = u + D_\tau$. The result for $\alpha = 0$ follows in an analogous way with S_τ replaced, for instance, by $A_\tau^{(1)}$. □

It is obviously possible to give results about higher moments of A_{τ_u} and B_{τ_u} by differentiating several times under the expectation in $E_0\left(L_{\tau_u}(\theta)\right)$. The result (10.2.9) is usually called *Wald's identity.*

Example 10.2.5 *Diffusion-type processes.* Consider the class of stochastic differential equations

$$dX_t = \left[a_t(X) + \theta b_t(X)\right] dt + c_t(X)\, dW_t, \quad t > 0, \tag{10.2.11}$$

with initial condition $X_0 = x_0$, and where all quantities are one-dimensional. The functionals a_t, b_t, and c_t depend on X only through $\{X_s : s \leq t\}$, and $c > 0$. As usual, W is a standard Wiener process. We assume that (10.2.11) has a solution for all $\theta \in \mathbb{R}$. This type of model was discussed in Section 3.5.

If a process that solves (10.2.11) for some value of θ is observed in the time interval $[0, t]$, the likelihood function is, under regularity conditions stated in Section 3.5, given by

$$L_t(\theta) = \exp\left(\theta A_t - \tfrac{1}{2}\theta^2 S_t\right), \tag{10.2.12}$$

where A is given by (3.5.5), while S is given by (3.5.2) (where it is called B).

Here $\Theta_{\alpha,\beta} = \{\theta : \alpha\theta + \beta > 0\}$, and since the canonical processes A and S are both continuous, (10.1.5) is satisfied for all $\theta \in \Theta_{\alpha,\beta}$.

Let us first consider the case $\alpha \neq 0$. The function γ is given by

$$\gamma(\theta) = -\tfrac{1}{2}\theta^2 - \theta\alpha^{-1}\beta,$$

so

$$\gamma^{-1}(x) = \alpha^{-1}\sqrt{\beta^2 - 2\alpha^2 x} - \alpha^{-1}\beta.$$

By Corollary 10.2.3 the Laplace transform of S_τ is

$$\exp\left\{u\alpha^{-1}\left(\theta+\alpha^{-1}\beta\right)\left[1-\sqrt{1-2s\left(\theta+\alpha^{-1}\beta\right)^{-2}}\right]\right\},$$

which is the Laplace transform of an inverse Gaussian distribution. The distribution on $(0,\infty)$ with density function

$$\tfrac{1}{2}\left(\psi/\chi\right)^{\lambda/2}K_\lambda\left(\sqrt{\chi\psi}\right)^{-1}x^{\lambda-1}\exp\left(-\tfrac{1}{2}\left(\chi x^{-1}+\psi x\right)\right),$$

where K_λ is the modified Bessel function of the third kind, is called a generalized inverse Gaussian distribution (Jørgensen, 1982) and denoted by $N^-(\lambda,\chi,\psi)$. For $\lambda=-\frac{1}{2}$, the usual inverse Gaussian distribution is obtained. We have proved that

$$S_\tau \sim N^-\left(-\tfrac{1}{2},u^2\alpha^{-2},\left(\theta+\alpha^{-1}\beta\right)^2\right).$$

The maximum likelihood estimator is

$$\hat{\theta}_\tau = u\alpha^{-1}S_\tau^{-1}-\alpha^{-1}\beta,$$

and the likelihood ratio test statistic Q_τ for the hypothesis $\theta=\theta_0$ is given by

$$-2\log Q_\tau = \left(\theta_0+\alpha^{-1}\beta S_\tau^{\frac{1}{2}}-u\alpha^{-1}S_\tau^{-\frac{1}{2}}\right)^2.$$

By standard results about the generalized inverse Gaussian distribution (see, e.g., Jørgensen, 1982), it follows that

$$\hat{\theta}_\tau+\alpha^{-1}\beta \sim N^-\left(\tfrac{1}{2},u\alpha^{-1}\left(\theta+\alpha^{-1}\beta\right)^2,u\alpha^{-1}\right)$$

and, provided that $\theta=\theta_0$, that

$$-2\log Q_\tau \sim \chi^2(1).$$

Similar exact distributional results can be found when $\alpha=0$ and $\beta>0$. By Corollary 10.2.3 the Laplace transform of A_τ under P_θ is

$$\exp\left(\theta u\beta^{-1}s+\tfrac{1}{2}u\beta^{-1}s^2\right),$$

so

$$A_\tau \sim N(\theta u\beta^{-1},u\beta^{-1}).$$

In this case,

$$\hat{\theta}_\tau = \beta u^{-1}A_\tau$$

and

$$-2\log Q_\tau = u\beta^{-1}\left(A_\tau-\theta_0 u\beta^{-1}\right)^2.$$

Hence,

$$\hat{\theta}_\tau \sim N(\theta, \beta u^{-1}),$$

and $-2\log Q_\tau$ has a non-central $\chi^2(1)$-distribution with non-centrality parameter $(\theta - \theta_0)\sqrt{u/\beta}$.

The distributional results derived here could also have been found by utilizing the fact that

$$\alpha A_t + \beta S_t = Y_{S_t},$$

where Y is a Brownian motion with drift $\alpha\theta + \beta$ and variance parameter α. If $\alpha \neq 0$, then S_τ is the first time Y hits the level u. □

Example 10.2.6 *Counting processes.* Consider the class of counting processes introduced in Section 3.4. Their intensity λ with respect to the filtration generated by the process has the form

$$\lambda_t = \kappa H_t, \tag{10.2.13}$$

where $\kappa > 0$ is a parameter and H is a given positive predictable stochastic process with finite expectation for all $t < \infty$. Let N denote the observed counting process, and assume that $N_0 = n_0$. Then the likelihood function corresponding to observation of N in $[0, t]$ is

$$L_t(\theta) = \exp\left(\theta A_t - \left[e^\theta - 1\right] S_t\right), \tag{10.2.14}$$

where $\theta = \log\kappa$, $A_t = N_t - n_0$, and

$$S_t = \int_0^t H_s ds.$$

For a model of this type, $\Theta_{\alpha,\beta} = \{\theta : \beta + \alpha e^\theta > 0\}$. When $\alpha \leq 0$ and $\beta > 0$, (10.1.5) is obviously satisfied. This is also the case if $\alpha = 1$, $\beta = 0$, and $u \in \mathbb{N}$. First we consider the case $\alpha \neq 0$. Here

$$\gamma(\theta) = 1 - e^\theta - \theta\alpha^{-1}\beta.$$

There is no explicit expression for the inverse of this function, except when $\beta = 0$. For $\alpha = 1$ and $\beta = 0$ we have $\Theta_{\alpha,\beta} = \mathbb{R}$ and $\gamma^{-1}(x) = \log(1 - x)$, $x < 1$. This stopping rule is simply observation until a prescribed number of counts is attained (often called inverse sampling). By Corollary 10.2.3 the Laplace transform of S_τ under P_θ is $(1 - s/\kappa)^{-u}$, so

$$S_\tau \sim \text{Gamma}(u, \kappa),$$

and since $\hat{\kappa}_\tau = u/S_\tau$,

$$\hat{\kappa}_\tau \sim N^-(-u, 2\kappa u, 0).$$

The likelihood ratio test statistic Q_τ for the hypothesis $\theta = \theta_0$ is given by

$$-2\log Q_\tau = 2\{u\log u - u + \kappa_0 S_\tau - u\log(\kappa_0 S_\tau)\}$$

$(\kappa_0 = \exp(\theta_0))$. When $\kappa = \kappa_0$, the Laplace transform of $\kappa_0 S_\tau - u \log(\kappa_0 S_\tau)$ is

$$\int_0^\infty \exp(s\kappa_0 x - su\log(\kappa_0 x))\, x^{u-1} e^{-\kappa_0 x} \kappa_0^u \Gamma(u)^{-1}\, dx$$
$$= \int_0^\infty x^{u(1-s)-1} e^{-\kappa_0(1-s)x} \kappa_0^{u(1-s)} \Gamma(u)^{-1}\, dx$$
$$= \Gamma(u(1-s))\, \Gamma(u)^{-1} (1-s)^{-u(1-s)}$$

for $s < 1$. Hence the Laplace transform of $-2\log Q_\tau$ is

$$(1-2s)^{u(2s-1)} e^{-2us} \Gamma(u(1-2s)) u^{2us} / \Gamma(u), \quad s < \tfrac{1}{2}.$$

These results hold also when u is not an integer provided that u is replaced by the smallest integer that is larger than or equal to u. We will denote this integer by $\bar{u}$. This is because when u is not an integer, $\tau_u(1,0) = \tau_{\bar{u}}(1,0)$. This result for a non-integer u is consistent with Theorem 10.2.2, as can be seen by noting that the overshoot is $\bar{u} - u$, so that $\varphi_\theta^u(s) = \exp[(\bar{u} - u)s]$. The expression for $-2\log Q_\tau$ holds too provied that u is replaced by $\bar{u}$, again because $D_\tau = \bar{u} - u$.

Next consider the case $\alpha = 0$ and $\beta > 0$. Also here $\Theta_{\alpha,\beta} = \mathbb{R}$, and by Theorem 10.2.2 the Laplace transform of A_τ under P_θ is

$$\exp\left[u\beta^{-1} e^\theta (e^s - 1)\right],$$

so

$$A_\tau \sim \text{Poisson}\left(u\beta^{-1}\kappa\right).$$

In this case $\hat{\kappa}_\tau = \beta u^{-1} A_\tau$.

These results could also have been obtained from the fact that there exists a Poisson process Y with intensity κ such that

$$N_t - n_0 = Y_{S_t}$$

(cf. Section 5.1) and that S_τ is the time at which Y hits the level u.

Let us finally briefly consider marked counting processes of the type defined in Section 3.4, i.e.,

$$X_t = \sum_{i=1}^{N_t} Y_i,$$

where N is as above with $n_0 = 0$. For simplicity we assume that the marks Y_i are mutually independent and independent of N. The likelihood function is

$$L_t(\theta, \alpha) = \exp\left(\theta A_t - \left(e^\theta - 1\right) S_t\right) \exp\left(\alpha^T \sum_{i=1}^{N_t} T(Y_i) - \gamma(\alpha)\right),$$

where the domain of α is independent of θ. For inference about θ the stopping times discussed above can still be used. When X is observed in $[0, \tau]$, N_τ is S-ancillary for α, so inference about α should be drawn conditionally on N_τ. This is identical to repeated sampling from a classical exponential family, and it does not matter which stopping rule we have used. □

Example 10.2.7 *The linear birth-and-death process.* The birth-and-death process with immigration is a Markov process with transition probabilities

$$P(X_{t+h} = k | X_t = i) = \begin{cases} \lambda ih + o(h) & k = i+1, \ i = 1, 2, \cdots \\ 1 - (\lambda + \mu) ih + o(h) & k = i, \ i = 1, 2, \cdots \\ \mu ih + o(h) & k = i-1, \ i = 1, 2, \cdots \\ \gamma h + o(h) & k = 1, \ i = 0 \\ 1 - \gamma h + o(h) & k = i = 0 \\ o(h) & \text{otherwise,} \end{cases}$$

where $(\lambda, \mu, \gamma) \in (0, \infty)^3$ and where $X_0 = x_0$. The likelihood function corresponding to observation of X in $[0, t]$ is

$$L_t(\theta) = \exp\left[\theta_1 A_t^{(1)} + \theta_2 A_t^{(2)} - \left(e^{\theta_1} + e^{\theta_2} - 2\right) S_t\right] \exp\left[\theta_3 A_t^{(3)} - e^{\theta_3} T_t\right], \tag{10.2.15}$$

where $\theta_1 = \log \lambda$, $\theta_2 = \log \mu$, and $\theta_3 = \log \gamma$, and where $A_t^{(1)}$ is the number of jumps from a state $i \geq 1$ to $i+1$ (births) in $[0, t]$, $A_t^{(2)}$ is the number of downward jumps (deaths), and $A_t^{(3)}$ is the number of jumps from 0 to 1 (immigrations). Finally, T_t is the total time where the population size was zero in $[0, t]$, and

$$S_t = \int_0^t X_s ds$$

is the total time lived by all members of the population in $[0, t]$. We are only interested in the parameters λ and μ and will only consider stopping times of the type

$$\tau_u = \inf\left\{t > 0 : \alpha_1 A_t^{(1)} + \alpha_2 A_t^{(2)} + \beta S_t > u\right\}. \tag{10.2.16}$$

In the following we assume that γ is known.

The parameter values for which (10.2.16) is finite are $\{(\lambda, \mu) : \alpha_1 \lambda + \alpha_2 \mu + \beta > u\}$. Condition (10.1.5) is satisfied when $\beta > 0$ and $\alpha_i \leq 0$, $i = 1, 2$. If $\beta = 0$ and $u \in \mathbb{N}$, it is satisfied when either $\alpha_1 = 1$ and $-\alpha_2 \in \mathbb{N} \cup \{0\}$, or $\alpha_2 = 1$ and $-\alpha_1 \in \mathbb{N} \cup \{0\}$. A final possibility where (10.1.5) holds is $u \in \mathbb{N}$, $\beta = 0$, and $\alpha_1 = \alpha_2 = 1$. The next results follow from Corollary 10.2.3.

If $\alpha_1 = \alpha_2 = 0$ and $\beta > 0$, the canonical statistic is $(A_\tau^{(1)}, A_\tau^{(2)})$, $A_\tau^{(1)}$ and $A_\tau^{(2)}$ are independent, and

$$A_\tau^{(1)} \sim \text{Poisson}\left(\lambda u \beta^{-1}\right), \quad A_t^{(2)} \sim \text{Poisson}\left(\mu u \beta^{-1}\right).$$

From now on let $\beta = 0$ and $u \in \mathbb{N}$. When $\alpha_1 = \alpha_2 = 1$ the canonical statistic is $(S_\tau, A_\tau^{(2)})$, $S_\tau$and A_τ are independent, and

$$S_\tau \sim \text{Gamma}\,(u, \lambda + \mu)\,, \quad A_\tau^{(2)} \sim \text{Binomial}\left(u, \mu\,(\lambda+\mu)^{-1}\right).$$

For $\alpha_1 = 1$ and $\alpha_2 = 0$ we find that

$$\begin{aligned} &S_\tau \sim \text{Gamma}\,(u, \lambda)\,, \\ &A_\tau^{(2)} | S_\tau = s \sim \text{Poisson}\,(\mu s)\,, \\ &A_\tau^{(2)} \sim \text{Negative binomial}\left(u, \mu\,(\lambda+\mu)^{-1}\right). \end{aligned}$$

For $\alpha_2 = 1$ and $\alpha_1 = 0$, symmetric results are obtained.

When $\alpha_1^{-1}\alpha_2 = -1$,

$$\gamma\,(\theta)^T = \left(2 - e^{\theta_1} - e^{\theta_2}, \theta_1 + \theta_2\right).$$

If $\alpha_1 = -1$ and $\alpha_2 = 1$, we must require that $\mu \geq \lambda$ in order that the stopping time (10.2.16) be finite. Note that $\tau_{x0}\,(-1,1,0)$ is the first time of extinction. When $\mu \geq \lambda$,

$$\gamma^{-1}\,(x)_1 = \log\left\{\tfrac{1}{2}\left(2 - x_1 - \sqrt{(2-x_1)^2 - 4e^{x_2}}\right)\right\},$$

so

$$\begin{aligned} &E_{\lambda,\mu}\left(\exp\left(s_1 S_\tau + s_2 A_\tau^{(2)}\right)\right) &(10.2.17) \\ &\quad = (2\lambda)^{-u}\left[(\lambda+\mu-s_1) - \sqrt{(\lambda+\mu-s_1)^2 - 4\lambda\mu e^{s_2}}\right]^u. \end{aligned}$$

The Laplace transform of S_τ is obtained for $s_2 = 0$, and using a formula in Roberts and Kaufman (1966) we find that the density function of S_τ is

$$f_{S_\tau}\,(x) = u\,(\mu/\lambda)^{u/2}\,e^{-(\lambda+\mu)x} I_u\left(x2\sqrt{\lambda\mu}\right)x^{-1}, \qquad (10.2.18)$$

where I_u denotes a modified Bessel function. The distribution with density (10.2.18) is the distribution of the first passage time to u of a random walk with jump times determined by a Poisson process; see Feller (1966, p. 414). Feller called (10.2.18) the Bessel function density.

The probability generating function (as a function of x) for $A_\tau^{(2)}$ is found from (10.2.17) by putting $s_1 = 0$ and replacing e^{s_2} by x. By differentiating the probability generating function with respect to x it is found that the distribution of $A_\tau^{(2)}$ equals the distribution of the sum of u independent random variables each of which has the point probabilities

$$\begin{aligned} p_1 &= \lambda\,(\lambda+\mu)^{-1} \\ p_i &= \frac{(2i-3)\,(2i-5)\cdots 1}{i!}\cdot\frac{\mu\,(2\lambda\mu)^i}{(\lambda+\mu)^{2i-1}}, \quad \text{for } i \geq 2. \end{aligned}$$

For $\alpha_1 = 1$ and $\alpha_2 = -1$, the condition $\lambda \geq \mu$ is needed to ensure a finite stopping rule. In that case

$$\gamma^{-1}(x)_1 = \log\left\{\tfrac{1}{2}\left(2 - x_1 + \sqrt{(2-x_1)^2 - 4e^{x_2}}\right)\right\},$$

so

$$E_{\lambda,\mu}\left[\exp\left(s_1 S_\tau + s_2 A_\tau^{(2)}\right)\right]$$
$$= (2\mu)^{-u} e^{-us_2}\left[\lambda + \mu - s_1 - \sqrt{(\lambda+\mu-s_1)^2 - 4\lambda\mu e^{s_2}}\right]^u.$$

Since $A_\tau^{(1)} = u + A_\tau^{(2)}$, the Laplace transform of $(S_\tau, A_\tau^{(1)})$ is equal to (10.2.17) with λ and μ interchanged. Thus for $\alpha_1 = 1$ and $\alpha_2 = -1$ the distribution of $(S_\tau, A_\tau^{(1)})$ equals that of $(S_\tau, A_\tau^{(2)})$ for $\alpha_1 = -1$ and $\alpha_2 = 1$.

As for the counting processes in the previous example, the results for $\beta = 0$ also hold when u is not an integer provided that u is replaced by $\bar{u}$, the smallest integer larger than or equal to u. The argument is exactly as in Example 10.2.6 □

Example 10.2.8 *Finite state space Markov processes.* A continuous time Markov process with finite state space $\{1, \cdots, m\}$ and intensity matrix $\{\lambda_{ij}\}$, where $\lambda_{ij} > 0$ for $i \neq j$, has, when observed in $[0, t]$, the likelihood function

$$L_t(\theta) = \exp\left\{\sum_{i=1}^{m}\left[\sum_{j\neq i}\theta_{ij}K_t^{(i,j)} - \left(\sum_{j\neq i}e^{\theta_{ij}} + 1 - m\right)S_t^{(i)}\right]\right\}, \tag{10.2.19}$$

provided that the initial state $X_0 = x_0$ is fixed. Here $\theta_{ij} = \log\lambda_{ij}$, the process $K_t^{(i,j)}$ is the number of transitions from state i to state j in the time interval $[0, t]$, and $S_t^{(i)}$ is the time the process has spent in state i before t. The assumption that $\lambda_{ij} > 0$ for all $i \neq j$ is made only to simplify the exposition. It is easy to modify the results to cover the case where some entries of the intensity matrix are zero.

If we are mainly interested in drawing inference about the ith row of the intensity matrix, we can use stopping rules of the form

$$\tau_u = \inf\left\{t > 0 : \sum_{j\neq i}\alpha_j K_t^{(i,j)} + \beta S_t^{(i)} \geq u\right\}. \tag{10.2.20}$$

Such a stopping rule is finite provided that

$$\sum_{j\neq i}\alpha_j\lambda_{ij} + \beta > 0$$

(Proposition 10.1.1 or Proposition 10.1.2). The condition (10.1.5) is satisfied if $\beta > 0$ and $\alpha_j \leq 0$, $j \neq i$. When $u \in \mathbb{N}$ and $\beta = 0$, it is also satisfied when some α_j's are equal to one and the rest are zero or a negative integer. The assumption that u is an integer is made only to simplify the exposition. As in the two previous examples, the following results also hold for a non-integer u if u is replaced by $\bar{u}$, the smallest integer that is larger than or equal to u.

The results stated below follow from Corollary 10.2.3. We fix i and let I denote the set of states different from i. If $\beta > 0$ and $\alpha_j = 0$, $j \in I$, the random variables $K_\tau^{(i,j)}$, $j \in I$ are independent and

$$K_\tau^{(i,j)} \sim \text{Poisson}\left(\lambda_{ij} u \beta^{-1}\right), \quad j \in I.$$

Next, suppose $\beta = 0$ and $u \in \mathbb{N}$, and let $I = I_1 \cup I_0 \cup I_{-1}$, where $\alpha_j = 1$ for $j \in I_1$, $\alpha_j = 0$ for $j \in I_0$, and $\alpha_j = -1$ for $j \in I_-$. We suppose that I_1 is non-empty. Without loss of generality we can assume that $1 \in I_1$. Define

$$\lambda = \sum_{j \in I_1} \lambda_{ij}, \quad \mu = \sum_{j \in I_{-1}} \lambda_{ij},$$

and

$$\begin{aligned} z_1(s) &= \lambda_{i1} + \sum_{j \in I_1 \setminus \{1\}} \lambda_{ij} e^{s_j}, \\ z_0(s) &= \sum_{j \in I} \lambda_{ij} - \sum_{j \in I_0} \lambda_{ij} e^{s_j}, \\ z_{-1}(s) &= \sum_{j \in I_{-1}} \lambda_{ij} e^{s_j}, \end{aligned}$$

where $s = (s_j : j \in I)$, and where a sum is zero if the index set is empty. Then

$$\begin{aligned} &E_\theta\left(\exp\left[s_1 S_\tau^{(i)} + \sum_{j \in I \setminus \{1\}} s_j K_\tau^{(i,j)}\right]\right) \qquad (10.2.21) \\ &= (2z_1(s))^u \left\{(z_0(s) - s_1) + \sqrt{(z_0(s) - s_1)^2 - 4z_1(s)z_{-1}(s)}\right\}^{-u}, \end{aligned}$$

provided that $\lambda > \mu$, which ensures that (10.2.20) is finite. Note that $z_1(0) = \lambda$, $z_0(0) = \lambda + \mu$, and $z_{-1}(0) = \mu$. Formula (10.2.21) can be rewritten as

$$(2z_{-1}(s))^{-u} \left\{(z_0(s) - s_1) - \sqrt{(z_0(s) - s_1)^2 - 4z_1(s)z_{-1}(s)}\right\}^u \qquad (10.2.22)$$

provided that I_{-1} is not empty. By putting $s_j = 0$ for $j \in I \setminus \{1\}$ we see that when $I_{-1} \neq \emptyset$, the distribution of S_τ is given by (10.2.18) with λ and μ interchanged.

When I_{-1} is empty, $z_{-1}(s) = 0$. We see from (10.2.21) that the two random vectors $(K_\tau^{(i,j)} : j \in I_1)$ and $(K_\tau^{(i,j)} : j \in I_0)$ are independent, and that

$$\left(K_\tau^{(i,j)} : j \in I_1\right) \sim \text{Multinomial}\left(u, \{\lambda_{ij}/\lambda : j \in I_1\}\right),$$

while $(K_\tau^{(i,j)} : j \in I_0)$ has a negative multinomial distribution, which can be described as follows. Conditional on $S_\tau^{(i)} = s$, the random variables $K_\tau^{(i,j)}$, $j \in I_0$, are independent and

$$K_\tau^{(i,j)} \sim \text{Poisson}\left(s\lambda_{ij}\right),$$

while

$$S_\tau^{(i)} \sim \text{Gamma}\left(u, \lambda\right).$$

Let us finally consider the situation where we are interested in more than one of the rows of the intensity matrix. In that case, stopping rules of the type (10.2.20) will not provide a non-curved exponential family. Note, however, that

$$\sum_{j \neq i} K_t^{(i,j)} + 1_{\{i\}}\left(X_t\right) = \sum_{j \neq i} K_t^{(j,i)} + 1_{\{i\}}\left(x_0\right) \tag{10.2.23}$$

for $i = 1, \cdots, m-1$. These $m-1$ equations imply a similar equation for $i = m$. The equations (10.2.23) show that if we choose a stopping time τ such that X_τ is fixed, then we get $m-1$ linear dependencies between the canonical statistics. To obtain a non-curved family we just need one more linear dependency, so a stopping time of the form

$$\tau_{u,r} = \inf\left\{t > 0 : \sum_{i \in J} \alpha_i K_t^{(i,r)} \geq u\right\}, \tag{10.2.24}$$

where $J \subseteq \{1, \cdots, m\}$ and $r \notin J$, will do the job, provided that $u \in \mathbb{N}$ and $\alpha_i \in \{1, 0, -1, -2, \cdots\}$. A stopping time of this form is finite under P_θ if

$$\sum_{i \in J} \alpha_i \lambda_{ir} c_i\left(\lambda\right) > 0, \tag{10.2.25}$$

where $c_i\left(\lambda\right) = \lambda_{i.}^{-1}\left(\sum_j \lambda_{j.}^{-1}\right)^{-1}$ and $\lambda_{i.} = \sum_{j \neq i} \lambda_{ij}$, cf. Proposition 10.1.2. Here we use that P_θ-almost surely $t^{-1} S_t^{(i)} \to c_i(\lambda)$.

Without loss of generality we can assume that $m \in J$ and $\alpha_m = 1$. Then we can write the likelihood function corresponding to observation of X in $[0, \tau_{u,r}]$ in the form

$$L_{\tau_{u,r}}\left(\theta\right) = \exp\left[\gamma\left(\theta\right)^T B_{\tau_{u,r}} + u\theta_{mr} + \left(u-1\right)\theta_{rm} + \theta_{x_0 m}\right]; \tag{10.2.26}$$

where $B_{\tau_{u,r}}$ is the $m(m-1)$-dimensional statistic with components

$$B_{\tau_{u,r}}^{(i,j)} = \begin{cases} K_\tau^{(i,j)} & i \neq m,\ j \neq i,\ j \neq m \text{ or } i = m,\ j \neq m,\ j \neq r \\ S_\tau^{(i)} & i \neq m,\ j = m \text{ or } i = m,\ j = r, \end{cases}$$

and where

$$\gamma(\theta)_{i,j} = \begin{cases} \theta_{ij} - \theta_{im} + \theta_{jm} & i \notin J,\ j \neq i,\ j \neq m \\ \theta_{ij} - \theta_{im} + \theta_{jm} & i \in J\backslash\{m\},\ j \neq i,\ j \neq m,\ j \neq r \\ \theta_{ir} - \theta_{im} + \theta_{rm} & i \in J\backslash\{m\},\ j = r \\ \quad -\alpha_i(\theta_{mr} + \theta_{rm}) & \\ \theta_{mj} + \theta_{jm} & i = m,\ j \neq m,\ j \neq r \\ -\sum_{j \neq i} \exp(\theta_{ij}) + 1 - m & i \neq m,\ j = m \text{ or } i = m,\ j = r. \end{cases}$$

Although (10.2.26) is not exactly of the form (10.2.1), it is of a very similar form. In fact, a result similar to Theorem 10.2.2 can be proved by arguments analogous to those in the proof of that theorem. Specifically, the Laplace transform of $B_{\tau_{u,r}}$ is given by

$$\begin{aligned} &E_\theta\left(\exp\left(s^T B_{\tau_{u,r}}\right)\right) \qquad (10.2.27) \\ &= \exp[u\left(\theta_{mr} + \theta_{rm} - \gamma^{-1}(s + \gamma(\theta))_{mr} - \gamma^{-1}(s + \gamma(\theta))_{rm}\right) \\ &\qquad + \theta_{x_0 m} - \theta_{rm} - \gamma^{-1}(s + \gamma(\theta))_{x_0 m} + \gamma^{-1}(s + \gamma(\theta))_{rm}]. \end{aligned}$$

Also the stopping times

$$\tau_{u,r}^* = \inf\left\{t > 0 : S_t^{(r)} > u\right\}, \qquad (10.2.28)$$

where $u > 0$, fix the state of X at the stopping time, so they too provide non-curved exponential families. Again in this case, the likelihood function has a form similar to (10.2.1), and an expression for the Laplace transform of B_τ can be found. □

10.3 Asymptotic likelihood theory

In this section we discuss some aspects of asymptotic likelihood theory for models with a likelihood function of the type (10.1.1) observed in a stochastic time interval $[0, \tau_u]$, where τ_u is of the form (10.1.3). The asymptotics are in the limit $u \to \infty$. In the following, θ_0 denotes the true value of θ.

When $\alpha \neq 0$ or when S is not continuous, the overshoot D_{τ_u} might well be non-zero. In order to cover this case we need a condition of the following type, which expresses to what extent D_{τ_u} is negligible as $u \to \infty$. The condition depends on a real number δ.

Condition C (δ) *There exists a* $\nu > 0$ *such that*

$$\sup_{(\theta,s)\in M_{\delta,\nu}} |\varphi_\theta^u(s) - 1| \to 0$$

as $u \to \infty$, *where*

$$M_{\delta,\nu} = \left\{(\theta, s) : \|\theta - \theta_0\| < \nu u^{-\delta}, \|s\| < \nu u^{-\delta}\right\}.$$

Note that if $\alpha = 0$ and S is continuous, then $D_\tau \equiv 0$ and Condition $C(\delta)$ is satisfied for all $\delta > 0$.

Theorem 10.3.1 *Suppose that* $\alpha = 0$ *and* $\beta > 0$, *and that Condition* $C(1)$ *is satisfied. Then under* P_{θ_0} *we have the following results on convergence in probability*

$$u^{-1}A_{\tau_u} \to \beta^{-1}\dot{\kappa}(\theta_0). \tag{10.3.1}$$

If the cumulant transform κ *is steep and* $\theta_0 \in \operatorname{int}\Theta$, *the maximum likelihood estimator* $\hat{\theta}_{\tau_u}$ *exists with a probability that tends to one as* $u \to \infty$ *and*

$$\hat{\theta}_{\tau_u} \to \theta_0. \tag{10.3.2}$$

Proof: The Laplace transform of $u^{-1}A_{\tau_u}$ is, by (10.2.5),

$$\begin{aligned} E_{\theta_0}\left(u^{-1}s^T A_{\tau_u}\right) &= \exp\left[u\beta^{-1}\left(\kappa\left(\theta + s/u\right) - \kappa\left(\theta\right)\right)\right] \\ &\quad \times \varphi_{\theta+s/u}^u\left[\beta^{-1}\left(\kappa\left(\theta + s/u\right) - \kappa\left(\theta\right)\right)\right], \end{aligned}$$

which for s in a neighbourhood of 0 converges to $\exp[\beta^{-1}\dot{\kappa}(\theta_0)^T s]$ as u tends to infinity. This proves (10.3.1). The other result follows immediately because $\dot{\kappa}(\hat{\theta}_\tau) = \beta A_\tau/(u + D_\tau)$, which has a solution provided that the righthand side belongs to $\operatorname{int} C$; cf. Theorem 10.2.1. □

Theorem 10.3.2 *Suppose that* $\alpha_1 \neq 0$ *and* $\theta_0 \in \Theta_{\alpha,\beta}$, *that the conditions of Proposition 10.1.1 or Proposition 10.1.2 hold, and that Condition* $C(1)$ *is satisfied. Then under* P_{θ_0},

$$u^{-1}B_{\tau_u} \to \left(\alpha^T \dot{\kappa}(\theta_0) + \beta\right)^{-1} \pi(\theta_0) \quad \textit{in probability} \tag{10.3.3}$$

as $u \to \infty$, *where* B *is defined by (10.2.2), and where*

$$\pi(\theta) = \left(1, \dot{\kappa}(\theta)_2, \cdots, \dot{\kappa}(\theta)_k\right)^T. \tag{10.3.4}$$

If the cumulant transform κ *is steep and* $\theta_0 \in \operatorname{int}\Theta$, *then*

$$\hat{\theta}_{\tau_u} \to \theta_0 \quad \textit{in probability}. \tag{10.3.5}$$

Proof: By (10.2.4),

$$E_{\theta_0}\left(\exp\left[u^{-1}s^T B_{\tau_u}\right]\right) = \exp\left\{u\alpha_1^{-1}\left(\theta_{0,1} - \gamma^{-1}\left(s/u + \gamma\left(\theta_0\right)\right)_1\right)\right\}$$
$$\times \varphi^u_{\gamma^{-1}(s/u+\gamma(\theta_0))}\left\{\alpha_1^{-1}\left(\theta_{0,1} - \gamma^{-1}\left(s/u + \gamma\left(\theta_1\right)\right)_1\right)\right\},$$

which for s in a neighbourhood of 0 converges to $\exp[(\alpha^T \dot\kappa(\theta_0) + \beta)^{-1} \pi(\theta_0)^T s]$ as u tends to infinity. This proves (10.3.3). The other result follows immediately because

$$\dot\kappa(\hat\theta_\tau) = A_\tau / S_\tau,$$

which has a solution provided that $A_\tau/S_\tau \in \operatorname{int} C$; cf. Theorem 10.2.1. In particular,

$$\dot\kappa(\hat\theta_\tau)_1 = \frac{u - \alpha_2 A_\tau^{(2)} - \cdots - \alpha_k A_\tau^{(k)}}{\alpha_1 S_\tau} - \alpha_1^{-1}\beta + O\left(u^{-1}D_\tau\right).$$

□

Theorem 10.3.3 *Suppose (10.1.5) is satisfied. Then, under the conditions of the respective theorems the convergence results (10.3.1), (10.3.2), (10.3.3), and (10.3.5) hold almost surely.*

Proof: When (10.1.5) holds, we have seen that an exponential family with non-empty interior is obtained. Therefore, $(B_{\tau_u})_u$ is a Lévy process with mean $(\alpha^T \dot\kappa(\theta) + \beta)^{-1}\pi(\theta)u$, which implies the results. □

Corollary 10.3.4 *Suppose there exists a function f and a random variable $V(\theta_0)$ such that*

$$S_t / f(t) \to V(\theta_0) \tag{10.3.6}$$

in P_{θ_0}-probability as $t \to \infty$. Then, under the conditions of Theorem 10.3.1 or Theorem 10.3.2,

$$u^{-1} f(\tau_u) \to \left(\alpha^T \dot\kappa(\theta_0) + \beta\right)^{-1} V(\theta_0)^{-1} \tag{10.3.7}$$

in P_{θ_0}-probability as $u \to \infty$. Here the lefthand side of (10.3.7) is interpreted as infinity if $V(\theta_0) = 0$. If the convergence in (10.3.6) is almost sure and if (10.1.5) holds, then the convergence (10.3.7) is almost sure too.

If the convergence in (10.3.6) is P_{θ_0}-almost sure and $\theta_0 \in \Theta_{\alpha,\beta}$, then we need only the assumption that S is continuous and strictly increasing or that $(\theta_0, \kappa(\theta_0)) \in \operatorname{int} \tilde\Gamma_t$ for all $t > 0$ to ensure that

$$\liminf_{u\to\infty} u^{-1} f(\tau_u) \geq \left(\alpha^T \dot\kappa(\theta_0) + \beta\right)^{-1} V(\theta_0)^{-1}.$$

Proof:

$$\frac{f(\tau_u)}{u} = \frac{f(\tau_u)/S_{\tau_u}}{u/S_{\tau_u}} \to \frac{V(\theta_0)^{-1}}{\left(\alpha^T \dot\kappa(\theta_0) + \beta\right)}$$

because $\tau_u \to \infty$ as $u \to \infty$. The last results follows because

$$u^{-1} f(\tau_u) \geq \left[S_{\tau_u}^{-1} f(\tau_u)\right] S_{\tau_u}\left(\alpha^T A_{\tau_u} + \beta S_{\tau_u}\right)^{-1},$$

where the righthand side converges P_{θ_0}-almost surely to $V(\theta_0)^{-1}(\alpha^T\dot{\kappa}(\theta_0)+\beta)^{-1}$ (see the proofs of Proposition 10.1.1 and Proposition 10.1.2). □

The result (10.3.7) can be used to evaluate τ_u under P_{θ_0}. For u large, $\tau_u \doteq f^{-1}\left(u\left(\alpha^T\dot{\kappa}(\theta_0)+\beta\right)^{-1}V(\theta_0)^{-1}\right)$.

Theorem 10.3.5 *Suppose that $\alpha_1 \neq 0$ and $\theta_0 \in \Theta_{\alpha,\beta}$, that the conditions of Proposition 10.1.1 hold, and that Condition $C(\frac{1}{2})$ is satisfied. Then the following results concerning weak convergence as $u \to \infty$ hold under P_{θ_0}.*

$$\begin{aligned} Z_u &= u^{-1/2}\left\{B_{\tau_u} - u\left(\alpha^T\dot{\kappa}(\theta_0)+\beta\right)^{-1}\pi(\theta_0)\right\} \qquad (10.3.8)\\ &\to N_k\left(0, \left(\alpha^T\dot{\kappa}(\theta_0)+\beta\right)^{-1}\left(\dot{\gamma}(\theta_0)^{-1}\right)^T\ddot{\kappa}(\theta_0)\dot{\gamma}(\theta_0)^{-1}\right), \end{aligned}$$

where

$$\dot{\gamma}(\theta_0)^{-1}_{ij} = \begin{cases} -\left(\alpha^T\dot{\kappa}(\theta_0)+\beta\right)^{-1}\alpha_i\pi(\theta_0)_j & \text{if } i\neq j \text{ or } i=j=1\\ -\left(\alpha^T\dot{\kappa}(\theta_0)+\beta\right)^{-1}\alpha_i\pi(\theta_0)_j+1 & \text{if } i=j\geq 2,\end{cases}$$

and

$$\sqrt{u}\left(A_{\tau_u}/S_{\tau_u} - \dot{\kappa}(\theta_0)\right) \to N_k\left(0, \left(\alpha^T\dot{\kappa}(\theta_0)+\beta\right)\ddot{\kappa}(\theta_0)\right). \qquad (10.3.9)$$

If $\theta_0 \in \operatorname{int}\Theta$, then

$$\sqrt{u}\left(\hat{\theta}_{\tau_u} - \theta_0\right) \to N_k\left(0, \left(\alpha^T\dot{\kappa}(\theta_0)+\beta\right)\ddot{\kappa}(\theta)^{-1}\right) \qquad (10.3.10)$$

and

$$-2\log Q_{\tau_u} \to \chi^2(k), \qquad (10.3.11)$$

where Q_{τ_u} is the likelihood ratio test statistic for the hypothesis $\theta = \theta_0$.

Proof: By (10.2.4) the Laplace transform of Z_u is

$$\begin{aligned} &\exp\left\{u\alpha_1^{-1}\left[\theta_{0,1} - \gamma^{-1}\left(u^{-1/2}s+\gamma(\theta_0)\right)_1\right] - u^{1/2}\left(\alpha^T\dot{\kappa}(\theta_0)+\beta\right)\pi(\theta_0)^T s\right\}\\ &\quad\times\varphi^u_{\gamma^{-1}(u^{-1/2}s+\gamma(\theta_0))}\left\{\alpha_1^{-1}\left[\theta_{0,1} - \gamma^{-1}\left(u^{-1/2}s+\gamma(\theta_0)\right)_1\right]\right\} =\\ &\exp\left\{\tfrac{1}{2}\left(\alpha^T\dot{\kappa}(\theta_0)+\beta\right)^{-1}s^T\left(\dot{\gamma}(\theta_0)^{-1}\right)^T\ddot{\kappa}(\theta_0)\dot{\gamma}(\theta_0)^{-1}+O\left(u^{-1}\right)\right\}\cdot O(1), \end{aligned}$$

which shows (10.3.8). Now,

$$\dot{\kappa}\left(\hat{\theta}_{\tau_u}\right) = A_{\tau_u}/S_{\tau_u}$$

and

$$\sqrt{u}\left(A_{\tau_u}/S_{\tau_u} - \dot{\kappa}(\theta_0)\right) = \dot{\gamma}(\theta)^T Z_u/\left(u^{-1}S_{\tau_u}\right) + O_p\left(u^{-\frac{1}{2}}D_\tau\right),$$

which by (10.3.8) and (10.3.3) converges weakly to $N_k(0, (\alpha^T \dot{\kappa}(\theta_0) + \beta) \ddot{\kappa}(\theta_0))$. Therefore, in view of (10.3.5), (10.3.10) follows via the delta method.

Finally, note that

$$-2 \log Q_{\tau_u} = S_{\tau_u} \left(\hat{\theta}_{\tau_u} - \theta_0 \right)^T \ddot{\kappa} \left(\hat{\theta}_{\tau_u} \right) \left(\hat{\theta}_{\tau_u} - \theta_0 \right) + o_p(1),$$

where $o_p(1) \to 0$ in probability. Hence (10.3.11) follows from (10.3.3), (10.3.5), and (10.3.10). □

Note in particular by considering the first coordinate of (10.3.8) that

$$u^{-1/2} \left(S_{\tau_u} - u \left(\alpha^T \dot{\kappa} (\theta_0) + \beta \right)^{-1} \right) \to N \left(0, \left(\alpha^T \dot{\kappa} (\theta_0) + \beta \right)^{-3} \alpha^T \ddot{\kappa} (\theta_0) \alpha \right). \tag{10.3.12}$$

The observed information is $\ddot{\kappa}(\theta) S_\tau$. As usual, we can normalize the maximum likelihood estimator by observed information instead of expected information. The results (10.3.3) and (10.3.10) imply that

$$S_{\tau_u}^{1/2} \left(\hat{\theta}_{\tau_u} - \theta_0 \right) \to N_k \left(0, \ddot{\kappa} (\theta_0) \right)$$

weakly under P_{θ_0}.

Theorem 10.3.6 *Suppose $\alpha = 0$ and $\beta > 0$, and that Condition $C(\frac{1}{2})$ is satisfied. Then under P_θ we have the following results about weak convergence as $u \to \infty$.*

$$u^{-\frac{1}{2}} \left(A_{\tau_u} - u\beta^{-1} \dot{\kappa} (\theta_0) \right) \to N \left(0, \beta^{-1} \ddot{\kappa} (\theta_0)^{-1} \right). \tag{10.3.13}$$

If $\theta_0 \in \operatorname{int} \Theta$,

$$u^{1/2} \left(\hat{\theta}_{\tau_u} - \theta_0 \right) \to N \left(0, \beta \ddot{\kappa} (\theta_0)^{-1} \right) \tag{10.3.14}$$

and

$$-2 \log Q_{\tau_u} \to \chi^2 (k), \tag{10.3.15}$$

where Q_{τ_u} is the likelihood ratio test statistic for the hypothesis that $\theta = \theta_0$.

Proof: Analogous to the proof of Theorem 10.3.5. □

Let us briefly discuss *Bartlett corrections* for the sampling schemes considered here. Suppose (10.1.5) holds. On series expansion we find by calculating the necessary moments using (10.2.6) and (10.2.7) that in the case $k = 1$,

$$E_{\theta_0} \left(-2 \log Q_{\tau_u} \right) = \tag{10.3.16}$$

$$1 + i_u (\theta_0)^{-1} \left\{ \frac{5\kappa^{(3)} (\theta_0)^2}{12\kappa^{(2)} (\theta_0)^2} - \frac{\kappa^{(4)} (\theta_0)}{4\kappa^{(2)} (\theta_0)} \right\} + O\left(i_u (\theta_0)^{-2} \right),$$

where $\kappa^{(k)}$ denotes the kth derivative of κ, and where $i_u(\theta)$ is the expected information (Fisher information)

$$i_u(\theta) = E_\theta\left(\ddot{\kappa}(\theta) S_{\tau_u}\right) = u\ddot{\kappa}(\theta)\left(\alpha\dot{\kappa}(\theta)+\beta\right)^{-1}. \tag{10.3.17}$$

We see that the Bartlett corrections are identical for different sampling rules of the type (10.1.3) provided that they give rise to the same expected information. Corollary 10.3.4 indicates that the waiting times until the sampling has been finished are similar for such stopping rules, and in Section 10.5 they will, under suitable regularity conditions, be shown to have the same approximate expectation for large values of u. However, in the next section we shall give a method by which stopping times with the same expected information can be compared.

When $D_{\tau_u} > 0$ with positive probability, (10.3.16) still holds provided that the contribution from the second factor of (10.2.4) or (10.2.5) is of a sufficiently small order of magnitude. For models where $S_t = t$ we see that there exist stopping times with the same Bartlett correction as fixed-time sampling. For such models we have earlier seen that $E_\theta(\tau_u) = u(\alpha^T\dot{\kappa}(\theta)+\beta)^{-1}$; cf. (10.2.8). Examples are the exponential families of Lévy processes discussed in Chapter 2 and exponential families of diffusion processes with proportional drift and diffusion coefficient. Particular examples are the Poisson processes, Brownian motion with drift, and geometric Brownian motion.

Corollary 10.3.7 *Suppose there exists a function f such that under P_{θ_0},*

$$S_t/f(t) \to 1, \quad f(t)^{\frac{1}{2}}\left(S_t/f(t) - 1\right) \to 0$$

in probability as $t \to \infty$. Then under the conditions of Theorem 10.3.5 or Theorem 10.3.6,

$$u^{-1/2}\left(f(\tau_u) - u\left(\alpha^T\dot{\kappa}(\theta_0)+\beta\right)^{-1}\right) \to N\left(0, \left(\alpha^T\dot{\kappa}(\theta)+\beta\right)^{-3}\alpha^T\ddot{\kappa}(\theta)\alpha\right) \tag{10.3.18}$$

weakly under P_{θ_0} as $u \to \infty$.

Proof:

$$\begin{aligned} & u^{-1/2}\left(f(\tau_u) - u\left(\alpha^T\dot{\kappa}(\theta_0)+\beta\right)^{-1}\right) \\ = \; & \frac{f(\tau_u)}{S_{\tau_u}} u^{-1/2}\left(S_{\tau_u} - u\left(\alpha^T\dot{\kappa}(\theta_0)+\beta\right)^{-1}\right) \\ & - \left(\alpha^T\dot{\kappa}(\theta_0)+\beta\right)^{-1}\left(\frac{u}{f(\tau_u)}\right)^{1/2}\frac{f(\tau_u)}{S_{\tau_u}} f(\tau_u)^{1/2}\left(S_{\tau_u}/f(\tau_u) - 1\right), \end{aligned}$$

where the last term tends to zero in probability. Now apply (10.3.12). □

In many cases there will be more randomness in S than assumed in the corollary. In such cases the last term in the proof will also contribute to the asymptotic distribution of $f(\tau_u)$.

The following lemma is sometimes useful for proving that Condition C(δ) is satisfied. We remind the reader that if two distributions with distribution functions F_1 and F_2 are given on the real line, then F_2 is said to be stochastically greater than F_1 if $F_1(x) \geq F_2(x)$ for all real x.

Lemma 10.3.8 *Let $\rho \in \mathbb{R}$ be given, and suppose that under all P_θ with θ in a neighbourhood of θ_0 and for all $t > 0$ the distributions of the random variables*

$$\|\alpha^T \Delta A_t + \beta \Delta S_t\| \cdot \|\alpha^T A_{t-} + \beta S_{t-}\|^{-\rho}$$

are stochastically smaller than a distribution on $[0, \infty)$ with distribution function F. Then Condition C(δ) is satisfied for all $\delta > \rho$, provided that zero belongs to the interior of the domain of the Laplace transform of F.

Proof: Let φ_F denote the Laplace transform of F. Since

$$\|D_{\tau_u}\| u^{-\rho} \leq \|D_{\tau_u}\| \cdot \|\alpha^T A_{\tau_u -} + \beta S_{\tau_u -}\|^{-\rho}$$

for all $u > 0$, it follows that

$$\varphi_F(-s_0) \leq E_\theta \left(e^{u^{-\rho} s^T D_{\tau_u}} \right) \leq \varphi_F(s_0)$$

for $\|s\| \leq s_0$, where $s_0 > 0$ belongs to the interior of dom φ_F. This implies the conclusion of the lemma. □

Example 10.3.9 *The Gaussian autoregression* The non-explosive Gaussian autoregression of order one is given by (3.2.6) with $|\theta| < 1$. We assume here that $X_0 = 0$. If the process is observed at the times $1, 2, \cdots, t$, the likelihood function is of the form (10.1.1) with

$$A_t = \sum_{s=1}^{t} X_s X_{s-1}, \quad S_t = \sum_{s=1}^{t} X_{s-1}^2, \quad \kappa(\theta) = \tfrac{1}{2}\theta^2;$$

cf. (3.2.7).

Obviously, when a stopping rule of the type (10.1.3) is used, the process $\alpha A_t + \beta S_t$ may jump far above u at the time of stopping. However,

$$\alpha(A_t - A_{t-1}) + \beta(S_t - S_{t-1}) = (\alpha\theta + \beta) X_{t-1}^2 + \alpha Z_t X_{t-1},$$

where $\alpha\theta + \beta > 0$ to ensure that τ_u is finite. The random variable

$$\tilde{X}_{t-1} = \left(\frac{1 - \theta^2}{1 - \theta^{2t}} \right)^{\frac{1}{2}} X_{t-1}$$

is N(0, 1)-distributed for all t and independent of Z_t, which is also standard normally distributed. For any interval (θ_-, θ_+) of (-1,1) there exist constants k_1 and k_2 such that

$$|\alpha(A_t - A_{t-1}) + \beta(S_t - S_{t-1})| \leq k_1 \tilde{X}_{t-1}^2 + k_2 |Z_t| |\tilde{X}_{t-1}|$$

for all $\theta \in (\theta_-, \theta_+)$ and for all $t > 1$. This proves that the conditions of Lemma 10.3.8 are satisfied for every $\theta_0 \in (-1, 1)$. Hence consistency and asymptotic normality of the maximum likelihood estimator follow. □

Example 10.3.10 *Finite state space Markov processes.* Let us return to the finite state Markov process considered in Example 10.2.8. There we studied stopping times for which a non-curved exponential family was obtained. Here we are interested in general stopping times of the type (10.1.12), i.e.,

$$\tau_u = \inf\left\{t > 0 : \sum_{i=1}^{m}\left(\beta_i S_t^{(i)} + \sum_{j\neq i}\alpha_{ij}K_t^{(i,j)}\right) \geq u\right\}, \tag{10.3.19}$$

where

$$\sum_{i=1}^{m}\left(\beta_i + \sum_{j\neq i}\alpha_{ij}\lambda_{ij}\right) c_i(\lambda) > 0,$$

with $c_i(\lambda)$ defined as in (10.2.25). By Proposition 10.1.2 the stopping time τ_u is almost surely finite under all P_θ.

Without loss of generality we can assume that $\alpha_{m1} + \alpha_{1m} \neq 0$. Then we can write the likelihood function corresponding to observation of X in $[0, \tau_u]$ in the form

$$L_{\tau_u}(\theta) = \exp\left[\gamma(\theta)^T B_{\tau_u} + u\mu(\theta) + D_{\tau_u}(\theta)\right], \tag{10.3.20}$$

where

$$\mu(\theta) = (\theta_{m1} + \theta_{1m})(\alpha_{m1} + \alpha_{1m})^{-1},$$

$$B_\tau^{(i,j)} = \begin{cases} K_{\tau_u}^{(i,j)} & i \neq m,\ j \neq i,\ j \neq m \text{ or } i = m,\ j \neq m,\ j \neq 1 \\ \\ S_{\tau_u}^{(i)} & i \neq m,\ j = m \text{ or } i = m,\ j = 1, \end{cases}$$

and

$$\gamma(\theta)_{ij} = \begin{cases} \theta_{ij} - \theta_{im} + \theta_{jm} & i \neq m,\ j \neq i,\ j \neq m \\ -\mu(\theta)(\alpha_{ij} - \alpha_{im} + \alpha_{jm}) & \\ \\ \theta_{mj} + \theta_{jm} & i = m,\ j \neq m,\ j \neq i \\ -\mu(\theta)(\alpha_{mj} + \alpha_{jm}) & \\ \\ -(\kappa_i(\theta) + \beta_i\mu(\theta)) & i \neq m,\ j = m \text{ or } i = m,\ j = 1. \end{cases}$$

The statistic $D_{\tau_u}(\theta)$ is bounded for all u:

$$|D_{\tau_u}(\theta)| \leq 2\sum_{i=1}^{m-1} |\theta_{im}| + 2m\mu(\theta) \max_{i,j} |\alpha_{ij}|, \quad \text{for all } u > 0.$$

The likelihood function (10.3.20) is not exactly of the form (10.2.1), so we cannot use the limit results given in this section. However, (10.3.20) is sufficiently similar to (10.2.1) that the proofs of Theorem 10.3.2 and Theorem 10.3.5 can be copied directly to show consistency and asymptotic normality of the maximum likelihood estimator and other similar limit results given above. □

10.4 Comparison of sampling times

When a process with fixed sampling time likelihood function given by (10.1.1) is observed from time zero to a stopping time of the type (10.1.3), the expected information (Fisher information) is

$$i_u(\alpha, \beta) = u\ddot{\kappa}(\theta)\left(\alpha^T \dot{\kappa}(\theta) + \beta\right)^{-1}, \tag{10.4.1}$$

provided (10.1.5) holds. Otherwise, the expected information is asymptotically equal to (10.4.1). In Section 10.3 it was noted that a whole sub-class of the stopping times of the type (10.1.3) has the same value of (10.4.10). When (10.1.5) holds, also the Bartlett correction for the log-likelihood ratio test of the point hypothesis $\theta = \theta_0$ is identical for stopping times in such a sub-class.

In the following we give a method of comparing such stopping times. By Corollary 10.3.4 they are asymptotically identical, but the following more refined analysis shows that some of the stopping times tend to be smaller than others. For some models the probability that one stopping time is smaller than another can be given explicitly. More generally, an Edgeworth expansion of this probability is given.

In the following we work under the additional assumption that S_t is a continuous function of t, and that either S is strictly increasing or that $(\theta, \kappa(\theta)) \in \text{int } \tilde{\Gamma}_t$ for all $t > 0$. As was the case with several results given above, the following results hold also when the fixed-time likelihood function is a product of a factor of the form (10.1.1) and another factor depending only on a parameter φ independent of θ.

Proposition 10.4.1 *Suppose $\alpha_1 \neq 0$, $\theta \in \Theta_{\alpha,\beta} \cap \Theta_{\tilde{\alpha},\tilde{\beta}}$ and that*

$$u\left(\alpha^T \dot{\kappa}(\theta) + \beta\right)^{-1} = \tilde{u}\left(\tilde{\alpha}^T \dot{\kappa}(\theta) + \tilde{\beta}\right)^{-1}. \tag{10.4.2}$$

Let $i(\theta)$ denote this common value in (10.4.2). Assume further that (10.1.5) is satisfied for (α, β) as well as for $(\tilde{\alpha}, \tilde{\beta})$ and that $\tilde{\alpha}^T A_t + \tilde{\beta} S_t$ is a strictly increasing process. Then

$$P_\theta\left(\tau_u(\alpha,\beta) \leq \tau_{\tilde{u}}\left(\tilde{\alpha},\tilde{\beta}\right)\right) = P_\theta\left(a^T B_{\tau_u(\alpha,\beta)} \leq i(\theta)\, a^T \pi(\theta)\right), \quad (10.4.3)$$

where $B_{\tau_u(\alpha,\beta)}$ is given by (10.2.2), $\pi(\theta)$ by (10.3.4), and where

$$a = \left(\tilde{\beta} - \tilde{\alpha}_1 \alpha_1^{-1}\beta, \tilde{\alpha}_2 - \tilde{\alpha}_1\alpha_1^{-1}\alpha_2, \cdots, \tilde{\alpha}_k - \tilde{\alpha}_1\alpha_1^{-1}\alpha_k\right)^T.$$

When $\tilde{\alpha} = 0$, it is neither necessary to assume that (10.1.5) holds for (α, β) nor that $\tilde{\alpha}^T A_t + \tilde{\beta} S_t$ is strictly increasing. In this case (10.4.3) simplifies to

$$P_\theta\left(\tau_u(\alpha,\beta) \leq \tau_{\tilde{u}}\left(0,\tilde{\beta}\right)\right) = P_\theta\left(S_{\tau_u(\alpha,\beta)} \leq i(\theta)\right). \quad (10.4.4)$$

Proof: Let τ denote $\tau_u(\alpha, \beta)$. When $\tilde{\alpha}^T A_t + \tilde{\beta} S_t$ is strictly increasing, we obviously have that

$$P_\theta\left(\tau_u(\alpha,\beta) \leq \tau_{\tilde{u}}\left(\tilde{\alpha},\tilde{\beta}\right)\right) = P_\theta\left(\tilde{\alpha}^T A_\tau + \tilde{\beta} S_\tau \leq \tilde{u}\right). \quad (10.4.5)$$

From this (10.4.3) follows, since

$$\tilde{\alpha}^T A_\tau + \tilde{\beta} S_\tau = \tilde{\alpha}_1 \alpha_1^{-1} u + a^T B_\tau,$$

and since by (10.4.2),

$$\tilde{u} - \tilde{\alpha}_1\alpha_1^{-1} u = \frac{u a^T \pi(\theta)}{\alpha^T \dot{\kappa}(\theta) + \beta}.$$

When $\tilde{\alpha} = 0$, (10.4.5) also holds in cases where S is not strictly increasing because under the conditions imposed on the exponential family the canonical process (A, S) is constant on subsets of the time axis where S is constant; cf. Section 5.1. When $\tilde{\alpha} = 0$, the righthand side of (10.4.5) is $P_\theta(\tilde{\beta} S_\tau \leq \tilde{u})$, which equals the righthand side of (10.4.4). □

Remarks: When $\tilde{\alpha}^T A_t + \tilde{\beta} S_t$ is not strictly increasing and $\tilde{\alpha} \neq 0$, we can only deduce that

$$P_\theta\left(\tau_u(\alpha,\beta) \leq \tau_{\tilde{u}}\left(\tilde{\alpha},\tilde{\beta}\right)\right) \leq P_\theta\left(a^T B_{\tau_u(\alpha,\beta)} \leq i(\theta)\, a^T \pi(\theta)\right).$$

This result also holds if (10.1.5) is not satisfied in situations where $\tilde{\alpha} \neq 0$.

Note also that $a = 0$ if and only if the two stopping times $\tau_{\tilde{u}}(\tilde{\alpha}, \tilde{\beta})$ and $\tau_u(\alpha, \beta)$ are identical, i.e., if there exists a constant $c \in \mathbb{R}$ such that $\tilde{\alpha} = c\alpha$, $\tilde{\beta} = c\beta$, and $\tilde{u} = cu$.

Finally, note that for $k = 1$, formula (10.4.3) simplifies to

$$P_\theta\left(\tau_u(\alpha,\beta) \le \tau_{\tilde{u}}\left(\tilde{\alpha},\tilde{\beta}\right)\right) = \begin{cases} P_\theta\left(S_{\tau_u(\alpha,\beta)} \le i(\theta)\right) & \text{if } \tilde{\beta} > \tilde{\alpha}\alpha^{-1}\beta \\ P_\theta\left(S_{\tau_u(\alpha,\beta)} \ge i(\theta)\right) & \text{if } \tilde{\beta} < \tilde{\alpha}\alpha^{-1}\beta. \end{cases}$$ □

When the distribution of $a^T B_{\tau_u(\alpha,\beta)}$ can be found explicitly from Theorem 10.2.2, equation (10.4.3) can be used directly to compare stopping times, as we shall see in the examples below. Otherwise, it is necessary to find an approximation to the righthand side of (10.4.3).

We shall need the first three moments of $a^T B_\tau$. They are given in the following corollary to Theorem 10.2.2.

Corollary 10.4.2 *Suppose $\alpha_1 \neq 0$ and that (10.1.5) is satisfied. Then for $a \in \mathbb{R}^k$,*

$$E_\theta\left(a^T B_{\tau_u(\alpha,\beta)}\right) = \frac{u}{\alpha^T \dot{\kappa}(\theta) + \beta} a^T \pi(\theta), \tag{10.4.6}$$

$$V_\theta\left(a^T B_{\tau_u(\alpha,\beta)}\right) = \frac{u}{\alpha^T \dot{\kappa}(\theta) + \beta} \tilde{a}(\theta)^T \ddot{\kappa}(\theta)\tilde{a}(\theta), \tag{10.4.7}$$

and the third central moment of $a^T B_{\tau_u(\alpha,\beta)}$ is given by

$$\frac{u}{\alpha^T \dot{\kappa}(\theta) + \beta} \sum_{i,j,l=1}^{k} \kappa_{ijl}^{(3)}(\theta)\tilde{a}(\theta)_i \tilde{a}(\theta)_j \tilde{a}(\theta)_l \tag{10.4.8}$$

$$- \frac{3u}{\left(\alpha^T \dot{\kappa}(\theta) + \beta\right)^2}\left(\tilde{a}(\theta)^T \ddot{\kappa}(\theta) a(\theta)\right)\left(\alpha^T \ddot{\kappa}(\theta)\tilde{a}(\theta)\right),$$

where $\pi(\theta)$ is given by (10.3.4),

$$\tilde{a}(\theta) = \dot{\gamma}(\theta)^{-1} a \tag{10.4.9}$$

with $\gamma(\theta)$ defined by (10.2.3) (see also Theorem 10.3.5), and

$$\kappa_{ijl}^{(3)}(\theta) = \frac{\partial^3}{\partial\theta_i \partial\theta_j \partial\theta_l}\kappa(\theta).$$

When $k = 1$,

$$E_\theta\left(B_{\tau_u(\alpha,\beta)}\right) = u\left(\alpha\dot{\kappa}(\theta) + \beta\right)^{-1}, \tag{10.4.10}$$

$$V_\theta\left(B_{\tau_u(\alpha,\beta)}\right) = u\alpha^2 \ddot{\kappa}(\theta)\left(\alpha\dot{\kappa}(\theta) + \beta\right)^{-3}, \tag{10.4.11}$$

and

$$E_\theta\left(\left[B_{\tau_u(\alpha,\beta)} - u\left(\alpha\dot{\kappa}(\theta) + \beta\right)^{-1}\right]^3\right) \tag{10.4.12}$$

$$= u\alpha^3\left(\alpha\dot{\kappa}(\theta) + \beta\right)^{-4}\left\{3\alpha\ddot{\kappa}(\theta)^2\left(\alpha\dot{\kappa}(\theta) + \beta\right)^{-1} - \kappa^{(3)}(\theta)\right\}.$$

Note that $i(\theta)a^T\pi(\theta)$ is the expectation of $a^T B_{\tau_u(\alpha,\beta)}$. It is not difficult to see from Theorem 10.2.2 that $\left[a^T B_{\tau_u(\alpha,\beta)} - i(\theta)\, a^T\mu(\theta)\right]$ normalized by the square root of (10.4.7) under P_θ converges in distribution to a standard normal distribution. This result tells us that asymptotically the probability (10.4.3) equals one half for stopping times with the same expected information, so it can also not distinguish between such stopping times. We will therefore approximate the righthand side of (10.4.3) by an Edgeworth expansion that includes the third moment of $a^T B_\tau$. This can easily be done because the Laplace transform of B_{τ_u} is known and has the form of the Laplace transform for a Lévy process (Corollary 10.2.3), so that classical results on Edgeworth expansions apply; see e.g., Feller (1966, Chapter 16). In fact, $\{B_{\tau_u} : u \geq 0\}$ is a Lévy process. This follows from Theorem 4.2.1.

Corollary 10.4.3 *Suppose the distribution of $B_{\tau_1(\alpha,\beta)}$ is not a lattice distribution. Then under the conditions of Proposition 10.4.1,*

$$P_\theta\left(\tau_u(\alpha,\beta) \leq \tau_{\tilde{u}}\left(\tilde{\alpha},\tilde{\beta}\right)\right) = \tfrac{1}{2} + i(\theta)^{-\frac{1}{2}}\frac{\mu_3}{6\sigma^3\sqrt{2\pi}} + o\left(i(\theta)^{-1/2}\right), \tag{10.4.13}$$

where $i(\theta) = u(\alpha^T\dot{\kappa}(\theta)+\beta)^{-1}$,

$$\sigma^2 = \tilde{a}(\theta)^T\ddot{\kappa}(\theta)\,\tilde{a}(\theta), \tag{10.4.14}$$

and

$$\begin{aligned}\mu_3 = &\sum_{i,j,l=1}^{k}\kappa_{ijl}^{(3)}(\theta)\,\tilde{a}(\theta)_i\,\tilde{a}(\theta)_j\,\tilde{a}(\theta)_l \\ &- \frac{3\tilde{a}(\theta)^T\ddot{\kappa}(\theta)\,\tilde{a}(\theta)}{\alpha^T\dot{\kappa}(\theta)+\beta}\left[\alpha^T\ddot{\kappa}(\theta)\,\tilde{a}(\theta) - a^T\pi(\theta)\right] \\ &+ \frac{\left(a^T\mu(\theta)\right)^3}{\left(\alpha^T\dot{\kappa}(\theta)+\beta\right)^2},\end{aligned} \tag{10.4.15}$$

with $\tilde{a}(\theta)$ given by (10.4.9).

When $k=1$, (10.4.13) holds with $\sigma^2 = \ddot{\kappa}(\theta)$ and

$$\mu_3 = \kappa^{(3)}(\theta) - 3\ddot{\kappa}(\theta)\left[\alpha\ddot{\kappa}(\theta)\left(\alpha\dot{\kappa}(\theta)+\beta\right)^{-1} + \alpha^{-1}\right] - \left(\alpha\dot{\kappa}(\theta)+\beta\right)\alpha^{-3},$$

provided that $\alpha\tilde{\beta} < \tilde{\alpha}\beta$. If $\alpha\tilde{\beta} > \tilde{\alpha}\beta$, the sign of μ_3 must be changed.

Example 10.4.4 *Diffusion-type processes.* Consider the class of stochastic differential equations of the form (10.2.11). The fixed-time likelihood function is given by (10.2.12), provided that the regularity conditions of Section 3.5 are satisifed. This model satisfies the conditions imposed above.

We will compare $\tau_u(\alpha,\beta)$ with $\tau_{\tilde{u}}(0,1)$, where $\alpha\theta+\beta>0$, $\alpha\neq 0$, and $\tilde{u} = u(\alpha\theta+\beta)^{-1}$. For all these sequential plans the same amount $i(\theta) =$

$u(\alpha\theta+\beta)^{-1}$ of expected information is obtained. Without loss of generality we can assume that $u = 1$. By Proposition 10.4.1,

$$P_\theta\left(\tau_1(\alpha,\beta) \le \tau_{\tilde{u}}(0,1)\right) = P_\theta\left(S_{\tau_1(\alpha,\beta)} \le i(\theta)\right),$$

and by Corollary 10.2.3 (see Example 10.2.5), $i(\theta)^{-1} S_{\tau_1(\alpha,\beta)}$ has an inverse Gaussian distribution with density

$$\frac{e^{\alpha^{-2}i(\theta)^{-1}}}{\sqrt{2\pi\alpha^2 i(\theta)}} x^{-3/2} \exp\left[-\tfrac{1}{2}\alpha^{-2} i(\theta)^{-1}\left(x+x^{-1}\right)\right] \tag{10.4.16}$$

and cumulative distribution function (cf. Johnson and Kotz, 1970, p.141),

$$\Phi\left((x-1)/\sqrt{i(\theta)\,\alpha^2 x}\right) + e^{2i(\theta)^{-1}\alpha^{-2}}\Phi\left(-(x+1)/\sqrt{i(\theta)\,\alpha^2 x}\right),$$

where Φ is the distribution function of the standard normal distribution. Hence

$$P_\theta\left(\tau_1(\alpha,\beta) \le \tau_{\tilde{u}}(0,1)\right) = \tfrac{1}{2} + e^{2\alpha^{-2}i(\theta)^{-1}}\Phi\left(-2/\sqrt{\alpha^2 i(\theta)}\right).$$

The righthand side is strictly larger than one half, so $\tau_1(\alpha,\beta)$ tends to be smaller than $\tau_{\tilde{u}}(0,1)$. When $\alpha \to \infty$, the righthand side tends to one. □

Example 10.4.5 *Counting processes.* Consider the class of counting processes that have a predictable intensity of the form (10.2.13) and a fixed-time likelihood function given by (10.2.14). This model satisfies the conditions imposed above.

We will study the class of stopping times $\tau_u(\alpha,\beta)$ for which $\alpha \neq 0$ and that have the same positive value of $i(\theta) = u(\alpha e^\theta + \beta)^{-1}$. For $\tilde{u} = i(\theta)$, we get by (10.4.4) that

$$P_\theta\left(\tau_u(\alpha,\beta) \le \tau_{\tilde{u}}(0,1)\right) = P_\theta\left(S_{\tau_u(\alpha,\beta)} \le i(\theta)\right).$$

First, supppose $\beta = 0$. Without loss of generality we can put $\alpha = 1$. If $u \in \mathbb{N}$, there is no overshoot, and by Corollary 10.2.3 the Laplace transform of $S_{\tau_u(1,0)}$ is $(1 - se^{-\theta})^{-u}$. If u is not an integer, it is obvious that $\tau_u(1,0) = \tau_{\bar{u}}(1,0)$, where $\bar{u}$ denotes the smallest integer that is larger than or equal to u. Hence for any $u > 0$, $S_{\tau_u(1,0)}$ is Gamma-distributed with shape parameter $\bar{u}$ and scale parameter e^θ. That this is consistent with Corollary 10.2.3 follows by noting that the overshoot is $\bar{u} - u$ and $\varphi_\theta^u(s) = \exp\left((\bar{u}-u)\,s\right)$. The cumulative distribution function for the distribution of $S_{\tau_u(1,0)}$ is (cf. Johnson and Kotz, 1970, p. 173)

$$1 - \exp\left(-xe^\theta\right)\sum_{j=0}^{\bar{u}-1}\frac{\left(xe^\theta\right)^j}{j!}.$$

Hence

$$P_\theta\left(\tau_u(1,0) \le \tau_{\tilde{u}}(0,1)\right) = 1 - e^{-\tilde{u}} \sum_{j=0}^{\tilde{u}-1} \frac{\tilde{u}^j}{j!}. \tag{10.4.17}$$

For $\tilde{u} = 1$ this probability is $1 - e^{-1} \doteq 0.63$. As $\tilde{u} \to \infty$ it decreases towards 0.5. For $\tilde{u} = 10$ it is 0.54, for $\tilde{u} = 25$ it is 0.53, and for $\tilde{u} = 50$ it is 0.52. For all u the plan with $\beta = 0$ (inverse sampling) tends to have a shorter sampling time than the plan with $\alpha = 0$, but for large values of u the difference is small.

When $\alpha < 0$ and $\beta > -\alpha e^\theta$, there is no overshoot; i.e., (10.1.5) is satisfied, and we can find the Laplace transform for $B_{\tau_u(\alpha,\beta)}$ under P_θ by Corollary 10.2.3. However, this Laplace transform cannot be explicitly inverted, so we apply Corollary 10.4.3 and obtain

$$\begin{aligned} &P_\theta\left(\tau_u(\alpha,\beta) \le \tau_{\tilde{u}}(0,1)\right) \\ &= \tfrac{1}{2} + \frac{e^{-\theta/2} - 3\left[\alpha e^{\theta/2}\left(\alpha e^\theta + \beta\right)^{-1} + \alpha^{-1} e^{-\theta/2}\right] - \left(\alpha e^\theta + \beta\right)\alpha^{-3} e^{-3\theta/2}}{6\sqrt{2\pi}\sqrt{i(\theta)}} \\ &\quad + o\left(i(\theta)^{-1/2}\right). \end{aligned}$$

Since $\alpha < 0$, the correction term is positive, so also in this case $\tau_u(\alpha,\beta)$ tends to be smaller than $\tau_{\tilde{u}}(0,1)$. □

Example 10.4.6 *Lévy processes.* Consider a natural exponential family of one-dimensional Lévy processes. For such models $S_t = t$, and the conditions imposed above are satisfied. We suppose further that A_t is increasing and that $\alpha < 0$ and $\beta > -\alpha\dot{\kappa}(\theta)$ (> 0). Thus (10.1.5) is satisfied, and by (10.4.13),

$$\begin{aligned} &P_\theta\left(\tau_u(\alpha,\beta) \le \tilde{u}\right) = \tfrac{1}{2} \\ &+ \tilde{u}^{-\frac{1}{2}} \frac{\kappa^{(3)}(\theta) - 3\ddot{\kappa}(\theta)\left[\alpha\ddot{\kappa}(\theta)\left(\alpha\dot{\kappa}(\theta) + \beta\right)^{-1} + \alpha^{-1}\right] - \left(\alpha\dot{\kappa}(\theta) + \beta\right)\alpha^{-3}}{6\ddot{\kappa}(\theta)^{3/2}\sqrt{2\pi}} \\ &+ o\left(\tilde{u}^{-\frac{1}{2}}\right), \end{aligned}$$

where $\tilde{u} = u(\alpha\dot{\kappa}(\theta) + \beta)^{-1}$. If the third cumulant, $\kappa^{(3)}(\theta)$, of A_1 is non-negative, i.e., if the distribution of A_1 has a non-negative skewness, $\tau_u(\alpha,\beta)$ tends to be smaller than $\tilde{u}$. If $\kappa^{(3)}(\theta) < 0$, there might exist α values for which $\tau_u(\alpha,\beta)$ tends to be larger than $\tilde{u}$. However, for α close to zero or for α numerically sufficiently large, the opposite is the case. If A_t is decreasing, we choose $\alpha > 0$ and $\beta > -\alpha\dot{\kappa}(\theta)$ (> 0), and the situation is reversed. □

10.5 Moments of the stopping times

In this section we consider models with a likelihood function of the general form (10.1.11) and stopping times of the type (10.1.12). We shall give results about moments of τ_u or of a function of τ_u, and results about L_1-convergence of τ_u. We shall work under the following conditions.

Condition A (θ)

(i) $(\theta, \varphi(\theta)) \in \operatorname{int} \tilde{\Gamma}_t$ *for all* $t > 0$.

(ii) *Every coordinate of* $\dot{\varphi}(\theta)\alpha + \beta$ *is positive.*

(iii) *There exist a strictly increasing function* f *and a constant vector* $c(\theta) \neq 0$ *such that* $f(0) = 0$ *and* $S_t / f(t) \to c(\theta)$ *in* P_θ*-probability as* $t \to \infty$.

When $(\theta, \varphi(\theta)) \in \operatorname{int} \tilde{\Gamma}_t$, we know from Chapter 8 that the score vector

$$U_t(\theta) = A_t - \dot{\varphi}(\theta)^T S_t, \tag{10.5.1}$$

where $\dot{\varphi}(\theta) = \{\partial \varphi_i(\theta)/\partial \theta_j\}$, is a square integrable P_θ-martingale and that the (i, j)th element of its quadratic characteristic is

$$\langle U(\theta)\rangle_t^{(i,j)} = -\frac{\partial^2}{\partial \theta_i \partial \theta_j} \varphi(\theta)^T S_t. \tag{10.5.2}$$

We assume that we can find an exponential representation, with $A_t = (A_t^c, A_t^d)^T$ where A_t^c is k_1-dimensional, and for which the corresponding decomposition of the score vector $U_t(\theta) = (U_t^c(\theta), U_t^d(\theta))$ has the property that $U_t^c(\theta)$ is a continuous martingale, whereas $U_t^d(\theta)$ is a purely discontinuous martingale. These and other concepts from stochastic calculus will be discussed more fully in Chapter 11 and in Appendix A. This assumption essentially means that the diffusion and the jump mechanism are parametrized separately. It implies that the quadratic characteristic of the martingale

$$M_t = \alpha^T U_t(\theta) \tag{10.5.3}$$

has the form

$$\langle M\rangle_t = \pi^c(\alpha)^T S_t + \pi^d(\alpha)^T S_t, \tag{10.5.4}$$

where the vectors $\pi^c(d)$ and $\pi^d(\alpha)$ are given by

$$\pi^c(\alpha) = \sum_{i,j=1}^{k_1} \alpha_i \alpha_j \frac{\partial^2}{\partial \theta_i \partial \theta_j} \varphi(\theta), \tag{10.5.5}$$

$$\pi^d(\alpha) = -\sum_{i,j=k_1+1}^{k} \alpha_i \alpha_j \frac{\partial^2}{\partial \theta_i \partial \theta_j} \varphi(\theta). \tag{10.5.6}$$

In (10.5.4), $\pi^c(\alpha)^T S_t$ is the quadratic characteristic of the continuous martingale part of M, while $\pi^d(\alpha)^T S_t$ is the quadratic characteristic of the purely discontinuous part of M. We shall impose the following condition on the martingale M.

Condition M (θ) *The martingale $M = \alpha^T U(\theta)$ is quasi-left-continuous and its jump characteristic ν under P_θ has the form*

$$\nu(\omega, dt, dx) = \sum_{i=1}^{m-k} K_i(dx)\, dS_t^{(i)}, \tag{10.5.7}$$

where K_i is a non-random measure satisfying

$$\int x^2 K_i(dx) < \infty, \quad i = 1, \cdots, m-k. \tag{10.5.8}$$

Under Condition M(θ) the quadratic characteristic of the purely discontinous martingale part of M is

$$\int_0^t \int_{\mathbb{R}\setminus\{0\}} x^2 \nu(ds, dx) = v(\alpha)^T S_t, \tag{10.5.9}$$

where

$$v_i(\alpha) = \int_{\mathbb{R}\setminus\{0\}} x^2 K_i(dx),$$

which is in accordance with (10.5.4). A condition implying (10.5.7) is $\nu(\omega, dt, dx) = K(dx) m(\omega, dt)$, where K is as in Condition $M(\theta)$. Note that (10.5.8) implies that

$$\int_0^t \int_{|x|>1} |x| \nu(ds, dx) < \infty, \quad t > 0.$$

Theorem 10.5.1 *Suppose Condition $A(\theta)$ and Condition $M(\theta)$ are satisfied. Assume further the Cramér condition that there exists $\lambda_0 > 0$ such that*

$$\int_{|x|>1} e^{\lambda x} K_i(dx) < \infty, \quad i = 1, \cdots, m-k, \quad \textit{for } \lambda \in (0, \lambda_0].$$

Then for $g(t) = f^{-1}(\kappa t)$, where $\kappa > [(\dot\varphi(\theta)\alpha + \beta)^T c]^{-1}$,

$$P_\theta(\tau_u > g(u)) \tag{10.5.10}$$
$$\le \inf_{\lambda \in \Lambda_a, q > 1} \exp\left[q^{-1}(q-1)\lambda u\right] \left\{ E_\theta \left(\exp\left[(q-1)\, a(\lambda)^T S_{g(u)} \right] \right) \right\}^{1/q},$$

where

$$a(\lambda) = \tfrac{1}{2}\lambda^2 \pi^c(\alpha) + h(\lambda) - \lambda(\dot\varphi(\theta)\alpha + \beta) \tag{10.5.11}$$

with

$$h_i(\lambda) = \int_{\mathbb{R}\setminus\{0\}} \left(e^{\lambda x} - 1 - \lambda x\right) K_i(dx)$$

and with $\pi^c(\alpha)$ *given by (10.5.5). The set*

$$\Lambda_a = \{\lambda \in (0, \lambda_0] : a(\lambda) < 0\} \tag{10.5.12}$$

is non-empty. By $a(\lambda) < 0$ *we mean that every coordinate of* $a(\lambda)$ *is negative.*

Remark: It is obvious from the proof that (10.5.10) holds for any positive and strictly increasing function g satisfying $f(t) - [(\dot{\varphi}(\theta)\alpha + \beta)^T c]^{-1} g^{-1}(t)$ is a positive increasing function. □

Proof:

$$P_\theta(\tau_u > g(u)) \leq P_\theta\left(K_{g(u)} < u\right),$$

where

$$K_t = \alpha^T A_t + \beta^T S_t = M_t - \tilde{B}_t + f(t)\left(\dot{\varphi}(\theta)\alpha + \beta\right)^T c$$

with

$$\tilde{B}_t = \left(\dot{\varphi}(\theta)\alpha + \beta\right)^T (cf(t) - S_t).$$

We will therefore apply the large deviation result in Liptser and Shiryaev (1989, Theorem 4.13.3) to the probability

$$P(K_t < \tilde{g}(t)) = P\left(X_t > \left(\dot{\varphi}(\theta)\alpha + \beta\right)^T cf(t) - \tilde{g}(t)\right),$$

where $X = \tilde{B} - M$. In order to do this, we need the local characteristics (see Chapter 11 and Appendix A) of the semimartingale X under P_θ. The jump characteristic and the continuous martingale characteristic of X equal those of M, i.e., ν and $\pi^c(\alpha)^T S_t$, respectively. The drift characteristic of X is $\tilde{B}$ minus the compensation of the jumps numerically larger than one, so our $\tilde{B}$ equals that in Liptser and Shiryaev's theorem. Now for $\lambda \in (0, \lambda_0]$ define

$$G_t(\lambda) = \lambda \tilde{B}_t + \tfrac{1}{2}\lambda^2 \pi^c(\alpha)^T S_t + \int_0^t \int_{\mathbb{R}\setminus\{0\}} \left(e^{\lambda x} - 1 - \lambda x\right) \nu(ds, dx).$$

Under Condition M(θ),

$$\int_0^t \int_{\mathbb{R}\setminus\{0\}} \left(e^{\lambda x} - 1 - \lambda x\right) \nu(ds, dx) = h(\alpha)^T S_t.$$

Thus

$$G_t(\lambda) = \lambda \left(\dot{\varphi}(\theta)\alpha + \beta\right)^T cf(t) + a(\lambda)^T S_t.$$

Let us next show that the set Λ_a is non-empty. First, note that

$$h_i'(\lambda) = \int_{\mathbb{R}\setminus\{0\}} x\left(e^{\lambda x} - 1\right) K_i(dx)$$

when $\lambda \in (0, \lambda_0)$. Hence $h_i'(0+) = 0$ and $a'(0+) = -(\dot{\varphi}(\theta)\alpha + \beta)$. Now, since $a(0) = 0$ and every coordinate of $a'(0+)$ is negative, we can find $\bar{\lambda} \in (0, \lambda_0)$ such that every coordinate of $a(\bar{\lambda})$ is negative.

We have now checked the conditions in Liptser and Shiryaev's theorem, and their formula (4.13.26) yields that for every positive and non-decreasing function ϕ_t,

$$P(X_t > \delta\phi_t) \leq \left\{E_\theta\left(\exp\left[-(q-1)\phi_t\left(\lambda\delta - \phi_t^{-1}G_t(\lambda)\right)\right]\right)\right\}^{1/q}$$

for every $\lambda \in \Lambda_a$, $q > 1$, and $\delta > 0$. By choosing $\delta = (\dot{\varphi}(\theta)\alpha + \beta)^T c$ and $\phi_t = f(t) - \delta^{-1}\tilde{g}(t)$ we obtain

$$P_\theta(X_t > \delta\phi_t) \leq \exp\left[q^{-1}(q-1)\lambda\tilde{g}(t)\right]\left\{E_\theta\left(\exp\left[(q-1)a(\lambda)^T S_t\right]\right)\right\}^{1/q},$$

provided that $\tilde{g}$ is a function such that $f - \delta^{-1}\tilde{g}$ is positive and non-decreasing.

For $t = g(u)$ and $\tilde{g}(t) = g^{-1}(t) = \kappa^{-1}f(t)$, we get that $\phi_t = f(t)(1 - (\delta\kappa)^{-1}) > 0$ and

$$P_\theta\left(K_{g(u)} < u\right) \leq \exp\left[q^{-1}(q-1)\lambda u\right]\left\{E_\theta\left(\exp\left[(q-1)a(\lambda)^T S_{g(u)}\right]\right)\right\}^{1/q},$$

which proves the theorem. □

With the notation

$$\psi_{\lambda,q}(t) = \left\{E_\theta\left(\exp\left[(q-1)a(\lambda)^T S_t\right]\right)\right\}^{1/q}$$

we have the following result.

Theorem 10.5.2 *Suppose Condition $A(\theta)$ and Condition $M(\theta)$ are satisfied and that there exist $q > 1$ and $\lambda \in \Lambda_a$ such that*

$$\int_0^\infty \psi_{\lambda,q}\left(f^{-1}(x)\right)x^{p-1}dx < \infty$$

for some $p > 0$. Then the pth moment of $f(\tau_u)$ under P_θ is finite.

Proof: The result follows because

$$E_\theta\left(\left(u^{-1}f(\tau_u)\right)^p\right) = \int_0^\infty P_\theta\left(\tau_u > f^{-1}\left(ux^{1/p}\right)\right)dx$$

and because the integrand for x sufficiently large is dominated by

$$\exp\left[q^{-1}(q-1)\lambda u\right]\psi_{\lambda,q}\left(f^{-1}\left(ux^{1/p}\right)\right),$$

which has been assumed integrable. □

Theorem 10.5.3 *Suppose Condition $A(\theta)$ and Condition $M(\theta)$ are satisfied and that there exist $q > 1$ and $\lambda \in \Lambda_a$ such that*

$$\psi_{\lambda,q}\left(f^{-1}(x)\right) = O\left(e^{-bx}x^{\rho}\right) \quad as \ x \to \infty,$$

where $b > 0$ and $\rho \geq 0$. Then the family of random variables $\{(u^{-1}f(\tau_u))^p : u \geq \epsilon > 0\}$ is uniformly integrable under P_θ for every $p > 0$ and $\epsilon > 0$. When S is one-dimensional, then under the conditions of Theorem 10.3.1 or Theorem 10.3.2,

$$u^{-1}f(\tau_u) \to \left(\alpha^T\dot{\varphi}(\theta) + \beta\right)^{-1} c(\theta)^{-1}$$

in L_p ($p > 0$) as $u \to \infty$ under P_θ.

Proof: With $\mu = q^{-1}(q-1)\lambda$ and with the constant k_1 suitably chosen, we have for c sufficiently large that

$$\begin{aligned}
&E_\theta\left[\left(u^{-1}f(\tau_u)\right)^p 1_{\{(u^{-1}f(\tau_u))^p > c\}}\right] \\
&= cP_\theta\left[\tau_u > f^{-1}\left(uc^{1/p}\right)\right] + \int_c^\infty P_\theta\left[\tau_u > f^{-1}\left(uy^{1/p}\right)\right] dy \\
&\leq ce^{\mu u}\psi_{\lambda,q}\left(f^{-1}\left(uc^{1/p}\right)\right) + e^{\mu u}\int_c^\infty \psi_{\lambda,q}\left(f^{-1}\left(uy^{1/p}\right)\right) dy \\
&\leq ce^{\mu u}\psi_{\lambda,q}\left(f^{-1}\left(uc^{1/p}\right)\right) + e^{\mu u}u^{-p}p\int_{uc^{1/p}}^\infty \psi\left(f^{-1}(z)\right) z^{p-1}dz \\
&\leq k_1u^\rho c^{\rho/p+1}e^{-u\left(bc^{1/p}-\mu\right)} + k_1e^{\mu u}u^{-p}p\int_{uc^{1/p}}^\infty e^{-bz}z^{\rho+p-1}dz \\
&= k_1u^\rho c^{\rho/p+1}e^{-u\left(bc^{1/p}-\mu\right)}\left\{1 + p\int_0^\infty e^{-buc^{1/p}y}(y+1)^{p+\rho-1}dy\right\} \\
&\leq k_1\epsilon^\rho c^{\rho/p+1}e^{-\epsilon\left(bc^{1/p}-\mu\right)}\left\{1 + p\int_0^\infty e^{-b\epsilon c^{1/p}y}(y+1)^{p+\rho-1}dy\right\},
\end{aligned}$$

for $c > ((\rho\epsilon^{-1}+\mu)/b)^p$. This expression goes to zero as $c \to \infty$, which proves the first claim. The result on L_p-convergence follows in view of Corollary 10.3.4. □

Example 10.5.4 *Diffusion-type processes.* Consider again the class of diffusion-type processes given by (10.2.11). Since a process of this type does not jump, Condition $M(\theta)$ and the Cramér condition are automatically satisfied. Thus we need only assume Condition $A(\theta)$ to obtain that

$$P_\theta(\tau_u > g(u)) \leq \inf_{\lambda\in\Lambda_a, q>1} \exp\left[q^{-1}(q-1)\lambda u\right]\psi_{\lambda,q}(g(u)). \qquad (10.5.13)$$

For diffusion-type processes, $a(\lambda)$ has the simple form

$$a(\lambda) = \tfrac{1}{2}\lambda^2\alpha^2 - \lambda(\alpha\theta + \beta),$$

and

$$\Lambda_a = \left(0, 2\alpha^{-2}(\alpha\theta + \beta)\right).$$

For the particular case of the Ornstein-Uhlenbeck process

$$dX_t = \theta X_t dt + dW_t, \quad X_0 = x_0,$$

Condition $A(\theta)$ is satisfied with $f(t) = t$ and $c(\theta) = -(2\theta)^{-1}$ when $\theta < 0$. From results in Section 7.6 it follows that

$$\psi_{\lambda,q}(t) = O\left(\exp\left[-\frac{t}{2q}\left\{\sqrt{\theta^2 - 2(q-1)a(\lambda)} + \theta\right\}\right]\right)$$

for all $q > 1$ and $\lambda \in \Lambda_a$, so the conditions of Theorem 10.5.3 are satisfied. Hence $\{(u^{-1}\tau_u)^p : u \geq \epsilon > 0\}$ is uniformly integrable for every $p > 0$ and

$$u^{-1}\tau_u \to -2\theta(\alpha\theta + \beta)^{-1}$$

in $L_p(P_\theta)$ as $u \to \infty$ for every $p > 0$. □

Example 10.5.5 *Finite state space Markov processes.* Let us briefly return to the Markov processes studied in the Examples 10.2.8 and 10.3.10. We consider a general stopping time of the form (10.3.19). The score vector is a P_θ-martingale, and Condition A(θ) (iii) is satisfied with $f(t) = t$ and $c_i(\theta) = \lambda_{i\cdot}(\theta)^{-1}\left[\sum_j \lambda_{j\cdot}(\theta)^{-1}\right]^{-1}$, where $\lambda_{i\cdot}(\theta) = \sum_{j \neq i} \exp(\theta_{ij})$. The jump characteristic of M has the form (10.5.7) with

$$K_i(dx) = \sum_{j \neq i} \alpha_{ij} e^{\theta_{ij}} \delta_{\{j\}}(dx),$$

where $\delta_{\{j\}}$ is the Dirac measure at j. The Cramér condition is obviously satisfied. Hence by Theorem 10.5.1,

$$P_\theta(\tau_u > u) \leq \inf_{\lambda \in \Lambda_a, q > 1} \exp\left[q^{-1}(q-1)\lambda u\right] \psi_{\lambda,q}(u).$$

For finite state space Markov processes,

$$a_i(\lambda) = \sum_{j \neq i} \alpha_{ij} e^{\theta_{ij}} \left(e^{j\lambda} - 1 - (j+1)\lambda\right) - \lambda\beta_i.$$

□

10.6 The sequential probability ratio test

In the following we consider continuous time processes only. We study models of the general type defined at the beginning of the chapter with the extra assumption that either S is strictly increasing or $(\theta, \kappa(\theta)) \in \operatorname{int} \tilde{\Gamma}_t$ for all

$t > 0$ when $\theta \in \operatorname{int}\Theta$. This extra condition ensures the existence of a Lévy process Z such that $A_t = Z_{S_t}$ and such that the cumulant transform of Z_1 under P_θ is $s \to \kappa(\theta + s) - \kappa(\theta)$.

Suppose we want to test sequentially the simple hypothesis $H_0 : \theta = \theta_0$ against the simple alternative $H_1 : \theta = \theta_1$ $(\theta_1 \neq \theta_0)$. This can be done by the following procedure. Observation starts at time zero and continues as long as $-b < R_t < a$, where

$$R_t = \log\{L_t(\theta_1)/L_t(\theta_0)\} = (\theta_1 - \theta_0)^T A_t - (\kappa(\theta_1) - \kappa(\theta_0)) S_t \quad (10.6.1)$$

is the logarithm of the likelihood ratio test statistic for H_1 against H_0, and where a and b are prescribed positive constants. Observation is terminated at the stopping time

$$\tau^* = \inf\{t > 0 : R_t \geq a \text{ or } R_t \leq -b\}. \quad (10.6.2)$$

If $R_{\tau^*} \geq a$, the hypothesis H_1 is accepted. Otherwise H_0 is accepted. Wald (1945) called this test the sequential probability ratio test.

Note that $\tau^* = \tau_1 \wedge \tau_2$, where τ_1 and τ_2 are stopping times of the type defined by (10.1.3). Specifically, τ_1 is obtained for $\alpha = \theta_1 - \theta_0$, $\beta = \kappa(\theta_0) - \kappa(\theta_1)$, and $u = a$, while τ_2 is obtained for $\alpha = \theta_0 - \theta_1$, $\beta = \kappa(\theta_1) - \kappa(\theta_0)$, and $u = b$. Now, $(\theta_1 - \theta_0)^T \dot{\kappa}(\theta_1) + \kappa(\theta_0) - \kappa(\theta_1) = E_{\theta_1}(Y_1)$, where $Y_t = (\theta_1 - \theta_0)^T Z_t - (\kappa(\theta_1) - \kappa(\theta_0))t$, and $-E_{\theta_1}(Y_1) < E_{\theta_1}(\exp(-Y_1)) - 1 = 0$, where the inequality is strict because the distribution of Y_1 is not concentrated in zero. In a similar way it follows that $(\theta_0 - \theta_1)^T \dot{\kappa}(\theta_0) + \kappa(\theta_1) - \kappa(\theta_0) > 0$. Therefore, the following result follows immediately from Proposition 10.1.1.

Proposition 10.6.1

$$P_{\theta_i}(\tau^* < \infty) = 1, \quad i = 0, 1, \quad (10.6.3)$$

and

$$E_{\theta_i}(S_{\tau^*}) < \infty, \quad i = 0, 1. \quad (10.6.4)$$

The function $\varphi^* : \{\tau^* < \infty\} \to \{0, 1\}$ given by

$$\varphi^* = 1_{(0,\infty)}(R_{\tau^*}) \quad (10.6.5)$$

is called the decision function associated with the sequential probability ratio test. In order to judge the performance of the test, the following error probabilities are studied:

$$q_1^* = E_{\theta_0}(\varphi^*), \quad q_0^* = E_{\theta_1}(1 - \varphi^*). \quad (10.6.6)$$

It is possible to give rather accurate bounds for q_i^*, $i = 0, 1$. To do this we need the random variable D given by

$$R_{\tau^*} = \begin{cases} a + D & if \ R_{\tau^*} \geq a \\ -b + D & if \ R_{\tau^*} \leq -b. \end{cases} \quad (10.6.7)$$

Proposition 10.6.2 *Suppose $P(|D| > x) \leq F(x)$ for some cumulative distribution function F. Then*

$$\sup_{M\geq 0} \frac{e^{-a-M}\left(F(M)-e^{-b}\right)}{1-e^{-a-b-M}} \leq q_1^* \leq \inf_{M\geq 0} \frac{e^{-a}\left(1-e^{-b-M}F(M)\right)}{1-e^{-a-b-M}} \tag{10.6.8}$$

and

$$\sup_{M\geq 0} \frac{e^{-b-M}\left(F(M)-e^{-a}\right)}{1-e^{-a-b-M}} \leq q_0^* \leq \inf_{M\geq 0} \frac{e^{-b}\left(1-e^{-a-M}F(M)\right)}{1-e^{-a-b-M}}. \tag{10.6.9}$$

Proof: Note that

$$\begin{aligned} q_1^* = E_{\theta_1}\left[\varphi^* \exp\left(-R_{\tau^*}\right)\right] &= e^{-a} E_{\theta_1}\left(\varphi^* e^{-D}\right) \leq e^{-a}\left(1-q_0^*\right), \\ q_0^* = E_{\theta_0}\left[\left(1-\varphi^*\right) \exp\left(R_{\tau^*}\right)\right] &= e^{-b} E_{\theta_0}\left(\left(1-\varphi^*\right) e^{D}\right) \leq e^{-b}\left(1-q_1^*\right), \\ E_{\theta_1}\left(\varphi^* e^{-D}\right) &\geq e^{-M} E_{\theta_1}\left(\varphi^* 1_{\{|D|\leq M\}}\right) \\ &\geq e^{-M}\left(1-q_0^*-\left(1-F(M)\right)\right), \end{aligned}$$

and that

$$E_{\theta_0}\left[\left(1-\varphi^*\right) e^{D}\right] \geq e^{-M}\left(F(M)-q_1^*\right).$$

Hence

$$e^{-b-M}\left(F(M)-e^{-a}\right) \leq q_0^*\left(1-e^{-b-a-M}\right) \leq e^{-b}\left(1-e^{-a-M}F(M)\right).$$

The result for q_1^* follows similarly. □

As is clear from the proof, one can give an exact expression for q_i^*, $i=0,1$ if D has a known fixed value. This is, for instance, the case for exponential families of counting processes and diffusions.

Note that the results in Proposition 10.6.2 neither depend on the exponential structure of the likelihood function nor on the assumption about continuous time. Thus Proposition 10.6.2 holds quite generally. Note also that $(e^a-1)/(e^{a+b}-1)$ and $(e^b-1)/(e^{a+b}-1)$ lie in the intervals (10.6.8) and (10.6.9), respectively ($M=0, F(M)=1$). These values for q_i^*, $i=0,1$ are often used as approximations (Wald's approximation). For models with a time-continuous likelihood function this approximation is exact. Such models are characterized in Section 11.3.

Proposition 10.6.3 *Suppose $\theta_i \in \operatorname{int}\Theta_{\alpha_i,\beta_i}$, $i=0,1$, where $\alpha_1 = \theta_1 - \theta_0$, $\beta_1 = \kappa(\theta_0) - \kappa(\theta_1)$, $\alpha_2 = -\alpha_1$, and $\beta_2 = -\beta_1$. Then A_{τ^*} and S_{τ^*} have moments of any order under P_{θ_i}, $i=0,1$. In particular,*

$$E_{\theta_i}\left(A_{\tau^*}\right) = \dot{\kappa}\left(\theta_i\right) E_{\theta_i}\left(S_{\tau^*}\right) \tag{10.6.10}$$

and

$$E_{\theta_i}\left(S_{\tau^*}\right) = \frac{E_{\theta_i}\left(R_{\tau^*}\right)}{\left(\theta_1-\theta_0\right)^T \dot{\kappa}\left(\theta_i\right) - \kappa\left(\theta_1\right) + \kappa\left(\theta_0\right)}, \tag{10.6.11}$$

$$E_{\theta_0}(S_{\tau^*}) = \frac{b-(a+b)q_1^* - E_{\theta_0}(D)}{(\theta_0-\theta_1)^T \dot{\kappa}(\theta_0) + \kappa(\theta_1) - \kappa(\theta_0)}, \qquad (10.6.12)$$

$$E_{\theta_1}(S_{\tau^*}) = \frac{a-(a+b)q_0^* + E_{\theta_1}(D)}{(\theta_1-\theta_0)^T \dot{\kappa}(\theta_1) + \kappa(\theta_0) - \kappa(\theta_1)}. \qquad (10.6.13)$$

Proof: As noted earlier, $\tau^* = \tau_a(\alpha_1, \beta_1) \wedge \tau_b(\alpha_2, \beta_2)$. From $\theta_1 \in \operatorname{int} \Theta_{\alpha_1,\beta_1}$, it follows that zero is an interior point in the Laplace transform of $S_{\tau_a(\alpha_1,\beta_1)}$ under P_{θ_1}, cf. the proof of Corollary 10.2.4, and since $S_{\tau^*} \leq S_{\tau_a(\alpha_1,\beta_1)}$, the same is true about S_{τ^*}. By arguments similar to the proofs of Proposition 7.3.1 and Corollary 7.3.3 it follows that $(\theta_1, \kappa(\theta_1))$ belongs to the interior of the canonical parameter set of the exponential family generated by A_{τ^*} and S_{τ^*}. Hence the first claim and (10.6.10) follow; cf. the proof of Corollary 10.2.4. Equation (10.6.11) is an immediate consequence of (10.6.10) and (10.6.1), and (10.6.13) holds because

$$\begin{aligned} E_{\theta_1}(R_{\tau^*}) &= E_{\theta_1}(a+D|R_{\tau^*} \geq a)(1-q_0^*) + E_{\theta_1}(-b+D|R_{\tau^*} \leq -b)\, q_0^* \\ &= a(1-q_0^*) - bq_0^* + E_{\theta_1}(D). \end{aligned}$$

The results for θ_0 follow in a similar way. □

An obvious approximation to $E_{\theta_i}(S_{\tau^*})$ follows from Proposition 10.6.3 by deleting $E_{\theta_i}(D)$. This approximation as well as (10.6.12) and (10.6.13) can, of course, be combined with (10.6.8) and (10.6.9) to give expressions which do not involve q_i^*, $i = 0, 1$. Equation (10.6.10) is usually called Wald's identity.

Note also that when A is a Lévy process, which is the case if and only if $S_t = t$, Proposition 10.6.3 gives a result about the mean sampling time. Otherwise, it is a result about what could be called the operational sampling time.

In the rest of this section we shall compare the sequential probability ratio test to other sequential test procedures. For a general stopping time τ with respect to $\{\mathcal{F}_t\}$ and an associated decision function $\varphi : \{\tau < \infty\} \to \{0, 1\}$, we define the following error probabilities:

$$q_1(\tau, \varphi) = \int_{\{\tau<\infty\}} \varphi dP_{\theta_0}, \quad q_0(\tau, \varphi) = \int_{\{\tau<\infty\}} (1-\varphi)\, dP_{\theta_1}. \qquad (10.6.14)$$

Sometimes we will omit (τ, φ) in the notation and just write q_i. We call (τ, φ) a sequential test. The sequential probability ratio test is optimal in the following sense.

Theorem 10.6.4 *Suppose $S_t = t$ for all $t \geq 0$ (or equivalently that A is a Lévy process), and let (τ, φ) be a sequential test such that*

$$q_i(\tau, \varphi) \leq q_i^*, \quad i = 0, 1. \qquad (10.6.15)$$

Then

$$E_{\theta_i}(\tau) \geq E_{\theta_i}(\tau^*), \quad i = 0, 1, \tag{10.6.16}$$

and both inequalities in (10.6.16) are strict if

$$(q_0(\tau, \varphi), q_1(\tau, \varphi)) \neq (q_0^*, q_1^*).$$

Corollary 10.6.5 *When S is of the general type studied in this section, the conclusion is*

$$E_{\theta_i}(S_\tau) \geq E_{\theta_i}(S_{\tau^*}), \quad i = 0, 1. \tag{10.6.17}$$

Also here, both inequalities are strict if $(q_0(\tau, \varphi), q_1(\tau, \varphi)) \neq (q_0^*, q_1^*)$.

Proof of Corollary 10.6.5: The corollary follows because S_τ is a stopping time with respect to the filtration $\{\mathcal{F}_{\sigma_u}\}$, where $\sigma_u = \inf\{t : S_t > u\}$ $(\{S_\tau \leq u\} = \{\sigma_u \geq \tau\} \in \mathcal{F}_{\sigma_u})$, and because $\{P_\theta : \theta \in \Theta\}$ is an exponential family with respect to $\{\mathcal{F}_{\sigma_u}\}$ with a likelihood function of the form (10.1.1) with $A_t = Z_t$ and $S_t = t$; cf. Chapter 5. Remember that S_{τ^*} is the first time Z crosses a or $-b$. □

When A is a Lévy process the sequential probability ratio test minimizes the mean sampling time. For a general exponential family we have proved only that the mean operational time $E_{\theta_i}(S_\tau)$ is minimized by τ^*. Of course, (10.6.17) does not in general imply (10.6.16).

In the rest of this section we will prove Theorem 10.6.4. To do this we need the following lemmas. Below we work under the condition of Theorem 10.6.4 that $S_t = t$.

Lemma 10.6.6 *For all $u, v \geq 0$,*

$$\inf_{(\tau,\varphi)} \{E_{\theta_0}(\tau) + uq_1(\tau, \varphi) + vq_0(\tau, \varphi)\} = \inf_\tau E_{\theta_0}[\tau + u \wedge (v \exp(R_\tau))], \tag{10.6.18}$$

$$\inf_{(\tau,\varphi)} \{E_{\theta_1}(\tau) + uq_1(\tau, \varphi) + vq_0(\tau, \varphi)\} = \inf_\tau E_{\theta_1}[\tau + (u \exp(-R_\tau)) \wedge v]. \tag{10.6.19}$$

Proof: Note that

$$\begin{aligned} uq_1(\tau, \varphi) + vq_0(\tau, \varphi) &= \int_{\{\tau<\infty\}} [u\varphi + (1-\varphi) v \exp(R_\tau)] \, dP_{\theta_0} \\ &\geq \int_{\{\tau<\infty\}} u \wedge (v \exp(R_\tau)) \, dP_{\theta_0}. \end{aligned}$$

Equality holds for the decision function

$$\hat{\varphi} = \begin{cases} 1 & \text{if } u < v\exp(R_\tau) \\ 0 & \text{if } u \geq v \exp(R_\tau). \end{cases}$$

This implies (10.6.18). A similar argument proves (10.6.19). □

Lemma 10.6.7 *Define the function*

$$R_0(u,v) = \inf_{\tau} E_{\theta_0}\left[\tau + u \wedge (v \exp(R_\tau))\right]. \tag{10.6.20}$$

Then the stopping time

$$\tau^0 = \inf\{t > 0 : u \wedge (v \exp(R_t)) = R_0(u, v \exp(R_t))\} \tag{10.6.21}$$

satisfies

$$E_{\theta_0}\left[\tau^0 + u \wedge (v \exp(R_{\tau^0}))\right] = R_0(u,v).$$

We shall not prove this lemma. We note only that since R_t is a Markov process, it follows from the general solution to a Markovian optimal stopping problem; cf. Shiryaev (1978). See also Irle and Schmitz (1984).

To simplify the notation put

$$c = \left\{(\theta_0 - \theta_1)^T \dot{\kappa}(\theta_0) + \kappa(\theta_1) - \kappa(\theta_0)\right\}^{-1}. \tag{10.6.22}$$

We have seen earlier that $c^{-1} = -E_{\theta_0}(R_1) > 0$.

Lemma 10.6.8 *For $u, v > 0$ satisfying*

$$vc^{-1} \geq e^{u/c} - 1 \quad or \quad vc^{-1} \leq 1 - e^{-u/c} \tag{10.6.23}$$

the following inequality holds:

$$R_0(u,v) \geq \min\{u, v, g(v)\},$$

where

$$g(v) = \begin{cases} v & \text{for } v \leq c \\ c + c\log(v/c) & \text{for } v > c. \end{cases} \tag{10.6.24}$$

Proof: Let (τ, φ) be a sequential test with $E_{\theta_0}(\tau) < \infty$.

(a) If $q_0 + q_1 \geq 1$, then $E_0(\tau) + uq_1 + vq_0 \geq u \vee v$.

(b) If $q_0 = 0$, then $\varphi = 1$, P_{θ_0}-almost surely on $\{\tau < \infty\}$, and, since $P^{\mathcal{T}}_{\theta_1} | \{\tau < \infty\} \sim P^{\mathcal{T}}_{\theta_2} | \{\tau < \infty\}$, $\varphi = 1$, P_{θ_0}-almost surely on $\{\tau < \infty\}$.

(c) For $q_0 + q_1 < 1$, $q_0 > 0$, we use the following inequality of Wald (1947, p. 197); see also Irle and Schmitz (1984):

$$E_{\theta_0}(\tau) \geq c\left\{(1 - q_1)\log\left(\frac{1 - q_1}{q_0}\right) + q_1 \log\left(\frac{q_1}{1 - q_0}\right)\right\}.$$

In Exercise 10.4 a proof is given for a special class of stopping times. Define f on $D = \{(x,y) : x > 0, y > 0, x + y < 1\}$ by

$$f(x,y) = c\left\{(1 - x)\log\left(\frac{1 - x}{y}\right) + x\log\left(\frac{x}{1 - y}\right)\right\} + ux + vy,$$

and extend f by continuity to $\bar{D} = \{(x,y) : 0 \le x < 1, 0 \le y \le 1, x+y \le 1\}$. Then

$$\inf_{(x,y)\in D} f(x,y) = \inf_{(x,y)\in \bar{D}\setminus D} f(x,y); \tag{10.6.25}$$

see Exercise 10.5. The lemma follows since

$$\inf_{(x,y):x+y=1} f(x,y) \ge u \wedge v$$

and

$$\inf_{(0,y):0\le y\le 1} f(x,y) = g(v).$$

□

Lemma 10.6.9 *The functions $U, V : [0,\infty) \to [0,\infty)$ defined by*

$$U(v) = \sup\{u : R_0(u,v) = u\}$$

$$V(u) = \sup\{v : R_0(u,v) = v\}$$

are concave, continuous, and non-decreasing. Moreover, they have the following properties:

(i)

$$\begin{aligned}
&U(0) = V(0) = 0,\\
&U(x) > 0,\ V(x) > 0 \quad \textit{for all } x > 0,\\
&R_0(u,v) = v \Leftrightarrow v \le V(u),\\
&R_0(u,v) < v \Leftrightarrow v > V(u),\\
&R_0(u,v) = u \Leftrightarrow u \le U(v),\\
&R_0(u,v) < u \Leftrightarrow u > U(v).
\end{aligned}$$

(ii) The function U is strictly increasing with $\lim_{v\to\infty} U(v) = \infty$, and

$$c\log(v/c+1) \le U(v) \le v \tag{10.6.26}$$

for all $v > 0$.

(iii) The function V is bounded from above, and

$$c\left(1 - e^{-u/c}\right) \le V(u) \le u \tag{10.6.27}$$

for all $u > 0$.

(iv)

$$\lim_{v\to 0} v^{-1} V(U(v)) = 1. \tag{10.6.28}$$

Proof: Obviously, $R_0(0,v) = R_0(u,0) = 0$ for all $u, v \geq 0$. First, let $u > 0$ be fixed. It is not difficult to see that $v \to R_0(u,v)$ is a continuous, concave, and non-decreasing function, and that $R_0(u,v) \leq u \wedge v$ (choose $\tau = 0$). For $v \leq c(1 - e^{-u/c})$ we have $v \leq u$, and by Lemma 10.6.8

$$v \geq R_0(u,v) \geq \min(u, v, g(v)) = u \wedge v = v.$$

Since $R_0(u,v) \leq u \wedge v < v$ for $v > 0$, it follows that

$$c\left(1 - e^{-u/c}\right) \leq V(u) \leq u,$$

and because the concavity of $v \to R_0(u,v)$ implies that $R_0(u,v) < v$ for $v > V(u)$, we have proved the results about V in (i).

From

$$\begin{aligned}\{(u,v) : v \leq V(u)\} &= \{(u,v) : v = R_0(u,v)\} \\ &= \bigcap_{(\tau,\varphi)} \{(u,v) : v(1 - q_0) \leq E_{\theta_0}(\tau) + uq_1\}\end{aligned}$$

it follows that V is a continuous, concave, and non-decreasing function of u. Moreover, for $\tilde{\tau} = \inf\{t : R_t \leq -1\}$ and $\tilde{\varphi} = 0$ on $\{\tilde{\tau} < \infty\}$,

$$E_{\theta_0}(\tilde{\tau}) < \infty, \quad q_1 = 0, \quad q_0 < 1$$

(because $P_{\theta_1}(\tilde{\tau} = \infty) > 0$), so

$$\{(u,v) : v \leq V(u)\} \subset \left\{(u,v) : v \leq \frac{E_{\theta_0}(\tilde{\tau})}{1 - q_0}\right\}.$$

Hence V is bounded.

Most of the results about U are proved analogously. In particular, for fixed $v > 0$, $e^{u/c} - 1 \leq v/c$ implies that $u \leq v$ and (by Lemma 10.6.8) that

$$u \geq R_0(u,v) \geq \min(u, v, g(v)) = u \wedge g(v) = u.$$

Since $R_0(u,v) \leq u \wedge v < u$ for $u > v$, we have proved that

$$c \log(v/c + 1) \leq U(v) \leq v.$$

The concavity of $u \to R_0(u,v)$ implies that $R_0(u,v) < u$ for $u > U(v)$.

The function U is continuous, concave, and non-decreasing, and (10.6.26) shows that $U(v) \to \infty$ as $v \to \infty$. These properties imply that U is strictly increasing.

Finally, note that

$$\begin{aligned}\frac{v}{v/c + 1} &= c\{1 - \exp[-\log(v/c + 1)]\} \\ &\leq V(c \log(v/c + 1)) \\ &\leq V(U(v)) \leq V(v) \leq v,\end{aligned}$$

which implies (10.6.28). □

Proof of Theorem 10.6.4. We consider only the mean value under P_{θ_0}. The proof of result for P_{θ_1} is symmetric.

It is sufficient to show the existence of $u, v > 0$ such that

$$E_{\theta_0}(\tau) + uq_1 + vq_0 \geq E_{\theta_0}(\tau^*) + uq_1^* + vq_0^*. \tag{10.6.29}$$

Consider the continuous function $h : (0, \infty) \to \mathbb{R}$ given by

$$h(v) = v^{-1} V(U(v)).$$

By Lemma 10.6.9, (iii) and (iv),

$$\begin{aligned} h(v) &\to 1 \quad \text{for } v \to 0 \\ h(v) &\to 0 \quad \text{for } v \to \infty. \end{aligned}$$

Therefore, there exists a $w_1 > 0$ such that $h(w_1) = \exp(-a - b)$. Define

$$u = U(w_1) \quad \text{and} \quad v = w_1 e^{-a}.$$

Then (Lemma 10.6.9)

$$e^{-b} v = V(u) \leq u, \qquad u = U(ve^a) \leq v.$$

Now, using Lemma 10.6.9 (i), we see that for $0 \leq w \leq u$,

$$R_0(u, w) = w \Leftrightarrow w \leq V(u) = e^{-b} v,$$

and for $w > v$

$$\begin{aligned} R_0(u, w) = u \quad &\Leftrightarrow \quad u \leq U(w) \\ &\Leftrightarrow \quad w \geq U^{-1}(u) = w_1 = e^a v. \end{aligned}$$

We have thus proved that

$$\{w \geq 0 : u \wedge w = R_0(u, w)\} = [0, e^{-b} v] \cup [e^a v, \infty]$$

and that

$$\tau^0 = \inf\{t > 0 : R_t \leq -b \text{ or } R_t \geq a\} = \tau^*,$$

where τ^0 is the optimal stopping time given by (10.6.21). Moreover, since $e^{-b} v \leq u \leq e^a v$,

$$\varphi^* = \begin{cases} 1 & \text{if } u < v \exp(R_\tau) \\ 0 & \text{if } u \geq v \exp(R_\tau), \end{cases}$$

so Lemma 10.6.6 shows that (10.6.28) holds.

Because $u, v > 0$, $(q_0, q_1) \neq (q_0^*, q_1^*)$ implies a strict inequality in (10.6.16). □

10.7 Exercises

10.1 Prove the distributional results in Example 10.2.7 in detail. In the same example, find the Laplace transform of $(S_\tau, A_\tau^{(1)})$ when $\beta = 0$, $\alpha_1 = 1$, $\alpha_2 = -2$, and $\lambda > 2\mu$.

10.2 Prove the formulae (10.2.21) and (10.2.22).

10.3 Give an explicit expression for the error probability q_1^* defined by (10.6.6) in the case of an exponential family of counting processes as defined in Example 10.2.6. Hint: See the proof of Proposition 10.6.2.

10.4 Consider a model with likelihood function of the form (10.1.1) with $S_t = t$, and let τ be a finite stopping time such that $\{R_{t\wedge\tau} : t > 0\}$ is uniformly integrable. Here R is given by (10.6.1). Show that

$$E_{\theta_0}(R_\tau) = -c^{-1} E_{\theta_0}(\tau),$$

with c given by (10.6.22).
Next show that

$$\begin{aligned} E_{\theta_0}(R_\tau) \;&\leq\; \log\left(E_{\theta_0}\left(\frac{L_\tau(\theta_1)}{L_\tau(\theta_0)}|\varphi = 1\right)\right) q_1(\tau,\varphi) \\ &\quad + \log\left(E_{\theta_0}\left(\frac{L_\tau(\theta_1)}{L_\tau(\theta_0)}|\varphi = 0\right)\right)(1 - q_1(\tau,\varphi)), \end{aligned}$$

where $q_i(\tau,\varphi)$ is defined by (10.6.14).
Finally, conclude that

$$\begin{aligned} E_{\theta_0}(\tau) \;&\geq\; c\left(q_1(\tau,\varphi)\log\left(\frac{q_1(\tau,\varphi)}{1 - q_0(\tau,\varphi)}\right)\right. \\ &\quad \left. + (1 - q_0(\tau,\varphi))\log\left(\frac{1 - q_1(\tau,\varphi)}{q_0(\tau,\varphi)}\right)\right). \end{aligned}$$

10.5 Prove (10.6.25). Hint: Give an indirect proof by assuming that

$$\inf_{(x,y)\in D} f(x,y) < \inf_{(x,y)\in \bar{D}\setminus D} f(x,y),$$

and consider the partial derivatives of f at the point $(\bar{x},\bar{y}) \in D$, at which f has a minimum.

10.6 Consider the class of ergodic discrete time Markov chains with finite state space $I = \{1,\cdots,m\}$ starting at $x_0 \in I$. Let $\{p_{ij}\}$ denote the transition matrix and use the notation in Section 3.2. In particular, the likelihood function is given by (3.2.5). Show that when the process is observed until the stopping time

$$\tau_{u,r} = \inf\left\{t > 0 : \sum_{i\in J} N_t^{(i,r)} \geq u\right\},$$

where $J \subseteq I$, $u \in \mathbb{N}$, and $\sum_{i \in J} p_{ir} > 0$, then the model is a non-curved exponential family. Find the Laplace transform of $N_{\tau_{u,r}}$.

10.7 Consider the same class of processes as in Exercise 10.6. Again use the notation of Section 3.2. Show that the likelihood function corresponding to observation of the process up to the stopping time

$$\tau_u = \inf \left\{ t > 0 : \sum_{i,j=1}^{m} \alpha_{ij} N_t^{(i,j)} \geq u \right\},$$

where $u > 0$ and $\sum_{i,j} \alpha_{ij} p_{ij} > 0$, has the form

$$L_{\tau_u}(\theta) = \exp \left[\gamma(\theta)^T B_{\tau_u} + u\mu(\theta) + D_{\tau_u}(\theta) \right],$$

where $\gamma(\theta)$ and B_{τ_u} are $m(m-1)$-dimensional, $\mu(\theta)$ is one-dimensional, and $D_{\tau_u}(\theta)$ is one-dimensional and uniformly bounded in u. Use this to show that B_{τ_u} is asymptotically normal distributed as $u \to \infty$.

10.8 Consider the two-dimensional diffusion process given by

$$\begin{aligned} dX_t^{(1)} &= (\theta_1 X_t^{(1)} - \theta_2 X_t^{(2)})dt + dW_t^{(1)}, \\ dX_t^{(2)} &= (\theta_1 X_t^{(2)} + \theta_2 X_t^{(1)})dt + dW_t^{(2)}, \end{aligned}$$

where $W^{(1)}$ and $W^{(2)}$ are independent standard Wiener processes and where $(\theta_1, \theta_2) \in \mathbb{R}^2$. We assume that $X_0^{(1)}$ and $X_0^{(2)}$ are non-random. Show that the likelihood function is of the form (10.1.1) with $\kappa(\theta_1, \theta_2) = \frac{1}{2}(\theta_1^2 + \theta_2^2)$,

$$S_t = \int_0^t |X_s|^2 ds,$$

$$A_t^{(1)} = \int_0^t X_s^T dX_s, \text{ and } A_t^{(2)} = \int_0^t X_s^T B dX_s,$$

where

$$B = \left\{ \begin{array}{cc} 0 & 1 \\ -1 & 0 \end{array} \right\}.$$

Prove that $i(\theta_1, \theta_2)^{-1} S_{\tau_u(\alpha_1, \alpha_2, \beta)}$, where the stopping time $\tau_u(\alpha_1, \alpha_2, \beta)$ is given by (10.1.3), has an inverse Gaussian distribution with density (10.4.16), where $i(\theta) = u/(\alpha_1\theta_1 + \alpha_2\theta_2 + \beta)$ and $\alpha = \alpha_1^2 + \alpha_2^2$, provided that $(\theta_1, \theta_2) \in \Theta_{\alpha_1, \alpha_2, \beta}$ defined in Section 10.2. Use this result to compare stopping times in analogy with Example 10.4.4.

10.9 Consider the birth-and-death process X with birth rates $\lambda_i = \lambda$, $i = 0, 1, \cdots$, and death rates $\mu_i = \mu$, $i = 1, 2, \cdots$, and $\mu_0 = 0$. Assume that $X_0 = 0$. This model is the same as the simple-single server queuing model (M/M/1 queue). Show that the model is an exponential family of stochastic processes. It is not of the type (10.1.1). Let $A_0(t)$ denote the sojourn time in the state 0 in the time interval $[0, t]$. Show that a non-curved exponential family is obtained when the process is observed in the random time interval $[0, \tau(u)]$, where $u > 1$ and $\tau(u) = \inf\{t > 1 : A_0(t) \geq u\}$, and find the Laplace transform of the canonical statistic.

10.8 Bibliographic notes

Much of the material in Sections 10.1, 10.2, and 10.3 is an updated and extended version of results in Sørensen (1986a), while Sections 10.4 and 10.5 are based on Sørensen (1994a,b). Material from several papers by other authors is included too. In particular, the last part of Example 10.2.8 and Example 10.3.10 are based on Stefanov (1991, 1995a), while Proposition 10.1.1 was first proved by Stefanov (1986b). Stopping times of the type (10.1.3) and (10.1.12) have been studied by numerous authors, mainly from the point of view of obtaining efficient sequential plans. The example in Exercise 10.9 is taken from Stefanov (1995b).

Moran (1951, 1953) and Kendall (1952) studied such stopping rules for the linear birth-and-death process, as did Franz (1982). More general birth-and-death processes were considered by Manjunath (1984). Continuous time branching processes were studied by Basawa and Becker (1983) and Franz (1985). Poisson processes were investigated by Trybuła (1968), compound Poisson processes with marks from a linear exponential family of order one by Stefanov (1982), and multivariate point processes by Pruscha (1987). The Ornstein-Uhlenbeck process with the level as parameter was studied by Różański (1980), diffusion branching processes by Brown and Hewitt (1975), and general exponential families of diffusions by Sørensen (1983, 1986a). Finite state space Markov processes were considered by Bai (1975), Trybuła (1982), Stefanov (1984, 1988, 1991), and Sørensen (1986a).

The stopping rules (10.1.3) were studied for the class of natural exponential families of Lévy processes by Magiera (1974, 1977), Franz and Winkler (1976), Franz (1977), Franz and Magiera (1978, 1990), Winkler and Franz (1979), Winkler (1980, 1985), Winkler, Franz, and Küchler (1982), and Stefanov (1985). They were studied for general exponential famlies of the form (10.1.1) by Stefanov (1986a, 1986b) and Sørensen (1986a); see also Franz (1986) and Stefanov (1995a). Franz and Winkler (1987) considered an even more general type of model.

Barndorff-Nielsen and Cox (1984) derived the stopping rules (10.1.3) for the one-dimensional Brownian motion with drift as the only stopping rules for which the Bartlett adjustment of $-2\log Q$ is zero, as it is for fixed time sampling. Here Q denotes the likelihood ratio test statistic for the hypothesis that the drift has a particular value. They also showed that the Bartlett adjustments for $\alpha = 0$ and $\beta = 0$ (inverse sampling) are equivalent for the class of Poisson processes and for diffusion type processes with proportional drift and diffusion coefficient. Höpfner (1987) proved that for continuous-time ergodic Markov chains the stopping time (10.1.3) has finite mean and that τ_u/u converges in L_p.

When $\alpha = 0$, the stopping rules (10.1.3) correspond essentially to observation until a prescribed amount of observed information has been obtained. This approach was investigated for exponential families of diffusions by Novikov (1972), Liptser and Shiryaev (1977), Sørensen (1983), and Le Breton and Musiela (1984), and for Gaussian autoregressions by Lai and Siegmund (1983). The same idea was investigated for other types of stochastic process models by Grambsch (1983) and Sellke and Siegmund (1983). Okamoto, Amari, and Takeuchi (1991) studied how the geometry of an exponential family is changed when fixed time sampling is replaced by sampling up to a stopping time (10.1.3).

The sequential probability ratio test has been studied in the context of exponential families of processes by Shiryaev (1978), Küchler (1979), Winkler, Franz, and Küchler (1981), Irle and Schmitz (1984), and Bhat (1988b). The proof of Theorem 10.6.4 is based on Irle and Schmitz (1984).

Readers wishing to study sequential analysis more thoroughly can, for instance, read the book by Sigmund (1985).

11

The Semimartingale Approach

11.1 The local characteristics of a semimartingale

Consider a general exponential family $\mathcal{P}$ on a filtered space with exponential representation (3.1.1). If the filtration is generated by observation of a stochastic process X that is a semimartingale under one of the measures in $\mathcal{P}$, then X is a semimartingale under all measures in the family, and we call $\mathcal{P}$ an *exponential family of semimartingales.*

A *semimartingale* is a stochastic process X that has a decomposition $X = A + M$, where A is a process of locally finite variation and M is a local martingale. We can loosely think of A as describing a systematic trend of the process and M as describing random fluctuations around this general trend. The semimartingales form a very broad class of stochastic processes covering most processes of practical interest. A comprehensive presentation of the semimartingale theory needed in this book is given in Appendix A. We call a d-dimensional stochastic process a d-dimensional semimartingale if every coordinate is a semimartingale. Usually we just refer to a multidimensional semimartingale as a semimartingale.

In Chapter 2 we studied the local behaviour of the observed process under the probability measures in a natural exponential family of Lévy processes by considering the Lévy characteristics of this process. In this chapter we shall use a similar method to study the local behaviour of the observed process for an exponential family of semimartingales. Specifically, we shall consider the so-called local characteristics.

We shall define exponential families of semimartingales by constructing suitable parametrized classes of local characteristics. This is a useful way of constructing parametric models of real-life processes. A problem in this approach is that the local characteristics do not, in general, determine the semimartingale uniquely. For an example of this see Appendix A. However, in many important special cases the local characteristics do determine the semimartingale uniquely.

We start by giving a brief introduction to the concept of the *local characteristics* of a semimartingale and illustrate it by some examples. A more comprehensive presentation of the theory can be found in Appendix A. First, we consider a one-dimensional semimartingale X. For simplicity of exposition we assume that $X_0 = x_0$, where x_0 is non-random. We define the following process associated with X (the big jumps)

$$X_t^1 = x_0 + \sum_{s \leq t} \Delta X_s 1_{\{|\Delta X_s| > 1\}}, \tag{11.1.1}$$

where $\Delta X_s = X_s - X_{s-}$. We also define a random measure μ on $\mathbb{R}_+ \times E$ associated with the jumps of X, E being $\mathbb{R} \backslash \{0\}$:

$$\mu(\omega; dt, dx) = \sum_{s \geq 0} 1_{\{\Delta X_s(\omega) \neq 0\}} \epsilon_{(s, \Delta X_s(\omega))}(dt, dx). \tag{11.1.2}$$

Here ϵ_a denotes the Dirac measure at a. There exists a unique predictable random measure ν on $\mathbb{R}_+ \times E$ such that $U * (\mu - \nu)$ is a local martingale under P. Here

$$U(\omega; t, x) = x 1_{[-1,1]}(x). \tag{11.1.3}$$

We use the convenient notation $Y * (\mu - \nu)$ for the stochastic integral of a predictable random function $Y : \mathbb{R}_+ \times E \mapsto \mathbb{R}$ with respect to $\mu - \nu$. The random measure ν is the so-called *compensator* of μ.

Now, there exists a continuous local martingale M^c and a predictable process α such that

$$X = X^1 + M^c + U * (\mu - \nu) + \alpha. \tag{11.1.4}$$

In analogy with (2.1.11) this equation shows how the local characteristics describe the local behaviour of X. The process $U * (\mu - \nu)$ represents the jumps smaller than one, while α can be thought of as the integrated drift. Let β denote the quadratic characteristic $\langle M^c \rangle$ of M^c. Then the triple (α, β, ν) is called the *local characteristics* of X with respect to P. We see that β generalizes the integrated diffusion coefficient, whereas ν can be throught of as a generalized jump intensity. The process β is continuous. The triplet (α, β, ν) is uniquely determined by X and P. Also the continuous local martingale M^c is uniquely determined by X and P. It is called the *continuous martingale part* of X and is denoted by X^c.

A Lévy process with Lévy characteristics (δ, σ^2, ν) is a semimartingale with local characteristics $(\alpha t, \sigma^2 t, \nu(dx)dt)$, where

$$\alpha = \gamma - \int_E y(1+y^2)^{-1}\nu(dy). \tag{11.1.5}$$

In particular, the Wiener process with drift θ can be written in the form

$$X_t = \theta t + W_t, \tag{11.1.6}$$

where W is a standard Wiener process, so the local characteristics are $(\theta t, t, 0)$. A Poisson process with intensity λ can be written in the form

$$X_t = \lambda t + (X_t - \lambda t), \tag{11.1.7}$$

so its local characteristics are $(\lambda t,\ 0,\ \lambda dt \epsilon_1(dx))$.

A counting process of the type considered in Section 3.4 with intensity (3.4.1) has a similar decomposition:

$$X_t = x_0 + \kappa \int_0^t H_s ds + \left(X_t - \kappa \int_0^t H_s ds \right). \tag{11.1.8}$$

The process in the parenthesis is a local martingale, so the local characteristics of X are

$$\left(\kappa \int_0^t H_s ds,\ 0,\ \kappa H_t dt\, \epsilon_1(dx) \right).$$

A one-dimensional diffusion-type process that solves the stochastic differential equation (3.5.1) satisfies

$$X_t = x_0 + \int_0^t [a_s(X) + b_s(X)\theta] ds + \int_0^t c_s(X) dW_s, \tag{11.1.9}$$

where the last term is a local martingale. We see that the local characteristics are

$$\left(\int_0^t a_s(X)ds + \int_0^t b_s(X)ds\, \theta,\ \int_0^t c_s(X)^2 ds,\ 0 \right).$$

For a d-dimensional semimartingale X we can also define the random measure μ on $\mathbb{R}_+ \times E^d$ associated with the jumps of X by (11.1.2). Let ν denote the unique predictable random measure on $\mathbb{R}_+ \times E^d$ which is the compensator of μ. If $X^{i,c}$ denotes the continuous martingale part of the ith coordinate of X, we define a $d \times d$-matrix of predictable processes β by

$$\beta^{ij} = \langle X^{i,c}, X^{j,c} \rangle. \tag{11.1.10}$$

Define the function U by

$$U(\omega; t, x) = x 1_{[-1,1]^d}(x). \tag{11.1.11}$$

Then X has a decomposition of the form

$$X = X^1 + X^c + U * (\mu - \nu) + \alpha, \tag{11.1.12}$$

where α is a unique predictable process, $X^c = (X^{1,c}, \cdots, X^{d,c})$, and

$$X_t^1 = x_0 + \sum_{s \le t} \Delta X_s 1_{\mathbb{R}^d \setminus [-1,1]^d}(\Delta X_s). \tag{11.1.13}$$

The processes α, β, and ν are called the local characteristics of X.

In order to calculate likelihood functions for semimartingale models, we need a result of the Girsanov type that gives Radon-Nikodym derivatives in terms of local characteristics. Results of this type are given in Appendix A. As mentioned above, the triple of local characteristics is uniquely determined by the probability measure P and the semimartingale X, whereas P is not necessarily uniquely determined by the triple; see Example A.9.4. Here we are mainly interested in the structure of exponential families of semimartingales in terms of their local characteristics. Therefore, we will, when necessary, assume that the uniqueness condition in Theorem A.10.4 is satisfied.

11.2 The natural exponential family generated by a semimartingale

In this section we discuss a natural way of embedding a semimartingale in an exponential family of semimartingales.

Let X be a d-dimensional semimartingale defined on the filtered probability space $(\Omega, \mathcal{F}, \{\mathcal{F}_t\}, P)$. We assume that $\{\mathcal{F}_t\}$ is the right-continuous filtration generated by X and that $\mathcal{F} = \sigma\{\mathcal{F}_t : t > 0\}$.

Consider the continuous process

$$Z_t(\theta) = \exp(\theta^T X_t^c - \tfrac{1}{2}\theta^T \beta_t \theta) \tag{11.2.1}$$

for $\theta \in \mathbb{R}^d$. As above, X^c denotes the continuous martingale part of X under P, and $\beta = \langle X^c \rangle$. Since $\langle \theta^T X^c \rangle = \theta^T \langle X^c \rangle \theta$, the process $Z(\theta)$ is the Doléans-Dade exponential of the local martingale $\theta^T X^c$. Therefore, $Z(\theta)$ is a local martingale for all $\theta \in \mathbb{R}^d$.

The process $Z(\theta)$ is our potential likelihood function. We shall now study to what extent $Z(\theta)$ has such an interpretation. Fix $\theta \in \mathbb{R}^d$, and define for every $u > 0$ the stopping time

$$\tau_u(\theta) = \inf\{t : \theta^T \beta_t \theta > u\}. \tag{11.2.2}$$

The process $\theta^T \beta_t \theta$ is generally a non-decreasing process. Let us for simplicity assume that $\theta^T \beta \theta$ is strictly increasing and that $\theta^T \beta_t \theta \to \infty$ as $t \to \infty$.

Then $\tau_u(\theta)$ is a strictly increasing continuous function of u, $\tau_u(\theta) < \infty$ and $\tau_u(\theta) \to \infty$ as $u \to \infty$. Moreover, $\tau_u(\theta)$ and $\theta^T \beta_t \theta$ are inverses of each other. It is well-known that $\theta^T X^c_{\tau_u(\theta)}$ is a standard Wiener process with respect to the filtration $\{\mathcal{F}_{\tau_u(\theta)}\}$ under P. As discussed in Chapter 2, there exists a class of probability measures $\{P_{\alpha,\theta} : \alpha \in \mathbb{R}\}$ on $(\Omega, \mathcal{F})$ such that $\theta^T X^c_{\tau_u(\theta)}$ is a Wiener process with drift α with respect to $\{\mathcal{F}_{\tau_u(\theta)}\}$ under $P_{\alpha,\theta}$. The restriction of $P_{\alpha,\theta}$ to $\mathcal{F}_{\tau_u(\theta)}$ is given by

$$\frac{dP^{\tau_u(\theta)}_{\alpha,\theta}}{dP^{\tau_u(\theta)}} = \exp\left((\alpha\theta)^T X^c_{\tau_u(\theta)} - \tfrac{1}{2}\alpha^2 u\right), \quad u > 0. \tag{11.2.3}$$

As we saw in Section 5.1, $\theta^T \beta_t \theta$ is a stopping time relative to the filtration $\{\mathcal{F}_{\tau_u(\theta)}\}$, and by Theorem B.1.1 in Appendix B we find that for $A \in \mathcal{F}_t$,

$$P_{\alpha,\theta}(A \cap \{\theta^T \beta_t \theta < \infty\}) = \int_{A \cap \{\theta^T \beta_t \theta < \infty\}} Z_t(\alpha\theta) dP. \tag{11.2.4}$$

Here we have used that $\mathcal{F}_t = \mathcal{F}_{\tau_u(\theta)}|_{u=\theta^T\beta_t\theta}$. Thus $P^t_{\alpha,\theta}$ is dominated by P on the set $\{\beta_t < \infty\}$ with Radon-Nikodym derivative $Z_t(\alpha\theta)$. This is also true without the assumption that $\theta^T \beta_t \theta$ is strictly increasing to infinity, because on intervals where $\theta^T \beta_t \theta$ is constant, $\theta^T X^c_t$ is also constant.

The considerations above can be made for every $\theta \in \mathbb{R}^d$, so we have defined a probability measure P_θ for every $\theta \in \mathbb{R}^d$ such that

$$P_\theta(A \cap \{\beta_t < \infty\}) = \int_A Z_t(\theta) dP, \quad A \in \mathcal{F}_t, \quad t > 0. \tag{11.2.5}$$

Unfortunately, it is quite possible that $P_\theta(\beta_t < \infty) < 1$ and that this probability varies with θ. For statistical purposes we must therefore restrict attention to values of θ for which

$$P_\theta(\beta_t < \infty) = E(Z_t(\theta)) = 1, \quad t > 0. \tag{11.2.6}$$

Therefore, we define the parameter set

$$\Theta = \left\{\theta \in \mathbb{R}^d : E(Z_t(\theta)) = 1, \ \forall t \geq 0\right\}. \tag{11.2.7}$$

The class $\{P_\theta : \theta \in \Theta\}$ is an exponential family with respect to $\{\mathcal{F}_t\}$. This construction is particularly important when X is a continuous semimartingale under P and hence under all P_θ. In this case we call the class $\{P_\theta : \theta \in \Theta\}$ the *natural exponential family* generated by X, and Θ we call the natural parameter set.

A result by Novikov (1973) shows that the set

$$\Theta_N = \left\{\theta \in \mathbb{R}^d : E\left[\exp(\tfrac{1}{2}\theta^T \langle X^c \rangle_t \theta)\right] < \infty, \ \forall t \geq 0\right\} \tag{11.2.8}$$

is included in the parameter set Θ given by (11.2.7). A somewhat stronger result by Kazamaki (1978) shows that also

$$\Theta_K = \left\{\theta \in \mathbb{R}^d : E\left[\exp(\tfrac{1}{2}\theta^T X_t^c)\right] < \infty, \ \forall t \geq 0\right\} \tag{11.2.9}$$

is a subset of Θ.

The following result is sometimes useful.

Lemma 11.2.1 *Suppose there exists a positive semi-definite $d \times d$-matrix of non-random functions $c(t)$ such that $c(t) - \langle X^c \rangle_t$ is P-a.s. positive semi-definite for all $t \geq 0$. Then $\Theta = \mathbb{R}^d$.*

Proof: Since $\theta^T \langle X^c \rangle \theta \leq \theta^T c(t)\theta$, we see that $\Theta_N = \mathbb{R}^d$. □

Example 11.2.2 Suppose $X = N + W$, where N is a Poisson process that is independent of the Wiener process W. Then $X^c = W = X - N$,

$$Z_t(\theta) = \exp(\theta(X_t - N_t) - \tfrac{1}{2}\theta^2 t), \tag{11.2.10}$$

and $\Theta = \mathbb{R}$. Under P_θ we can express X as

$$X_t = N_t + W_t^\theta + \theta t, \tag{11.2.11}$$

where $W_t^\theta = W_t - \theta t$ is a standard Wiener process under P_θ. □

Let (α, β, ν) denote the local characteristics of the semimartingale X under P. Under P_θ the local characteristics are $(\alpha^\theta, \beta, \nu)$, where

$$\alpha^\theta = \alpha + \beta\theta. \tag{11.2.12}$$

This follows from Corollary 11.3.2 below.

Example 11.2.3 Suppose the d-dimensional process X under P is the unique strong solution of the stochastic differential equation

$$dX_t = a(t, X_t)dt + \sigma(t, X_t)dW_t, \tag{11.2.13}$$

where W is an m-dimensional standard Wiener process under P, and where σ is a $d \times m$-matrix. In particular, we assume that P-almost surely

$$\int_0^t a(s, X_s)ds < \infty \quad \text{and} \quad \int_0^t \sigma(s, X_s)\sigma^T(s, X_s)ds < \infty$$

for all $t > 0$. Then $X^c = X - \int a(s, X_s)ds$,

$$Z_t(\theta) = \exp\left(\theta^T X_t - \theta^T \int_0^t a(s, X_s)ds - \tfrac{1}{2}\theta^T \int_0^t \sigma(s, X_s)\sigma(s, X_s)^T ds\theta\right), \tag{11.2.14}$$

and $\Theta = \mathbb{R}^d$. Under P_θ the process X satisfies the equation

$$dX_t = a(t, X_t)dt + \sigma(t, X_t)\sigma(t, X_t)^T\theta dt + \sigma(t, X_t)dW_t^\theta, \tag{11.2.15}$$

where W^θ is a standard Wiener process under P_θ. □

Example 11.2.4 Let X be a standard Wiener process under P. In particular, $X_0 = 0$. The continuous martingale part of $\frac{1}{2}X^2$ is $\{\frac{1}{2}(X_t^2 - t)\}$ with quadratic characteristic

$$\beta_t = \int_0^t X_s^2 ds.$$

Hence the natural exponential family generated by $\frac{1}{2}X^2$ is given by

$$Z_t(\theta) = \exp\left[\theta\tfrac{1}{2}(X_t^2 - t) - \tfrac{1}{2}\theta^2 \int_0^t X_s^2 ds\right]$$

with parameter set $\Theta = \mathbb{R}$. Under P_θ the process X is an Ornstein-Uhlenbeck process with drift θx. □

Comparison of (11.2.1) to the construction of the natural exponential family generated by a Lévy process in Chapter 2 reveals that (11.2.1) is not the most general way of embedding a non-continuous semimartingale in an exponential family of semimartingales. Let us therefore define $Z(\theta)$ in another way.

Let (α, β, ν) denote the local characteristics of X under P, and define U by (11.1.11). The process

$$B_t(\theta) = \int_{E^d} \{e^{\theta^T x} - 1 - \theta^T U(x)\}\nu([0,t] \times dx) \qquad (11.2.16)$$

is well-defined and finite for all $t > 0$ if θ belongs to the set

$$\Theta_0 = \left\{\theta \in \mathbb{R}^d : \int_{\{|x|>1\}} e^{\theta^T x}\nu([0,t] \times dx) < \infty, \quad t > 0, \quad P\text{-a.s.}\right\}. \qquad (11.2.17)$$

This is because near zero the integrand in (11.2.16) behaves like $|x|^2$, and the random measure ν satisfies $(|x|^2 \wedge 1) * \nu \in \mathcal{A}_{\mathrm{loc}}(P)$. For $\theta \in \Theta_0$ we can therefore define $Z(\theta)$ by

$$Z_t(\theta) = \exp[\theta^T(X_t - X_0) - \tfrac{1}{2}\theta^T\beta_t\theta - \log\ \mathcal{E}(A(\theta))_t], \qquad (11.2.18)$$

where $\mathcal{E}(A(\theta))$ denotes the Doléan-Dade exponential of the process

$$A_t(\theta) = \theta^T\alpha_t + B_t(\theta). \qquad (11.2.19)$$

Since $A(\theta)$ has finite variation,

$$\mathcal{E}(A(\theta))_t = \exp[A_t(\theta)] \prod_{s \le t} [1 + \Delta A_s(\theta)]e^{-\Delta A_s(\theta)}. \qquad (11.2.20)$$

The local characteristics satisfy

$$\Delta\alpha_t = \int_{E^d} U(x)\nu(\{t\} \times dx), \qquad (11.2.21)$$

where the integral is finite, and since

$$\Delta B_t(\theta) = \int_{E^d} [e^{\theta^T x} - 1 - \theta^T U(x)]\nu(\{t\} \times dx), \tag{11.2.22}$$

it follows that

$$\Delta A_t(\theta) = \int_{E^d} [e^{\theta^T x} - 1]\nu(\{t\} \times dx) > -1. \tag{11.2.23}$$

Therefore, $\mathcal{E}(A(\theta))_t > 0$, and (11.2.18) is well-defined. By substituting (11.2.20) into (11.2.18) we obtain

$$\begin{aligned} Z_t(\theta) &= \exp\Bigg\{\theta^T(X_t - X_0 - \alpha_t) - \tfrac{1}{2}\theta^T\beta_t\theta - B_t(\theta) \\ &\qquad - \sum_{s \le t}[\log(1 + \Delta A_t(\theta)) - \Delta A_t(\theta)]\Bigg\}. \end{aligned} \tag{11.2.24}$$

The process $Z(\theta)$ is a local martignale for $\theta \in \Theta_0$; see Exercise 11.5 or Liptser and Shiryaev (1989, Chapter 4). The set

$$\Theta = \{\theta \in \Theta_0 : E(Z_t(\theta)) = 1, \;\; t \ge 0\}, \tag{11.2.25}$$

is exactly the set of values of θ for which $Z(\theta)$ is a martingale. This is because a non-negative local martingale by Fatou's lemma is a supermartingale. Thus the class $\{P_\theta^{(t)} : t \ge 0\}$ of probability measures on $\mathcal{F}_t$ defined by

$$\frac{dP_\theta^{(t)}}{dP^t} = Z_t(\theta) \tag{11.2.26}$$

is consistent for every $\theta \in \Theta$. Under weak regularity conditions on the stochastic basis, see Ikeda and Watanabe (1981), we can for each $\theta \in \Theta$ find a probability measure P_θ on $\mathcal{F}$ such that $P_\theta^t = P_\theta^{(t)}$. If $Z_t(\theta)$ is of exponential form, i.e., of the form (3.1.1), for every $t > 0$, we call $\mathcal{P} = \{P_\theta : \theta \in \Theta\}$ the *natural exponential family* generated by X, and we call Θ the natural parameter set.

Obviously, $Z_t(\theta)$ is not always of exponential form. In the particular case where ν does not depend on ω, $B_t(\theta)$ and $\Delta A_t(\theta)$ are deterministic functions of θ, cf. (11.2.23), and thus $\mathcal{P}$ is an exponential family. We shall find other examples.

Let us consider the case where X is *quasi-left-continuous* in more detail. In this case $\nu(\{t\} \times dx) = 0$ for all $t \ge 0$, so by (11.2.23) $\Delta A_t(\theta) = 0$, $t \ge 0$, and

$$Z_t(\theta) = \exp[\theta^T(X_t - X_0 - \alpha_t) - \tfrac{1}{2}\theta^T\beta_t\theta - B_t(\theta)]. \tag{11.2.27}$$

If we assume that ν has the particular form

$$\nu(\omega; dt, dx) = \tilde{\nu}(\omega; dt)m(dx), \tag{11.2.28}$$

where m is a non-random measure on $(E^d, \mathcal{B}(E^d))$, we have

$$Z_t(\theta) = \exp[\theta^T(X_t - X_0 - \alpha_t) - \tfrac{1}{2}\theta^T\beta_t\theta - \kappa(\theta)S_t], \tag{11.2.29}$$

where

$$\kappa(\theta) = \int_{E^d} [e^{\theta^T x} - 1 - \theta^T U(x)]m(dx) \tag{11.2.30}$$

and

$$S_t(\omega) = \tilde{\nu}(\omega; [0, t]). \tag{11.2.31}$$

We see that in this case $\mathcal{P}$ is an exponential family. Note that the process S is continuous and increasing. In this situation Θ_0 has the simple form

$$\Theta_0 = \left\{ \theta \in \mathbb{R}^d : \int_{\{|x|>1\}} e^{\theta^T x} m(dx) < \infty \right\}. \tag{11.2.32}$$

Note that if X is a Lévy process under P, then (11.2.29) equals (2.1.23) and (11.2.32) equals the domain of the Laplace transform of X_1 under P. In situations where m is a probability measure, we find that

$$Z_t(\theta) = \exp[\theta^T X_t^c - \tfrac{1}{2}\theta^T\beta_t\theta + \theta^T X_t^d - (\varphi(\theta) - 1)S_t], \tag{11.2.33}$$

where X^c is the continuous martingale part of X under P,

$$X_t^d = X_t - X_0 - X_t^c - \alpha_t + S_t \int_{E^d} U(x)m(dx) \tag{11.2.34}$$

is the sum of the jumps of X, and φ is the Laplace transform of m

$$\varphi(\theta) = \int_{E^d} e^{\theta^T x} m(dx). \tag{11.2.35}$$

Note that if X is a continuous semimartingale, then (11.2.33) is identical to (11.2.1).

The problem of determining Θ is, of course, not easier here than it was for $Z(\theta)$ defined by (11.2.1). A discussion involving stochastic time transformations similar to the one above can be given here. If X^c and X^d are independent under P, we can use (11.2.8) or (11.2.9) to study the factor of $Z(\theta)$ involving only X^c. Note that if $\theta \in \Theta$, the process X^d can be written as $X_t^d = Y_{S_t}$, where Y is a compound Poisson process whose jump-distribution under P_θ has Laplace transform $s \mapsto \varphi(\theta + s)/\varphi(\theta)$.

Example 11.2.5 We consider the one-dimensional process $X = V + W$, where W is a standard Wiener process that is independent of the marked counting process

$$V_t = \sum_{i=1}^{N_t} Y_i, \quad V_0 = 0. \tag{11.2.36}$$

Here N is a counting process with intensity λ, and the Y_i's are independent of N and are mutually independent identically distributed with density f and Laplace transform φ. Thus $\beta_t = t$, and $\nu(\omega; dt, dx) = \lambda_t(\omega) f(x) dt dx$ satisfies (11.2.28). By (11.2.33),

$$Z_t(\theta) = \exp[\theta X_t - \tfrac{1}{2}\theta^2 t - (\varphi(\theta) - 1)\Lambda_t], \tag{11.2.37}$$

where

$$\Lambda_t = \int_0^t \lambda_s ds. \tag{11.2.38}$$

This is the likelihood function for the model where X is of the form above, but where W is a Wiener process with drift θ, N is a counting process with intensity $\varphi(\theta)\lambda_t$, and the distribution of Y_i has density

$$f_\theta(x) = \exp[\theta x - \log(\varphi(\theta))] f(x), \quad \theta \in \operatorname{dom} \varphi. \tag{11.2.39}$$

Thus $\Theta = \operatorname{dom} \varphi = \Theta_0$. Note that under P_θ, the third local characteristic of X is $\nu^\theta(dt, dx) = e^{\theta x} \nu(dt, dx)$. □

When X is quasi-left-continuous, both $B_t(\theta)$ given by (11.2.16) and the local characteristic α_t are continuous, so we see from (11.2.27) that

$$\Delta Z_t(\theta) = Z_{t-}(\theta)(e^{\theta^T \Delta X_t} - 1). \tag{11.2.40}$$

From Ito's formula it follows that

$$Z(\theta)^c = Z_-(\theta) \cdot (\theta^T X^c) \tag{11.2.41}$$

(we have here used the short dot-notation for stochastic integrals; see Appendix A), so that

$$\langle Z(\theta)^c, X^c \rangle = Z_- \cdot \beta\theta. \tag{11.2.42}$$

If we represent β in the form (A.9.6), i.e., $\beta = c \cdot A$ where A is an increasing predictable process and c is a matrix of predictable processes, it follows from Theorem A.10.1 that the local characteristics of X under P_θ are

$$\alpha^\theta = \alpha + \beta\theta + [U(e^{\theta^T x} - 1)] * \nu, \tag{11.2.43}$$

$$\beta^\theta = \beta, \tag{11.2.44}$$

$$\nu^\theta = e^{\theta^T x} \nu. \tag{11.2.45}$$

Example 11.2.6 Let us finally consider a d-dimensional discrete time stochastic process X. If $Y_i = X_i - X_{i-1}$ and if

$$K_i(\theta) = \log E(\exp(\theta^T Y_i)|\mathcal{F}_{i-1}) \tag{11.2.46}$$

is the conditional cumulant transform of Y_i given the σ-field $\mathcal{F}_{i-1} = \sigma(X_0, \cdots, X_{i-1})$, then the process $Z_n(\theta)$ given by (11.2.24) here takes the simple form

$$Z_n(\theta) = \exp\left(\theta^T(X_n - X_0) - \sum_{i=1}^{n} K_i(\theta)\right). \tag{11.2.47}$$

For $\theta \in \Theta = \cap_{i=1}^{\infty} \operatorname{dom} K_i$, a simple calculation shows that $Z(\theta)$ is a martingale with mean one. If we assume that the conditional cumulant transform has the product structure

$$K_i(\theta) = \kappa(\theta) V_i, \tag{11.2.48}$$

then

$$Z_n(\theta) = \exp[\theta^T(X_n - X_0) - \kappa(\theta) S_n], \tag{11.2.49}$$

where

$$S_n = \sum_{i=1}^{n} V_i, \tag{11.2.50}$$

has exponential structure. As above, we can find a class $\{P_\theta : \theta \in \Theta\}$ of probability measures on $\mathcal{F}$ such that $dP_\theta^n/dP^n = Z_n(\theta)$. The Radon-Nikodym derivative of the conditional distribution of Y_i given $\mathcal{F}_{i-1}$ under P_θ with respect to that under P is

$$\frac{dP_\theta.Y_i}{dP.Y_i} = \exp[\theta^T Y_i - K_i(\theta)], \tag{11.2.51}$$

so the natural exponential family generated by X is a model of the type discussed in Section 3.3. If X is a Markov process under P, it is also Markovian under P_θ. The decomposition

$$X_n = X_0 + M_n(\theta) + \dot{\kappa}(\theta) S_n, \tag{11.2.52}$$

where

$$M_n(\theta) = X_n - X_0 - \dot{\kappa}(\theta) S_n \tag{11.2.53}$$

is a zero-mean martingale under P_θ, shows how the drift of X changes with θ. □

It should be noted that the natural exponential family generated by a semimartingale is of the form (7.1.7). Note also that a model of the type (5.1.1) considered in Chapter 5 is the natural exponential family generated by the process A.

11.3 Exponential families with a time-continuous likelihood function

We begin this section by discussing conditions under which the likelihood function of a statistical semimartingale model is time-continuous. Let X be a d-dimensional semimartingale defined on the filtered probability space $(\Omega, \mathcal{F}, \{\mathcal{F}_t\}, P)$ and let (α, β, ν) be the local characteristics of X under P. As usual, we write β in the form (A.9.6), i.e., $\beta = c \cdot A$, where A is an increasing predictable process. Suppose a class of probability measures $\{P_\theta : \theta \in \Theta\}$, $\Theta \subseteq \mathbb{R}^k$, is given on $\mathcal{F}$ such that

$$P_\theta^t \ll P^t, \ \ \theta \in \Theta, \ \ t \geq 0,$$

where P_θ^t denotes the restriction of P_θ to $\mathcal{F}_t$. Then by Theorem A.10.1 the process X is also a semimartingale under P_θ, and the local characteristics of X under P_θ are given by

$$\nu^\theta = Y^\theta \nu, \ \ \beta^\theta = \beta, \tag{11.3.1}$$

$$\alpha^\theta = \alpha + (cz^\theta) \cdot A + [U(Y^\theta - 1)] * \nu \tag{11.3.2}$$

for some predictable random function $Y^\theta : \mathbb{R}_+ \times E^d \mapsto \mathbb{R}_+$ and some predictable process z^θ.

Under the conditions of Theorem A.10.2 or Theorem A.10.4, the likelihood function corresponding to observation of X in $[0, t]$ is given by

$$L_t(\theta) = \exp\{N_t^\theta - \tfrac{1}{2}[(z^\theta)^T \cdot cz^\theta] \cdot A_t\} \prod_{s \leq t} \{(1 + \Delta N_s^\theta) \exp(-\Delta N_s^\theta)\}, \tag{11.3.3}$$

where

$$N^\theta = z^\theta \cdot X^c + W^\theta * (\mu - \nu) \tag{11.3.4}$$

and

$$\begin{aligned} W^\theta(s, x) &= Y^\theta(s, x) - 1 \\ &+ 1_{\{\nu(\{s\} \times E^d) \neq 1\}} (1 - \nu(\{s\} \times E^d))^{-1} \int_E (Y^\theta(s, x) - 1) \nu(\{s\}, dx). \end{aligned} \tag{11.3.5}$$

Here X^c is, as usual, the continuous martingale part of X under P.

Since β is continuous, $L_t(\theta)$ is a continuous function of t if and only if N_t^θ is continuous. We see that a necessary and sufficient condition that the likelihood function is time-continuous is that $Y^\theta = 1$, i.e., that the jump mechanism is the same for all $\theta \in \Theta$ ($\nu^\theta = \nu$). Under this condition the local characteristics of X under P_θ are $(\alpha^\theta, \beta, \nu)$, where

$$\alpha^\theta = \alpha + (cz^\theta) \cdot A \tag{11.3.6}$$

and the likelihood function is

$$L_t(\theta) = \exp\{z^\theta \cdot X_t^c - \tfrac{1}{2}[(z^\theta)^T c z^\theta] \cdot A_t\}. \tag{11.3.7}$$

Note that only the drift α^θ depends on the parameter θ. The model considered in Example 11.2.2 is an example of a model where the observed process is discontinuous while the likelihood function is time-continuous.

In the following we will derive the structure of a time-homogeneous exponential family with a time-continuous likelihood function. Let $\mathcal{P} = \{P_\theta : \theta \in \Theta\}$, $\Theta \subseteq \mathbb{R}^k$, be a time-homogeneous exponential family on the filtered space $(\Omega, \mathcal{F}, \{\mathcal{F}_t\})$ with an exponential representation

$$L_t(\theta) = \frac{dP_\theta^t}{dP^t} = \exp[\gamma(\theta)^T B_t - \phi_t(\theta)], \quad t \geq 0, \quad \theta \in \Theta. \tag{11.3.8}$$

As usual, we assume that $P_\theta^0 = P^0$, so that $L_0(\theta) = 1$, and we can assume that $B_0 = \phi_0(\theta) = 0$.

Theorem 11.3.1 *Suppose the function $t \mapsto \phi_t(\theta)$ is continuous and of locally finite variation for all $\theta \in \Theta$, and that the canonical process B is a continuous semimartingale. Then there exists an exponential representation of the form*

$$L_t(\theta) = \exp[\gamma(\theta)^T B_t - \tfrac{1}{2}\gamma(\theta)^T \langle \tilde{B} \rangle_t \gamma(\theta)], \quad t \geq 0, \ \theta \in \Theta, \tag{11.3.9}$$

where $\tilde{B}$ is the continuous local martingale part of B under P, and where $\langle \tilde{B} \rangle$ denotes the $m \times m$-matrix $\{\langle \tilde{B}^{(i)}, \tilde{B}^{(j)} \rangle\}$.

Remark: Thus, $\mathcal{P}$ is the natural exponential family generated by the continuous semimartingale $\tilde{B}$ or a subfamily thereof. □

Proof: For fixed $\theta \in \Theta$ consider the process $l_t(\theta) = \log L_t(\theta)$. Under the dominating measure P, the process $L_t(\theta)$ is a continuous (and hence locally square integrable) martingale. By Ito's formula,

$$l_t(\theta) = \int_0^t L_s(\theta)^{-1} dL_s(\theta) - \tfrac{1}{2} \int_0^t L_s(\theta)^{-2} d\langle L(\theta) \rangle_s, \tag{11.3.10}$$

where the process

$$M_t(\theta) = \int_0^t L_s(\theta)^{-1} dL_s(\theta) \tag{11.3.11}$$

is a continuous locally square integrable martingale with quadratic characteristic

$$\langle M(\theta) \rangle_t = \int_0^t L_s(\theta)^{-2} d\langle L(\theta) \rangle_s. \tag{11.3.12}$$

Hence (11.3.10) can be written

$$\gamma(\theta)^T B_t - \phi_t(\theta) = M_t(\theta) - \tfrac{1}{2}\langle M(\theta) \rangle_t, \tag{11.3.13}$$

where we have used (11.3.8) on the lefthand side.

Since $B^{(i)}$ is a continuous semimartingale, it has a decomposition of the form

$$B_t^{(i)} = \tilde{B}_t^{(i)} + \bar{B}_t^{(i)}, \quad i = 1, \cdots, m, \tag{11.3.14}$$

where $\tilde{B}^{(i)}$ is a continuous local martingale and where $\bar{B}^{(i)}$ is a continuous process of locally finite variation. We can assume that $\tilde{B}_0 = \bar{B}_0 = 0$. By combining (11.3.13) and (11.3.14) we get that

$$M_t(\theta) - \gamma(\theta)^T \tilde{B}_t = \gamma(\theta)^T \bar{B}_t + \tfrac{1}{2}\langle M(\theta)\rangle_t - \phi_t(\theta). \tag{11.3.15}$$

The lefthand side of (11.3.15) is a continuous local martingale, while the righthand side is of locally finite variation. Therefore, the local martingale on the lefthand side must be constant. Since $M_0(\theta) = \tilde{B}_0 = 0$, we find that

$$M_t(\theta) = \gamma(\theta)^T \tilde{B}_t \tag{11.3.16}$$

and

$$\gamma(\theta)^T \bar{B}_t = \phi_t(\theta) - \tfrac{1}{2}\langle M(\theta)\rangle_t. \tag{11.3.17}$$

From (11.3.16) it follows that

$$\langle M(\theta)\rangle_t = \gamma(\theta)^T \langle \tilde{B}\rangle_t \gamma(\theta). \tag{11.3.18}$$

The theorem follows by inserting (11.3.16) and (11.3.18) in (11.3.13). □

Corollary 11.3.2 *Let X be a semimartingale under P with local characteristics (α, β, ν). Then the local characteristics of X under P_θ are $(\alpha^\theta, \beta, \nu)$, where*

$$\begin{aligned} \alpha_t^\theta &= \alpha_t + \sum_{i=1}^m \gamma_i(\theta)\langle X^c, \tilde{B}^{(i)}\rangle_t \\ &= \alpha_t + \langle X^c, \gamma(\theta)^T \tilde{B}\rangle_t. \end{aligned} \tag{11.3.19}$$

Remark: Corollary 11.3.2 holds in particular when the filtration $\{\mathcal{F}_t\}$ is generated by observation of the semimartingale X. Note that for $X = B$ we obtain the result (11.2.12). □

Proof: The result follows from Theorem A.10.1. Since $L_t(\theta)$ is continuous, it follows from (A.10.6) that the random function Y in that theorem is equal to one, implying that the jump mechanism is the same under P_θ as under P, i.e., $\nu^\theta = \nu$. By Ito's formula we find that

$$\langle X^c, L(\theta)\rangle_t = \sum_{i=1}^m \theta_i \int_0^t z_s(\theta) d\langle X^c, \tilde{B}^{(i)}\rangle_t,$$

which implies (11.3.19). □

The following example shows that Theorem 11.3.1 does not hold for exponential families that are not time-homogeneous.

Example 11.3.3 Consider the class of diffusion processes given by

$$dX_t = h(\theta, t)X_t dt + dW_t, \quad X_0 = 0, \tag{11.3.20}$$

where $\theta = (\theta_1, \theta_2)$, $\theta_2 \geq 0$, $-\theta_2 \leq \theta_1 < \theta_2 \coth(\theta_2)$, and

$$h(\theta, t) = \theta_1 + \frac{\theta_1^2 - \theta_2^2}{\theta_2 \coth\{(1-t)\theta_2\} - \theta_1}. \tag{11.3.21}$$

The function h tends to θ_2 as $t \to \infty$. For $\theta_2 = |\theta_1|$ the process X is an ordinary Ornstein-Uhlenbeck process. This model was considered in Section 3.5 and Section 7.6.

The likelihood function corresponding to observing X in $[0, t]$ is

$$L_t(\theta) = \exp\left(\tfrac{1}{2}\left[h(\theta, t)X_t^2 - \theta_2^2 \int_0^t X_s^2 ds + m(\theta, t)\right]\right), \tag{11.3.22}$$

where m is defined by (3.5.13).

This exponential family is not time-homogeneous and does not have a representation of the form (11.3.9). Of course, it has a representation of the form (11.3.13), viz., $\exp(M_t(\theta) - \frac{1}{2}\langle M(\theta)\rangle_t)$. Specifically, we see by Ito's formula that

$$M_t(\theta) = \int_0^t h(\theta, s)X_s dX_s \tag{11.3.23}$$

and

$$\langle M(\theta)\rangle_t = \int_0^t [(\dot{h}(\theta, s) - \theta_2^2)X_s^2 + h(\theta, s)]ds + m(\theta, t), \tag{11.3.24}$$

where $\dot{h}(\theta, s) = \frac{\partial}{\partial s}h(\theta, s)$. □

The exponential representation (11.3.9) need not be minimal. First of all, some of the coordinates of $\tilde{B}$ might be zero. Without loss of generality we can assume that $\tilde{B}^{(i)} = 0$, $i = m^* + 1, \cdots, m$, while the first m^* coordinates are not identically equal to zero. Note that necessarily $m^* > 0$. Note also that the sum in (11.3.19) is from one to m^* only. Define $A = (\tilde{B}^{(1)}, \cdots, \tilde{B}^{(m^*)})^T$ and $\tilde{\gamma}(\theta) = (\gamma_1(\theta), \cdots, \gamma_{m^*}(\theta))^T$. Then we have the exponential representation

$$L_t(\theta) = \exp[\tilde{\gamma}(\theta)^T A_t - \tfrac{1}{2}\tilde{\gamma}(\theta)^T \langle A\rangle_t \tilde{\gamma}(\theta)]. \tag{11.3.25}$$

This representation might still not be minimal because some elements of A or of $\langle A\rangle$ might be affinely dependent. However, the representation (11.3.25) is a very natural one, and we call the parametrization $\tilde{\gamma}(\theta)$ the *natural parametrization.*

In the following we study the connection between (11.3.25) and the original representation (11.3.8). From (11.3.17) it is obvious that

$$\phi_t(\theta) = \tfrac{1}{2}\tilde{\gamma}(\theta)^T \langle A\rangle_t \tilde{\gamma}(\theta) + \gamma(\theta)^T \bar{B}_t. \tag{11.3.26}$$

Since $\phi_t(\theta)$ is deterministic, random variables in the sums $\frac{1}{2}\tilde{\gamma}(\theta)^T\langle A\rangle_t\tilde{\gamma}(\theta)$ and $\gamma(\theta)^T\bar{B}_t$ must cancel each other. This can be expressed in a precise way. To do this, we assume that $\gamma_1(\theta),\cdots,\gamma_m(\theta)$ are linearly independent functions of θ. This is not a restriction, because we can always find a representation of the form (11.3.8) with this property. The linear independence implies that we can find $\theta_1,\cdots,\theta_m$ such that the matrix $\Gamma=(\gamma(\theta_1),\cdots,\gamma(\theta_m))^T$ is invertible. From (11.3.17) we find that

$$\bar{B}_t=\varphi_t-\sum_{i=1}^{m^*}\sum_{j=1}^{i}\Delta(i,j)\langle A^{(i)},A^{(j)}\rangle_t, \tag{11.3.27}$$

where $\varphi_t=\Gamma^{-1}(\phi_t(\theta_1),\cdots,\phi_t(\theta_m))^T$ and

$$\Delta(i,j)=\begin{cases}\Gamma^{-1}(\gamma_i(\theta_1)\gamma_j(\theta_1),\cdots,\gamma_i(\theta_m)\gamma_j(\theta_m))^T & \text{for } i\neq j\\ \frac{1}{2}\Gamma^{-1}(\gamma_i(\theta_1)^2,\cdots,\gamma_i(\theta_m)^2)^T & \text{for } i=j.\end{cases} \tag{11.3.28}$$

Some components of $\langle A\rangle$ might be non-random processes. This happens, for instance, if some coordinates of A have independent increments. Let I_1 denote the subset of $I=\{(i,j):1\le j\le i\le m^*\}$ for which $\langle A^{(i)},A^{(j)}\rangle_t$ is non-random for all $t\ge 0$, and define $I_2=I\backslash I_1$ and $J=\{(i,i):1\le i\le m^*\}$. Using (11.3.27) we can then rewrite (11.3.26) as

$$\begin{aligned}\phi_t(\theta)-\gamma(\theta)^T\varphi_t-&\sum_{(i,i)\in I_1}[\tfrac{1}{2}\gamma_i(\theta)^2-\gamma(\theta)^T\Delta(i,i)]\langle A^{(i)}\rangle_t \qquad (11.3.29)\\ -&\sum_{(i,j)\in I_1\backslash J}[\gamma_i(\theta)\gamma_j(\theta)-\gamma(\theta)^T\Delta(i,j)]\langle A^{(i)},A^{(j)}\rangle_t\\ =&\sum_{(i,i)\in I_2}[\tfrac{1}{2}\gamma_i(\theta)^2-\gamma(\theta)^T\Delta(i,i)]\langle A^{(i)}\rangle_t\\ &+\sum_{(i,j)\in I_2\backslash J}[\gamma_i(\theta)\gamma_j(\theta)-\gamma(\theta)^T\Delta(i,j)]\langle A^{(i)},A^{(j)}\rangle_t.\end{aligned}$$

In order to continue, we should now collect affinely dependent terms on the righthand side. However, to avoid this trivial complication, we assume that for some fixed $t>0$, all random variables $\langle A^{(i)},A^{(j)}\rangle_t$, $(i,j)\in I_2$, are affinely independent. Since the lefthand side of (11.3.29) is deterministic, it follows that for all $\theta\in\Theta$,

$$\gamma_i(\theta)\gamma_j(\theta)=\begin{cases}\gamma(\theta)^T\Delta(i,j) & \text{for } (i,j)\in I_2\backslash J\\ 2\gamma(\theta)^T\Delta(i,i) & \text{for } (i,j)\in I_2\cap J,\end{cases} \tag{11.3.30}$$

and that

$$\phi_t(\theta)=\gamma(\theta)^T\left\{\varphi_t-\sum_{(i,j)\in I_1}\Delta(i,j)\langle A^{(i)},A^{(j)}\rangle_t\right\} \tag{11.3.31}$$

$$+\tfrac{1}{2}\sum_{(i,i)\in I_1}\gamma_i(\theta)^2\langle A^{(i)}\rangle_t+\sum_{(i,j)\in I_1\backslash J}\gamma_i(\theta)\gamma_j(\theta)\langle A^{(i)},A^{(j)}\rangle_t.$$

By inserting (11.3.30) in (11.3.27) we find that

$$\begin{aligned}\gamma^T(\theta)\bar{B}_t &= \gamma^T(\theta)\left\{\varphi_t-\sum_{(i,j)\in I_1}\Delta(i,j)\langle A^{(i)},A^{(j)}\rangle_t\right\} \qquad (11.3.32)\\ &\quad -\tfrac{1}{2}\sum_{(i,i)\in I_2}\gamma_i(\theta)^2\langle A^{(i)}\rangle_t-\sum_{(i,j)\in I_2\backslash J}\gamma_i(\theta)\gamma_j(\theta)\langle A^{(i)},A^{(j)}\rangle_t.\end{aligned}$$

Example 11.3.4 Suppose the representation (11.3.8) is minimal, that B is two-dimensional, and that $B^{(2)}$ is of locally finite variation. Then $A=B_t^{(1)}$ ($m^*=1$), and there exist deterministic functions $\varphi_t^{(i)}$ and constants $\Delta^{(i)}$ ($i=1,2$) such that

$$\begin{aligned}\bar{B}_t^{(i)} &= \varphi_t^{(i)}-\Delta^{(2)}\langle A\rangle_t,\\ \phi_t(\theta) &= \gamma_1(\theta)\varphi_t^{(1)}+\gamma_2(\theta)\varphi_t^{(2)},\end{aligned}$$

and

$$\Delta^{(2)}\gamma_2(\theta)=\tfrac{1}{2}\gamma_1(\theta)^2-\Delta^{(1)}\gamma_1(\theta);$$

cf. (11.3.27), (11.3.30), and (11.3.31). Note that the assumption that (11.3.8) is minimal implies that $\langle A\rangle_t$ is random for some $t>0$, and that $\Delta^{(2)}\neq 0$ (otherwise $\gamma_1(\theta)$ is constant). The most compact representation is obtained for $\varphi_t^{(1)}=\varphi_t^{(2)}=\Delta^{(1)}=0$ and $\Delta^{(2)}=1$. Then $\gamma_2(\theta)=\frac{1}{2}\gamma_1(\theta)^2$, $\bar{B}^{(1)}=0$, $\bar{B}^{(2)}=-\langle A\rangle$, and $\phi_t(\theta)=0$.

These conclusions do not hold without the assumption that $B^{(2)}$ is of locally finite variation. For instance, if X under P_θ solves the equation

$$dX_t=(\theta+\theta^3\sqrt{t})dt+dW_t,\;\; X_0=0,$$

then the likelihood function is

$$\frac{dP_\theta^t}{dP_0^t}=\exp\left(\theta X_t+\theta^3\int_0^t\sqrt{s}dX_s-\tfrac{1}{2}[\theta^2t+\tfrac{4}{3}\theta^4t^{3/2}+\tfrac{1}{2}\theta^6t^2]\right).$$

Here $B^{(2)}$ is not of locally finite variation. In fact, $\bar{B}^{(1)}=\bar{B}^{(2)}=0$, and obviously $\gamma_2(\theta)=\frac{1}{2}\gamma_1(\theta)^2$ does not hold. □

11.4 Other types of exponential families of semimartingales

In this section we assume that X is a d-dimensional quasi-left-continuous semimartingale with local characteristics (α,β,ν) under P. We assume that

$\beta = c \cdot A$, where A is an increasing predictable process and c is a symmetric non-negative definite $d \times d$-matrix of predictable processes.

Let $z^{(i)}$, $i = 1, \cdots, n$, be d-dimensional predictable processes satisfying

$$((z^{(i)})^T c z^{(j)}) \cdot A \in \mathcal{A}_{\text{loc}}(P), \quad i, j = 1, \cdots, n, \tag{11.4.1}$$

and let $\{Y_\kappa : \kappa \in K\}$, $K \subseteq \mathbb{R}^k$, be a family of strictly positive predictable random functions from $\mathbb{R}_+ \times E^d$ into $\mathbb{R}_+$ such that

$$y^\kappa = (1 - \sqrt{Y_\kappa})^2 * \nu \in \mathcal{A}^+_{\text{loc}}(P), \quad \kappa \in K, \tag{11.4.2}$$

and $Y_\kappa(\omega; s, 0) \equiv 1$, P-a.s.

Suppose $\{P_{\theta,\kappa} : (\theta, \kappa) \in \Theta \times K\}$ is a family of probability measures on $(\Omega, \mathcal{F})$ such that X is a semimartingale with local characteristics $(\alpha^{\theta,\kappa}, \beta, \nu^\kappa)$ under $P_{\theta,\kappa}$, where

$$\nu^\kappa = Y_\kappa \nu, \quad \alpha^{\theta,\kappa} = \alpha + \sum_{i=1}^n \theta_i (c z^{(i)}) \cdot A + [U(Y_\kappa - 1)] * \nu. \tag{11.4.3}$$

As usual, U is defined by (11.1.11). Then it follows from Theorem A.10.4 that the condition

$$P_{\theta,\kappa}\left(\sum_{i=1}^n |((z^{(i)})^T c z^{(j)}) \cdot A_t| + y_t^\kappa < \infty\right) = 1, \quad t > 0, \tag{11.4.4}$$

for all $(\theta, \kappa) \in \Theta \times K$, implies that $P^t_{\theta,\kappa} << P^t$ for all $t > 0$ and $(\theta, \kappa) \in \Theta \times K$, and that

$$\begin{aligned} L_t(\theta, \kappa) = \frac{dP^t_{\theta,\kappa}}{dP^t} &= \exp\{\theta^T z \cdot X^c - \tfrac{1}{2}\theta^T (z c z^T) \cdot A\theta \\ &\quad + (Y_\kappa - 1) * (\mu - \nu)_t + [\log(Y_\kappa) - Y_\kappa + 1] * \mu_t\}, \end{aligned} \tag{11.4.5}$$

where z is the $n \times d$-matrix of predictable functions the ith row of which is $z^{(i)}$.

In order to obtain a likelihood function of exponential structure, we assume that Y_κ is non-random and of the form

$$Y_\kappa(\omega; s, x) = \exp\left(\sum_{i=1}^k \kappa_i \varphi_i(s, x)\right) = \exp(\kappa^T \varphi). \tag{11.4.6}$$

Here φ_i, $i = 1, \cdots, k$, are nonrandom functions satisfying $\varphi_i(s, 0) = 0$, $i = 1, \cdots, k$.

In the rest of this section we shall work under the following condition.

Condition 11.4.1 *For all $t > 0$ it holds that*

$$
\begin{aligned}
|\varphi_i|1_A * (\mu - \nu)_t &< \infty \ P\text{-a.s.}, \ i = 1, \cdots, k, &(11.4.7)\\
|e^{\kappa^T\varphi} - \kappa^T\varphi - 1|1_A * \nu_t &< \infty \ P\text{-a.s.}, \ \kappa \in K, &(11.4.8)\\
e^{\kappa^T\varphi}1_{A^c} * \nu_t &< \infty \ P\text{-a.s.}, \ \kappa \in K, &(11.4.9)
\end{aligned}
$$

where $A = [-1, 1]^d$.

Note that if φ_i is continuous in x for every $i = 1, \ldots, k$, then (11.4.8) can be replaced by

$$|\varphi_i\varphi_j|1_A * \nu_t < \infty \ P\text{-a.s.}, \ \kappa \in K. \tag{11.4.10}$$

This is because $|e^{\kappa^T\theta} - \kappa^T\varphi - 1| \le C(\kappa^T\varphi)^2$ for x in a neighbourhood of zero and for some constant C.

Under Condition 11.4.1 we can rewrite the part of the likelihood function (11.4.5) related to the jumps as follows.

$$
\begin{aligned}
&(Y_\kappa - 1) * (\mu - \nu)_t + [\log(Y_\kappa) - Y_\kappa + 1] * \mu_t &(11.4.11)\\
&= (e^{\kappa^T\varphi} - 1) * (\mu - \nu)_t - (e^{\kappa^T\varphi} - \kappa^T\varphi - 1) * \mu_t\\
&= \kappa^T B_t - g(\kappa, t),
\end{aligned}
$$

where $B_t = (B_t^{(1)}, \cdots, B_t^{(k)})^T$,

$$B_t^{(i)} = \varphi_i 1_A * (\mu - \nu)_t + \varphi_i 1_{A^c} * \mu_t, \ \ i = 1, \cdots, k,$$

and

$$g(\kappa, t) = (e^{\kappa^T\varphi} - 1 - \kappa^T\varphi)1_A * \nu_t + (e^{\kappa^T\varphi} - 1)1_{A^c} * \nu_t.$$

To see this, note that (11.4.9) implies that

$$|e^{\kappa^T\varphi} - 1|1_{A^c} * \nu_t < \infty,$$

because ν, being a local characteristic, satisfies $(|x|^2 \wedge 1) * \nu < \infty$. Moreover, by (11.4.8) and (11.4.7),

$$
\begin{aligned}
&(e^{\kappa^T\varphi} - 1)1_A * (\mu - \nu)_t - (e^{\kappa^T\varphi} - 1 - \kappa^T\varphi)1_A * \mu_t\\
&= \kappa^T(\varphi 1_A) * (\mu \dot{-} \nu)_t - (e^{\kappa^T\varphi} - 1 - \kappa^T\varphi)1_A * \nu_t.
\end{aligned}
$$

Finally, we have used that X has only finitely many jumps with $\Delta X_s \in A^c$ in a finite time interval.

We now consider five types of exponential families of semimartingales.

(I) If $g(\kappa, t)$ is non-random, we obtain an exponential family. This obviously happens when ν is non-random. An example is the natural exponential family generated by a Lévy process considered in Chapter 2. We shall study more general examples in the next section.

(II) We shall now impose the following further condition:

Condition 11.4.2 *For all $t > 0$ it holds that*

$$|\varphi_i| 1_A * \nu_t < \infty \;\; P\text{-a.s.}, \;\; i = 1, \cdots, k. \tag{11.4.12}$$

Under this condition, the part of the likelihood function (11.4.5) related to the jumps takes the form

$$\begin{aligned}(Y_\kappa - 1) * (\mu - \nu)_t + [\log(Y_\kappa) - Y_\kappa + 1] * \mu_t \\ = \kappa^T(\varphi * \mu_t) - (e^{\kappa^T\varphi} - 1) * \nu_t.\end{aligned} \tag{11.4.13}$$

When ν is random, the last term depends on κ as well as ω, so we have not in general obtained an exponential family. To do this we make the further assumptions that

$$\nu(\omega; ds, dx) = F(dx)\tilde{\nu}(\omega; ds) \tag{11.4.14}$$

and that the functions φ_i, $i = 1, \ldots, k$, depend on x only. Then

$$(e^{\kappa^T\varphi} - 1) * \nu = f(\kappa) \int_0^t \tilde{\nu}(ds), \tag{11.4.15}$$

where

$$f(\kappa) = \int_{E^d} (e^{\kappa^T\varphi(x)} - 1) F(dx), \tag{11.4.16}$$

so that a time-homogeneous exponential family is obtained. Note that the part of the likelihood function related to the jumps is of the type studied in Sections 5.1 and 5.2. An example of a model for which (11.4.14) is satisfied is a marked counting process with a mark distribution that is independent of the past; cf. Section 3.4.

(III) Another condition on ν and φ for which an exponential family is obtained is that

$$\nu(\omega; ds, dx) = F(ds)\tilde{\nu}(\omega, dx), \tag{11.4.17}$$

and that φ depends on s only. In this case

$$(e^{\kappa^T\varphi} - 1) * \nu = c(\kappa, t) \int_{E^d} \tilde{\nu}(dx), \tag{11.4.18}$$

where

$$c(\kappa, t) = \int_0^t e^{\kappa^T\varphi(s)} F(ds), \tag{11.4.19}$$

so an exponential family that is not time-homogeneous is obtained. Since ν is predictable, it follows that $\tilde{\nu}$ must be $\mathcal{F}_0$-measurable. Therefore, this type of model is included in type (I) if we assume that $\mathcal{F}_0$ is trivial. An example is an exponential family of compound Poisson processes with jump size distribution depending on $\mathcal{F}_0$; cf. Section 3.4.

(IV) Instead of assuming that ν has a product structure, we can suppose that φ is constant. This is essentially the same as assuming that $k = 1$ and $\varphi_1 \equiv 1$. Under this assumption, (11.4.9) and (11.4.12) imply that

$$1 * \nu_t < \infty, \quad P\text{-a.s.}, \quad t > 0, \tag{11.4.20}$$

and we find

$$(e^{\kappa\varphi} - 1) * \nu_t = e^{\kappa}(1 * \nu_t) - 1 * \nu_t. \tag{11.4.21}$$

The statistic $\varphi_1 * \mu_t = 1 * \mu_t$ is the number of jumps made by X before time t, which by (11.4.20) is finite. Also in this case, the part of the likelihood function related to the jumps is of the type considered in the first sections of Chapter 5. An example is an exponential family of counting processes; see Section 3.4.

(V) Finally, we shall consider a more general form of Y_κ. Assume that

$$Y_\kappa(\omega, s, x) = \sum_{i=1}^{m} \exp\left(\sum_{j=1}^{k_i} \kappa_{ij}\varphi_{ij}(x, s)\right) 1_{A_i}(\omega, s, x), \tag{11.4.22}$$

where A_i, $i = 1, \cdots, m$, are subsets of $\Omega \times [0, \infty) \times E$ such that $A_i \in \tilde{\mathcal{P}}$ and $A_i \cap A_j = \emptyset$, $i \neq j$. Then the likelihood function is a product of factors each of which corresponds to a particular A_i. We can now make each of these factors an exponential family by imposing conditions on ν, φ_{ij}, and A_i of the types discussed above. It is rather obvious how to do this, so we shall not go into details. An example of this situation is provided by the class of finite state space Markov processes; see Section 3.4. Let the state space be $\{1, \cdots, m\}$, and let $\{\lambda_{ij}\}$ denote the intensity matrix, $\lambda_{ij} > 0$, $i \neq j$. Then

$$\nu(\omega, ds, dx) = \sum_{i,j(i \neq j)} 1_{\{i\}}(X_{s-}(\omega))\epsilon_{\{j-i\}}(dx)ds,$$

where ϵ_a is the Dirac measure in a. In (11.4.22), $k_i = m - 1$, $i = 1, \cdots, m$,

$$\begin{aligned} \kappa_{ij} &= \log(\lambda_{ij}), \quad i, j = 1, \cdots, m, \quad i \neq j, \\ \varphi_{ij}(x, s) &= 1_{\{j-i\}}(x), \quad i, j = 1, \cdots, m, \quad i \neq j, \end{aligned}$$

and

$$A_i = \{(\omega, s, x) : X_{s-}(\omega) = i, \ \ x \in \mathbf{Z} \cap [1 - i, n - i]\}, \quad i = 1, \cdots, m.$$

11.5 General exponential families of Lévy processes

In Chapter 2 we considered natural exponential families of Lévy processes. We shall now study more general exponential families of Lévy processes defined by means of results from the preceding section. Let X be a one-dimensional Lévy process defined on $(\Omega, \mathcal{F}, P)$ with $X_0 = 0$ and Lévy characteristics (δ, σ^2, ν).

We define a new probability measure $P_{\alpha,\theta}$, $\theta = (\theta_1, \cdots, \theta_k)$, $\alpha \in \mathbb{R}$, on $\mathcal{F}$ by requiring that under $P_{\alpha,\theta}$ the Lévy characteristics of X be

$$\nu_\theta(dx) = e^{\theta^T \varphi(x)} \nu(dx), \tag{11.5.1}$$

$$\sigma_\theta^2 = \sigma^2, \tag{11.5.2}$$

$$\delta_{\alpha,\theta} = \delta + \alpha\sigma^2 + \int_E \frac{x}{1+x^2}(e^{\theta^T \varphi(x)} - 1)\nu(dx), \tag{11.5.3}$$

where φ is a Borel function $\mathbb{R} \mapsto \mathbb{R}^k$. This can be done provided that the following conditions are satisfied:

$$\int_E \frac{x^2}{1+x^2} e^{\theta^T \varphi(x)} \nu(dx) < \infty, \tag{11.5.4}$$

$$\int_E \frac{x}{1+x^2}(e^{\theta^T \varphi(x)} - 1)\nu(dx) < \infty. \tag{11.5.5}$$

We shall, moreover, assume that

$$\int_E [e^{\frac{1}{2}\theta^T \varphi(x)} - 1]^2 \nu(dx) < \infty, \tag{11.5.6}$$

$$\int_{0<|x|\leq 1} \varphi_i(x)^2 \nu(dx) < \infty, \quad i = 1, \cdots, k, \tag{11.5.7}$$

$$\int_{0<|x|\leq 1} |e^{\theta^T \varphi(x)} - 1 - \theta^T \varphi(x)|\nu(dx) < \infty. \tag{11.5.8}$$

Condition (11.5.4) ensures that ν_θ is a Lévy measure and (11.5.5) that $\delta_{\alpha,\theta}$ is well-defined. Condition (11.5.6) implies (11.4.2) and ensures that the probability measures $P_{\alpha,\theta}$ and P are equivalent, while (11.5.7) implies (11.4.7). Condition (11.5.8) is identical to (11.4.8), and (11.4.9) follows from (11.5.4), so Condition 11.4.1 is satisfied, and it follows that we are in situation (I) of the preceding section. Thus, the Radon-Nikodym derivative of $P_{\alpha,\theta}$ with respect to P has exponential structure.

For a given function φ, let Θ_φ denote the subset of $\mathbb{R}^k$ for which conditions (11.5.4)–(11.5.8) are satisfied for all $\theta \in \Theta_\varphi$. The likelihood function for the model $\mathcal{P}_\varphi = \{P_{\alpha,\theta} : \alpha \in \mathbb{R}, \theta \in \Theta_\varphi\}$ is

$$L_t(\alpha, \theta) = \frac{dP_{\alpha,\theta}^t}{dP^t} \tag{11.5.9}$$

$$= \exp[\alpha X_t^c - \tfrac{1}{2}\alpha^2\sigma^2 t + \theta^T B_t - t\phi(\theta)].$$

Here X^c denotes the continuous martingale part of X under P and is given by

$$X_t^c = X_t - \delta t - U_t - \sum_{s \le t} \Delta X_s 1_{\{|\Delta X_s| > 1\}}, \tag{11.5.10}$$

where U_t is the P-almost sure limit as $n \to \infty$ of

$$U_n(t) = \sum_{s \le t} \Delta X_s 1_{\{1/n \le |\Delta X_s| \le 1\}} - t \int_{1/n}^{1} x \nu(dx),$$

which exists uniformly for t in compact intervals; see, e.g., Ito (1969, p. 1.7.4). Moreover,

$$B_t^{(i)} = V_t^{(i)} + \sum_{s \le t} \varphi_i(\Delta X_s) 1_{\{|\Delta X_s| > 1\}}, \tag{11.5.11}$$

where $V_t^{(i)}$ is the limit of

$$V_n^{(i)}(t) = \sum_{s \le t} \varphi_i(\Delta X_s) 1_{\{1/n \le |\Delta X_s| \le 1\}} - t \int_{1/n}^{1} \varphi_i(x) \nu(dx)$$

as $n \to \infty$, and

$$\begin{aligned} \phi(\theta) &= \int_{|x| \le 1} (e^{\theta^T \varphi(x)} - 1 - \theta^T \varphi(x)) \nu(dx) \\ &\quad + \int_{|x| > 1} (e^{\theta^T \varphi(x)} - 1) \nu(dx). \end{aligned}$$

We call $\mathcal{P}_\varphi$ the exponential family generated by φ and X.

Let us consider the conditions (11.5.4)–(11.5.8) more closely. First, assume that there exist an i and $\delta, \epsilon > 0$ such that $|\varphi_i(x)| > \delta$ for $0 < |x| < \epsilon$. Then (11.5.7) implies that ν must be a finite measure, and hence that X is a compound Poisson process under P as well as under $P_{\alpha,\theta}$. In this case the likelihood function (11.5.9) simplifies considerably. Specifically, we can set

$$B_t^{(i)} = \sum_{s \le t} \varphi_i(\Delta X_s) 1_{\{\Delta X_s \ne 0\}}$$

and

$$\phi(\theta) = \int (e^{\theta^T \varphi(x)} - 1) \nu(dx).$$

Next, suppose there exist $K, \epsilon, \delta > 0$ such that it holds for all i that $|\varphi_i(x)| > \delta$ for $0 < |x| < \epsilon$, and $|\varphi_i(x)| < K$ for $|x| \le 1$. Again, ν is necessarily finite, so that (11.5.7) and (11.5.8) are satisfied for all θ in $\mathbb{R}^k$. Moreover, (11.5.4) implies (11.5.5) and (11.5.6). For a φ of this form, we see that

$$\Theta_\varphi = \left\{ \theta \in \mathbb{R}^k : \int_E \frac{x^2}{1+x^2} e^{\theta^T \varphi(x)} \nu(dx) < \infty \right\}, \tag{11.5.12}$$

i.e., the set of θ's for which ν_θ is a Lévy measure. If $\varphi_i(0+) = \infty$ or $\varphi_i(0-) = \infty$ for some i, conditions (11.5.4)–(11.5.8) must of course be checked.

Finally, suppose $\varphi_i(0+) = \varphi_i(0-) = 0$ for all $i = 1, \cdots, k$, and that φ is bounded for $|x| \leq 1$. Let $\beta \in [0, 2]$ be chosen such that

$$\int_{0<|x|\leq 1} x^\beta \nu(dx) < \infty, \tag{11.5.13}$$

which can always be done; cf. (11.5.9). Then (11.5.7) is satisfied provided that

$$\varphi_i(x) = O(x^{\frac{1}{2}\beta}) \tag{11.5.14}$$

for all i. This and (11.5.4) imply (11.5.5)–(11.5.8). Thus, the parameter set Θ_φ is also given by (11.5.12) when φ satisfies (11.5.14). We can always let α vary freely in $\mathbb{R}$. The natural exponential family generated by X is obtained for $k = 1$, $\alpha = \theta$, and $\varphi(x) = x$. This choice of φ obviously satisfies (11.5.14) with $\beta = 2$.

Example 11.5.1 Let X be a symmetric stable process with index $\alpha \in (0, 2)$ under P. In this case $\nu(dx) = |x|^{-(\alpha+1)}dx$, $x \in \mathbb{R}\backslash\{0\}$. By the considerations above, the parameter space for the exponential family generated by $\varphi(x) = x$ and X is given by (11.5.12). Hence $\Theta_\varphi = \{0\}$. For $\varphi(x) = x^2$ we find by (11.5.12) that $\Theta_\varphi = (-\infty, 0]$. The likelihood function is

$$L_t(\theta) = \exp\left[\theta \sum_{s\leq t} (\Delta X_s)^2 - t\int_E (e^{\theta x^2} - 1)|x|^{-(\alpha+1)}dx\right].$$

Consider finally $\varphi(x) = x/(1 + x)$. Again (11.5.14) is satisfied with $\beta = 2$, and $\Theta_\varphi = \mathbb{R}$ by (11.5.12). For $\alpha < 1$ the likelihood function is

$$L_t(\theta) = \exp\left[\theta \sum_{s\leq t} \frac{\Delta X_s}{1 + \Delta X_s} - t\int_E (e^{\theta x/(1+x)} - 1)|x|^{-(\alpha+1)}dx\right].$$

For $\alpha \geq 1$ the likelihood function is given by the more complicated expressions (11.5.9) and (11.5.11). □

For the general exponential families of Lévy processes, maximum likelihood estimation follows the classical pattern, as was the case for natural exponential families of Lévy processes. Consider the situation where the parameter set Θ_φ is given by (11.5.12). By the Hölder inequality, Θ_φ is convex, and without loss of generality we can assume that $\operatorname{int}\Theta_\varphi \neq \emptyset$. It is obvious from their construction that the canonical processes X^c and B have independent stationary increments. This also follows from Theorem 4.2.1. The cumulant transform of (X_1^c, B_1) under P is $(u, v) \mapsto \frac{1}{2}\sigma^2 u^2 + \phi(v)$, where $u \in \mathbb{R}$ and $v \in \Theta_\varphi$. We will assume that the coordinates of φ are affinely independent. The following theorem is proved in the same way as Theorem 2.2.1.

Theorem 11.5.2 *The maximum likelihood estimator*

$$\hat{\alpha}_t = X_t^c/(\sigma^2 t)$$

exists uniquely for all $t > 0$. If the cumulant transform ϕ is steep, the maximum likelihood estimator

$$\hat{\theta}_t = \dot{\phi}^{-1}(B_t/t) \tag{11.5.15}$$

exists and is unique for t sufficiently large. Let C denote the closed convex support of B_t/t, which does not depend on t. If $P(B_t/t \in \operatorname{bd} C) = 0$, then the maximum likelihood estimator $\hat{\theta}_t$ exists and is uniquely given by (11.5.15).

Suppose $\theta \in \operatorname{int} \Theta_\varphi$. Then under $P_{\alpha,\theta}$,

$$\hat{\alpha}_t \to \alpha \text{ and } \hat{\theta}_t \to \theta \text{ almost surely,}$$

$$\sqrt{t}(\hat{\theta}_t - \theta) \to N(0, \ddot{\phi}(\theta)^{-1}) \text{ weakly,}$$

and

$$-2 \log Q_t \to \chi^2(k - l) \text{ weakly}$$

as $t \to \infty$. Here Q_t is the likelihood ratio test statistic for a hypothesis of the type considered in Theorem 2.2.1 with l free parameters. Finally, we have the exact result

$$\sigma\sqrt{t}(\hat{\alpha}_t - \alpha) \sim N(0, 1).$$

This is also the likelihood ratio test for the hypothesis that α is the true parameter value.

11.6 Exponential families constructed by stochastic time transformation

In this section we discuss how to construct exponential families of processes with a more complicated behaviour from families of Lévy processes by stochastic time transformations. In particular, we study the probabilistic properties of the processes that arise by a given time transformation. Let a natural exponential family of d-dimensional Lévy processes with likelihood function (2.1.23) be defined as in Chapter 2. The right-continuous filtration $\{\mathcal{F}_t\}$ is generated by X.

Let $u \mapsto \tau_u$ $(u \geq 0)$ be a right-continuous increasing process such that for each u the random variable τ_u is a stopping time with respect to $\{\mathcal{F}_t\}$.

Suppose, moreover, that τ_u is finite P_θ-almost surely for all $u \geq 0$ and all $\theta \in \Theta$, and that $\tau_u \to \infty$ as $u \to \infty$ P_θ-almost surely for all $\theta \in \Theta$. Then the family of stopping times $\{\tau_u\}$ defines a *stochatic time transformation.* The family $\{P_\theta : \theta \in \Theta\}$ is by Corollary B.1.2 of Appendix B also an exponential family with respect to the filtration $\{\mathcal{F}_{\tau_u}\}$ with likelihood function

$$L_{\tau_u}(\theta) = \exp[\theta^T X_{\tau_u} - \tau_u \kappa(\theta)]. \tag{11.6.1}$$

This is an exponential family of the type studied in Section 5.1, except that here we have not assumed that $u \to \tau_u$ is continuous.

The family $\{P_\theta : \theta \in \Theta\}$ is, in fact, an exponential family with respect to the filtration generated by the process $\{(Y_u, \tau_u) : u \geq 0\}$, where $Y_u = X_{\tau_u}$; in some cases even with respect to the filtration generated by Y only, so that (11.6.1) is the likelihood function for the observation $\{Y_s : 0 \leq s \leq u\}$. It is therefore of considerable interest to study the behaviour of the process Y. This process is in general a semimartingale (see Jacod, 1979), so its local behaviour under P_θ is determined by its local characteristics $(\alpha_\theta, \beta_\theta, \nu_\theta)$. These can be expressed in terms of the Lévy characteristics (δ, Σ, ν) of X under P.

Proposition 11.6.1 *Suppose τ is X-continuous, i.e., that X is constant on all intervals $[\tau_{u-}, \tau_u]$, $u > 0$. Then the local characteristics of the process Y under P_θ are given by*

$$\begin{aligned} \nu_\theta(du, dx) &= d\tau_{u-} e^{\theta^T x} \nu(dx), & (11.6.2)\\ \beta(u) &= \Sigma \tau_{u-}, & (11.6.3)\\ \alpha_\theta(u) &= \left(\delta + \Sigma\theta - \int_{E^k} \mu(x)(1 - e^{\theta^T x})\nu(dx)\right)\tau_{u-}, & (11.6.4) \end{aligned}$$

where

$$\mu_i(x) = x_i/(1 + x_i^2), \;\; i = 1, \cdots, k. \tag{11.6.5}$$

Proof: The local characteristics of the process X under P_θ are essentially given by (2.1.24). By results in Chapter 10.1 in Jacod (1979), the local characteristics of Y are, under the assumption of the proposition, found from those of X by applying the random time transformation $\{\tau_{u-}\}$. □

Note that in comparison to (2.1.24), the drift α_θ, the quadratic characteristic β, and the intensity with which jumps occur have been changed, whereas the distribution of the jump size is preserved under the random time transformation.

Example 11.6.2 Let X be a Poisson process, i.e., $\kappa(s) = e^s - 1$, and suppose we want a family of counting processes with intensity proportional to the state of the process. We assume that $X_0 = 1$. From (11.6.2) we see that we can obtain this if we can find a family of stopping times satisfying

$$d\tau_u = X_{\tau_u} du. \tag{11.6.6}$$

To do this, define

$$Z_t = \int_0^t X_s^{-1} ds = \sum_{i=1}^{X_t - 1} i^{-1}(T_i - T_{i-1}) + X_t^{-1}(t - T_{X_t - 1}),$$

which is a strictly increasing continuous $\{\mathcal{F}_t\}$-adapted process. The random variables $T_1, T_2, \cdots$ ($T_0 = 0$) are the jump times of the Poisson process. Now let τ_u be the inverse of Z_t. Then obviously (11.6.6) holds, and since

$$\{\tau_u \leq t\} = \{u \leq Z_t\} \in \mathcal{F}_t,$$

τ_u is an $\{\mathcal{F}_t\}$-stopping time. Not surprisingly, the process X_{τ_u} is a pure birth process with intensity $e^{\theta} X_{\tau_u}$ under P_θ. □

Next, we consider again briefly the type of exponential families studied in Section 5.1. We assume that the same regularity conditions hold as in that section and let A and S be the processes appearing in the exponential representation (5.1.1). We denote the Lévy characteristics corresponding to κ, cf. (2.1.25), by (δ, Σ, ν). Then we have the following corollary.

Corollary 11.6.3 *Suppose S_t is strictly increasing. Then the local characteristics of the semimartingale A under P_θ are*

$$\nu_\theta(dt, dx) = dS_t e^{\theta^T x} \nu(dx), \tag{11.6.7}$$

$$\beta(t) = \Sigma S_t \tag{11.6.8}$$

$$\alpha_\theta(t) = \left(\delta + \Sigma\theta - \int_{E^k} \mu(x)(1 - e^{\theta^T x})\nu(dx)\right) S_t, \tag{11.6.9}$$

with $\mu(x)$ given by (11.6.5).

Proof: Under the condition of the corollary the semimartingale A is obtained from a Lévy process B by the continuous time transformation S_t (see Section 5.1). The cumulant transform of B_1 under P_θ is $s \mapsto \kappa(\theta + s) - \kappa(\theta)$. Hence the corollary follows from Proposition 11.6.1. □

Example 11.6.4 Consider the situation studied in Section 5.4. The family of stopping times $\{\tau_u\}$ defined by (5.4.1) is not X-continuous, and the local characteristics of $X_{\tau_u} = u$ are obviously not given by (11.6.3)–(11.6.5). This illustrates that Proposition 11.6.1 is not true without the condition that τ is X-continuous. □

11.7 Likelihood theory

In this section we shall see how some of the likelihood quantities considered in Chapter 8 can be expressed in terms of the local characteristics, and how some of the results simplify in the semimartingale models studied in Section 11.4.

Consider a semimartingale model of the general type defined in that section with likelihood function (11.4.5) and with Y_κ defined by (11.4.6). The parameters θ and κ are L-independent, i.e., the likelihood function is of the product form

$$L_t(\theta, \kappa) = L_t^{(1)}(\theta) L_t^{(2)}(\kappa), \quad (\theta, \kappa) \in \Theta \times K. \tag{11.7.1}$$

We can therefore study maximum likelihood estimation of θ and κ separately. The same is true of classical tests such as likelihood ratio tests, score tests, or Wald's tests for hypotheses about θ only or κ only. The L-independence of θ and κ implies that θ and κ are observed orthogonal, and hence also orthogonal, with respect to the incremental expected information and the expected information. We shall see below that θ and κ are also orthogonal with respect to the incremental observed information. The two parameters θ and κ are said to be orthogonal with respect to an information matrix $\mathcal{I}$ if $\mathcal{I}$ is a block diagonal matrix with one block corresponding to θ and the other to κ. The statistical consequences of parameter orthogonality are discussed in Barndorff-Nielsen and Cox (1994).

Under Condition 11.4.1 the log-likelihood function is

$$\log L_t(\theta, \kappa) = \theta^T H_t - \tfrac{1}{2}\theta^T \langle H \rangle_t \theta + \kappa^T B_t - g(\kappa, t), \tag{11.7.2}$$

where

$$H_t = z \cdot X_t^c, \quad \langle H \rangle_t = (zcz^T) \cdot A_t, \tag{11.7.3}$$

$$B_t^{(i)} = \varphi_i 1_A * (\mu - \nu)_t + \varphi_i 1_{A^c} * \mu_t, \quad i = 1, \cdots, k, \tag{11.7.4}$$

and

$$g(\kappa, t) = (e^{\kappa^T \varphi} - 1 - \kappa^T \varphi) 1_A * \nu_t + (e^{\kappa^T \varphi} - 1) 1_{A^c} * \nu_t \tag{11.7.5}$$

with $A = [-1, 1]^d$. We do not assume here that g is non-random, so the model need not be an exponential family in κ. The total score function is

$$U_t(\theta, \kappa) = \nabla \log L_t(\theta, \kappa) = U_t^{(1)}(\theta) + U_t^{(2)}(\kappa), \tag{11.7.6}$$

where

$$U_t^{(1)}(\theta) = \nabla_\theta \log L_t^{(1)}(\theta) = H_t - \langle H \rangle_t \theta \tag{11.7.7}$$

and

$$U_t^{(2)}(\kappa) = \nabla_\kappa \log L_t^{(2)}(\kappa) = B_t - \nabla_\kappa g(\kappa, t). \tag{11.7.8}$$

Note that $L_t^{(1)}(\theta)$ has the form of the general time-homogeneous exponential families with time-continuous likelihood function considered in Section

11.3. It follows from Corollary 11.3.2 that $U_t^{(1)}(\theta)$ is a continuous local martingale with quadratic characteristic $\langle H \rangle$ under $P_{\theta,\kappa}$. This is because under P the local characteristics of H are $(0, \langle H \rangle, 0)$, so under $P_{\theta,\kappa}$ they are $(\langle H \rangle \theta, \langle H \rangle, 0)$. The incremental expected information matrix $I_t(\theta)$ for θ is the quadratic characteristic of the score martingale $U^{(1)}(\theta)$, so

$$I_t(\theta) = \langle U^{(1)}(\theta) \rangle_t = \langle H \rangle_t. \tag{11.7.9}$$

Note that this matrix does not depend on θ. The observed information for θ is

$$j_t = -\nabla_\theta^2 \log L_t^{(1)}(\theta) = \langle H \rangle_t = I_t. \tag{11.7.10}$$

It is not surprising that j_t and I_t coincide, because I_t is generally the compensator of the matrix submartingale j_t, and in this case j_t is predictable. Also the incremental observed information $J_t(\theta)$, defined as the quadratic variation of the score martingale, equals I_t. This is because $U_t^{(1)}(\theta)$ is a continuous process. Note that for fixed κ, condition (11.4.4) requires that all entries of the information matrix I_t be finite.

The matrix $\langle H \rangle_t$ is symmetric and positive semi-definite. If it is strictly positive definite, the maximum likelihood estimator is given by

$$\hat{\theta}_t = \langle H \rangle_t^{-1} H_t, \tag{11.7.11}$$

provided that $\hat{\theta}_t$ belongs to the parameter domain. We assume that the two conditions hold for t large enough. Suppose $\langle H \rangle_t$ is integrable, so that $U_t^{(1)}(\theta)$ is square integrable. Then it follows from the central limit theorem and the weak law of large numbers for multivariate martingales (see Theorem A.7.7 and Corollary A.7.8 in Appendix A) applied to the martingale $U_t^{(1)}(\theta)$, that under $P_{\theta,\kappa}$ we have as $t \to \infty$ that

$$\hat{\theta}_t = \theta + I_t^{-1} U_t^{(1)}(\theta) \to \theta \tag{11.7.12}$$

in probability and that

$$(\hat{\theta}_t - \theta)^T j_t (\hat{\theta}_t - \theta) = U_t^{(1)}(\theta)^T I_t U_t^{(1)}(\theta) \to \chi^2(k) \tag{11.7.13}$$

weakly. A condition ensuring this is that there exists a family of invertible non-random $k \times k$-matrices $\{D_t(\theta, \kappa) : t > 0\}$ such that as $t \to \infty$,

$$D_t(\theta, \kappa) \to 0, \tag{11.7.14}$$

$$D_t(\theta, \kappa) \langle H \rangle_t D_t(\theta, \kappa)^T \to W(\theta, \kappa) \tag{11.7.15}$$

in $P_{\theta,\kappa}$-probability, where $W(\theta, \kappa)$ is a (possibly random) positive definite matrix, and

$$D_t(\theta, \kappa) E_\theta(\langle H \rangle_t) D_t(\theta, \kappa)^T \to \Sigma(\theta, \kappa), \tag{11.7.16}$$

where $\Sigma(\theta, \kappa)$ is a positive definite matrix. If in (11.7.13) we normalize the maximum likelihood estimator by the expected information

$$i_t(\theta, \kappa) = E_{\theta,\kappa}(\langle H \rangle_t) \tag{11.7.17}$$

instead of the observed information j_t, the limiting distribution will be the square of a normal variance-mixture when W is random; see the discussion in Chapter 8. If W is only positive semi-definite, the result (11.7.13) still holds conditionally on W being positive definite.

Strong consistency of the maximum likelihood estimator can be obtained via the strong law of large numbers for multivariate martingales if more restrictive conditions are imposed; see Theorem A.7.6. For $n = 1$ it follows without any further conditions from Lepingle's law of large numbers for martingales (Corollary A.7.5 in Appendix A) that (11.7.12) holds almost surely on the event $\{\langle H \rangle_\infty = \infty\}$.

The equality in (11.7.13) shows that the score test statistic and the Wald test statistic based on the observed information matrix for the point hypothesis $\theta = \theta_0$ coincide. Elementary calculations show that they also equal the likelihood ratio test statistic for the hypothesis $\theta = \theta_0$. The asymptotic distribution is given by (11.7.13).

If the conditions (11.7.14)–(11.7.16) hold for all values of the natural parameter θ, the model is obviously *locally asymptotically mixed normal* for κ fixed. For a definition of this concept see Section 8.5. If, moreover, the matrix W is non-random, it is locally asymptotically normal. To see this, note that

$$\begin{aligned} l_t(\theta + D_t(\theta)^T u) - l_t(\theta) &= u^T D_t(\theta) U_t^{(1)}(\theta) \\ &\quad - \tfrac{1}{2} u^T D_t(\theta) \langle H \rangle_t D_t(\theta)^T u, \end{aligned} \tag{11.7.18}$$

where u is a k-dimensional vector and $l_t(\theta) = \log L_t^{(1)}(\theta)$. Under the conditions (11.7.14)–(11.7.16) it follows from Theorem A.7.7 that under $P_{\theta,\kappa}$,

$$(D_t(\theta) U_t^{(1)}(\theta),\ D_t(\theta) \langle A \rangle_t D_t(\theta)^T) \to (W^{1/2} Z, W)$$

weakly as $t \to \infty$, where Z is a k-dimensional standard normally distributed random vector independent of W.

Let us finish the discussion of $L_t^{(1)}(\theta)$ by pointing out the connection between the observed information and the *Hellinger process* $h(\alpha)$ of order $\alpha \in (0, 1)$ between $P_{\theta_1,\kappa}$ and $P_{\theta_2,\kappa}$. From (11.4.5) it follows that

$$\begin{aligned} Z_t &= \frac{dP^t_{\theta_1,\kappa}}{dP^t_{\theta_2,\kappa}} \\ &= \exp[(\theta_1 - \theta_2)^T H_t - \tfrac{1}{2}\theta_1^T \langle H \rangle_t \theta_1 + \tfrac{1}{2}\theta_2^T \langle H \rangle_t \theta_2]. \end{aligned} \tag{11.7.19}$$

By Ito's formula,

$$Z_t = Z_0 + \sum_{i=1}^{m} (\theta_1^{(i)} - \theta_2^{(i)}) \int_0^t Z_s dH_s^{(i)}(\theta_2), \tag{11.7.20}$$

where

$$H_t^{(i)}(\theta_2) = H_t^{(i)} - \sum_{j=1}^{m} \theta_2^{(i)} \langle H^{(i)}, H^{(j)} \rangle_t, \quad i = 1, \cdots, n, \tag{11.7.21}$$

is a local $P_{\theta_2,\kappa}$-martingale by (11.3.19). Hence

$$\langle Z \rangle_t = \sum_{i,j=1}^{m} (\theta_1^{(i)} - \theta_2^{(i)})(\theta_1^{(j)} - \theta_2^{(j)}) \int_0^t Z_s^2 d\langle H^{(i)}, H^{(j)} \rangle_s, \tag{11.7.22}$$

and by Jacod and Shiryaev (1987, Corollary IV.1.37) the Hellinger process of order α between $P_{\theta_1,\kappa}$ and $P_{\theta_2,\kappa}$ is given by

$$h_t(\alpha) = \tfrac{1}{2}\alpha(1-\alpha)(\theta_1 - \theta_2)^T \langle H \rangle_t (\theta_1 - \theta_2). \tag{11.7.23}$$

Since the observed information is equal to $\langle H \rangle_t$, the Hellinger process of order α between $P_{\theta_1,\kappa}$ and $P_{\theta_2,\kappa}$ is proportional to the squared distance between θ_1 and θ_2 in the metric defined by the observed information. Remember that observed information equals incremental observed information as well as incremental expected information.

In Jacod (1989) a generalized Hellinger process for parametric statistical models is defined for every $\alpha = (\alpha(\theta) : \theta \in \Theta)$ for which $\alpha(\theta)$ is non-zero for only finitely many values of θ and $\sum_\theta \alpha(\theta) = 1$. By Theorem 4.3 in Jacod (1989) the generalized Hellinger process of order α is for a time-continuous exponential family of processes given by

$$h(\alpha) = \tfrac{1}{4} \sum_{\theta_1,\theta_2} \alpha(\theta_1)\alpha(\theta_2)(\theta_1 - \theta_2)^T \langle H \rangle_t (\theta_1 - \theta_2). \tag{11.7.24}$$

In the rest of this section we discuss inference about κ. We shall work under conditions that allow us to interchange differentiation with respect to κ and integration with respect to ν in the expression (11.7.5) for $g(\kappa, t)$. For the situations (II)–(IV) in Section 11.4 such conditions are easy to give, and if ν is non-random (case (I)), we can assume that the functions

$$\varphi_i(e^{\kappa^T \varphi} - 1)1_A, \quad \varphi_i e^{\kappa^T \varphi} 1_{A^c}, \quad i = 1, \cdots, k,$$

are locally dominated integrable with respect to ν. Under such conditions we find that

$$U_t^{(2)}(\kappa) = \varphi * (\mu - e^{\kappa^T \varphi} \nu)_t. \tag{11.7.25}$$

If

$$(\varphi_i^2 e^{\kappa^T \varphi}) * \nu \in \mathcal{A}_{\text{loc}}^+(P), \tag{11.7.26}$$

it follows (Theorem A.8.2) that $U_t^{(2)}(\kappa)$ is a purely discontinuous locally square integrable martingale under $P_{\theta,\kappa}$ and that the incremental observed information is given by

$$J_t(\kappa) = [U^{(2)}(\kappa)]_t = (\varphi\varphi^T) * \mu_t. \tag{11.7.27}$$

Note that J_t is independent of κ. Moreover, the incremental expected information is

$$I_t(\kappa) = \langle U^{(2)}(\kappa)\rangle_t = \left((\varphi\varphi^T)e^{\kappa^T\varphi}\right) * \nu_t. \tag{11.7.28}$$

Since $U^{(1)}(\theta)$ is continuous and $U^{(2)}(\kappa)$ is a purely discontinuous local martingale, we have that

$$[U^{(1)}(\theta), U^{(2)}(\kappa)] = \langle U^{(1)}(\theta), U^{(2)}(\kappa)\rangle = 0, \tag{11.7.29}$$

which shows the orthogonality of θ and κ with respect to the incremental observed information and the incremental expected information. The last result also follows directly from the L-independence of θ and κ; see Barndorff-Nielsen and Sørensen (1993).

Under further conditions that allow another interchange of differentiation and integration with respect to ν, we find that the observed information equals the incremental expected information

$$j_t(\kappa) = \nabla_\kappa^2 \log L_t^{(2)}(\kappa) = \left((\varphi\varphi^T)e^{\kappa^T\varphi}\right) * \nu_t. \tag{11.7.30}$$

Asymptotic results can be obtained by the methods used in Chapter 8. In some cases the situation simplifies considerably, as illustrated by the following example.

Example 11.7.1 Consider an exponential family of the type (IV) in Section 11.4 with $\kappa \in \mathbb{R}$. In this situation,

$$\log L_t^{(2)}(\kappa) = \kappa(1 * \mu_t) - e^\kappa(1 * \nu_t),$$

so we find immediately that

$$U_t^{(2)}(\kappa) = 1 * \mu_t - e^\kappa(1 * \nu_t)$$

and

$$j_t(\kappa) = -\nabla_\kappa^2 \log L_t^{(2)}(\kappa) = e^\kappa(1 * \nu_t).$$

The maximum likelihood estimator of κ is given by

$$\hat{\kappa}_t = \log\left(\frac{1 * \mu_t}{1 * \nu_t}\right).$$

Since $U^{(2)}(\kappa)$ is a locally square integrable martingale with $\langle U^{(2)}(\kappa)\rangle = e^\kappa(1*\nu)$ under $P_{\theta,\kappa}$, it follows by Lepingle's law of large numbers (Corollary A.7.5) that

$$\exp(\hat{\kappa}_t) = e^\kappa + e^\kappa\langle U^{(2)}(\kappa)\rangle_t^{-1} U_t^{(2)}(\kappa) \to e^\kappa$$

almost surely on $\{1*\nu_\infty = \infty\}$ under $P_{\theta,\kappa}$ as $t \to \infty$. Suppose, furthermore, that

$$[U^{(2)}(\kappa)]_t = 1 * \mu_t \to \infty$$

almost surely as $t \to \infty$ and that there exists a non-random increasing function $c_t > 0$ and a non-negative random variable W such that

$$1 * \mu_t / c_t \to W$$

in $P_{\theta,\kappa}$-probability as $t \to \infty$. Then it follows from the central limit theorem for martingales (Theorem A.7.7) that under $P_{\theta,\kappa}$,

$$(1 * \mu_t)^{1/2}(\hat{\kappa}_t - \kappa) \to N(0,1)$$

and

$$(1 * \mu_t)^{-1/2} U_t^{(2)}(\kappa) \to N(0,1)$$

weakly as $t \to \infty$. Note that $J_t = 1 * \mu_t$ is the number of observed jumps.

A particular case of the exponential families considered in this example is the type of counting process model presented in Section 3.4. There $1 * \mu_t$ equals the observed counting process. □

11.8 Exercises

11.1 Find the local characteristics of the process X given by

$$dX_t = -X_{t-}dt + dW_t^{(1)} + \sin(X_{t-})dN_t, \quad X_0 = x_0,$$

where $W^{(1)}$ is a standard Wiener process and where N is a counting process with intensity X_{t-}. Find the local characteristics of the two-dimensional process $(X_t, \exp(X_t - Y_t))$, where Y is given by

$$dY_t = Y_t dt + Y_t dW_t^{(2)}, \quad Y_0 = y_0.$$

Here $W^{(2)}$ is a standard Wiener process independent of $W^{(1)}$.

11.2 Prove that (11.2.37) is the likelihood function of the model described below the formula.

11.3 Prove (11.2.47) and that $Z(\theta)$ is a martingale with mean one.

11.4 Let A be a process of finite variation satisfying $A_0 = 0$ and that $\Delta A_t > -1$ for all $t \geq 0$. Show (Ito's formula) that the process

$$Y_t = \exp(A_t) \prod_{s \leq t} (1 + \Delta A_s) \exp(-\Delta A_s)$$

satisfies the equation

$$dY_t = Y_{t-} dA_t.$$

This shows that Y equals the Doléan-Dade exponential of A, i.e., $Y = \mathcal{E}(A)$. Show also that Y_t^{-1} satisfies the equation

$$dY_t^{-1} = Y_{t-}^{-1} d\tilde{A}_t,$$

where

$$\tilde{A}_t = -\int_0^t (1 + \Delta A_s)^{-1} dA_s.$$

Hint: Apply Ito's formula and use that $Y_t^{-1} = Y_{t-}^{-1}(1 + \Delta A_t)^{-1}$.

11.5 Let X be a one-dimensional semimartingale with local characteristics (α, β, ν) and $X_0 = 0$, and let $A(\theta) = \theta\alpha_t + B_t(\theta) + \frac{1}{2}\theta^2\beta_t$, where $B_t(\theta)$ is the process defined by (11.2.16). Define the following processes:

$$H_t(\theta) = \exp(\theta X_t),$$

$$Y_t(\theta) = \mathcal{E}_t(A(\theta)),$$

$$L_t(\theta) = H_t(\theta) Y_t(\theta)^{-1}.$$

Here $\theta \in \Theta_0 \subseteq \mathbb{R}$, which is defined by (11.2.17). Show that

$$\begin{aligned} H_t(\theta) &= 1 + \int_0^t H_{s-}(\theta) dA_s(\theta) \\ &\quad + \theta \int_0^t H_{s-}(\theta) dX_s^c + \int_0^t \int_{E^d} H_{s-}(\theta)(e^{\theta x} - 1) d(\mu - \nu) \end{aligned}$$

and that

$$\begin{aligned} L_t(\theta) &= 1 + \theta \int_0^t L_{s-}(\theta) dX_s^c \\ &\quad + \int_0^t \int_{E^d} L_{s-}(\theta) \frac{e^{\theta x} - 1}{1 + \Delta A_s(\theta)} d(\mu - \nu). \end{aligned} \qquad (11.8.1)$$

The result (11.8.1) shows that $L(\theta)$ is a local martingale under P. Hint: Ito's formula and Exercise 11.4.

11.9 Bibliographic notes

Chapter 11 is based on work by the authors. Section 11.2 presents results from Küchler and Sørensen (1994b) and Sørensen (1993), while Section 11.3 is based on Küchler and Sørensen (1994b). The material in Section 11.4 was first published in Küchler and Sørensen (1989). Section 11.5 draws on the same paper, as well as on Küchler and Sørensen (1994a). Also Section 11.6 is based on results from Küchler and Sørensen (1994a), while finally, Section 11.7 presents an updated version of material from Küchler and

Sørensen (1989) with a bit from Küchler and Sørensen (1994b) too. For an application of results in Section 11.2 to risk theory see Sørensen (1996a).

Not all exponential families of stochastic processes are families of semimartingales. An important example is the exponential family of fractional Brownian motions with drift, which was studied by Norros, Valkeila and Virtamo (1996).

12

Alternative Definitions

Definitions of an exponential family of stochastic processes other than the one on which this book is based have been given in the literature. In this chapter we shall review some of these.

The starting point for some interesting definitions is the following fact. Consider a parametric statistical model for independent, identically distributed observations $X_1, X_2, \cdots$, and define $S_n = \sum_{i=1}^{n} X_i$, $n \geq 1$. Then the statement that the class of marginal distributions of S_n is an exponential family for every $n \geq 1$ is equivalent to the statement that the distributions of the vector $(S_1, \cdots, S_n)$, or equivalently $(X_1, \cdots, X_n)$, form an exponential family for all n. One could therefore define an exponential family of processes as a statistical model for a stochastic process X under which the marginal distribution of X_t belongs to the same exponential family for all $t > 0$. Models of this type have been studied by B. Ycart and will be studied in Section 12.1.

In the i.i.d. situation there is also equivalence between the statements that the marginal distribution of X belongs to an exponential family and the statement that for the observations $(X_1, \cdots, X_n)$ there exists a non-trivial sufficient reduction with the same dimension for all n. Based on this fact, one could define an exponential family of stochastic processes as a statistical model for a stochastic process X for which there exists a non-trivial sufficient reduction of the observations $\{X_s : s \leq t\}$ that has the same dimension for all $t > 0$. This approach has been investigated by U. Küchler and will be the subject of Section 12.2.

12.1 Exponential marginal distributions

Let $\mathcal{P} = \{ P_\theta : \theta \in \Theta\}$, $\Theta \subseteq \mathbb{R}^k$, be an exponential family of probability distributions on $E \subseteq \mathbb{R}^d$, and let X be a stochastic process with state space E.

Definition 12.1.1 *The stochastic process X is said to be* stable *with respect to the exponential family $\mathcal{P}$ if there exist an open interval $I \subseteq (0, \infty)$ and an open mapping $\theta : I \mapsto \Theta$ such that for every $t \in I$ the distribution of X_t is $P_{\theta(t)}$.*

We call the mapping θ the *dynamic parameter function.* Simple examples are that a Poisson process with parameter λ is stable with respect to the exponential family of Poisson processes with dynamic parameter function $\theta(t) = \lambda t$, where we have used the mean value parametrization, and that a Brownian motion with drift μ and diffusion coefficient σ^2 is stable with respect to the exponential family of normal distributions. In the latter case the dynamic parameter function is $\theta(t) = (\mu t, \sigma^2 t)$, where we have used the usual parametrization of the normal distributions.

A mapping is called open if the image of every open set is itself open. This condition is imposed in order to exclude the case of stationary processes. A class of stationary processes for which the class of marginal distributions forms an exponential family is obviously interesting in some applications, but here we follow B. Ycart in avoiding this, from a structural point of view, relatively simple situation. Without the condtion that $\theta(t)$ is open, very many stationary processes would be stable with respect to an exponential family. In fact, whenever X is a stationary process for which the Laplace transform of the distribution F of X_t has a non-trivial domain, then the distribution of X belongs to the natural exponential family generated by F.

The question of when a Markov process with countable state space is stable with respect to an exponential family has been studied thoroughly by B. Ycart. We shall now review his results. We remind the reader that if μ is a measure on $\mathbb{N} = \{ 0, 1, 2, \cdots\}$, then the natural exponential family generated by μ is the family $\mathcal{P} = \{P_\theta : \theta \in \Theta\}$, where

$$P_\theta(n) = \frac{e^{\theta n}\mu(n)}{\sum_{m=0}^{\infty} e^{\theta m}\mu(m)} \tag{12.1.1}$$

and

$$\Theta = \{\theta \in \mathbb{R} : \sum_{m=0}^{\infty} e^{\theta m}\mu(m) < \infty\},$$

provided that $\operatorname{int}\Theta \neq \emptyset$.

Let X be a continuous time Markov process with state space $\mathbb{N}$ and intensity matrix $A = (a_{i,j})_{i,j \in \mathbb{N}}$. (To be quite precise, X is assumed to be a Feller process.) Moreover, let μ be a measure on $\mathbb{N}$ with support $S = \{n \in \mathbb{N} : \mu(n) \neq 0\}$. Without loss of generality we can in the following discussion assume that $0 \in S$. We seek conditions under which X is stable with respect to the natural exponential family generated by μ. Obviously, the state space of X must equal S, so $a_{ji} = a_{ij} = 0$ for all $j \in \mathbb{N}$ if $i \notin S$. We denote by s the smallest positive element in S, and define for all $i \in \mathbb{N}$

$$\alpha_{ij} = \begin{cases} a_{ij}\mu(i)/\mu(j) & \text{if } j \in S \\ 0 & \text{if } j \notin S. \end{cases} \tag{12.1.2}$$

Finally, we define the generating functions

$$g(z) = \sum_{n=0}^{\infty} \mu(n) z^n, \tag{12.1.3}$$

$$g_0(z) = \sum_{n=0}^{\infty} \alpha_{n,0} z^n, \tag{12.1.4}$$

$$g_s(z) = \sum_{n=0}^{\infty} \alpha_{n,s} z^{n-s}. \tag{12.1.5}$$

Theorem 12.1.2 *Suppose X is stable with respect to the natural exponential family generated by μ. Then*

(i) $\alpha_{ij} = 0$ *if* $j > i + s$.

(ii) For all $i \in S$ and $h \geq -s$,

$$s\alpha_{h+i,i} = i\alpha_{h+s,s} - (i - s)\alpha_{h,0}$$

with $\alpha_{h,0} = 0$ for all $h < 0$.

(iii)

$$zg'(z)[g_s(z) - g_0(z)] + sg(z)g_0(z) = 0 \tag{12.1.6}$$

whenever the series (12.1.3)–(12.1.5) converge.

(iv) The dynamic parameter function $\theta(t)$ is a solution of the differential equation

$$\theta'(t) = s^{-1}[g_s(e^{\theta(t)}) - g_0(e^{\theta(t)})]. \tag{12.1.7}$$

Conversely, (i), (ii), and (iii) imply that X is stable with respect to the natural exponential family generated by μ with $\theta(t)$ given by (12.1.7).

Remarks: Note that the problem of finding a stable Markov process by Theorem 12.1.2 is reduced to choosing explicit values of $\alpha_{h+s,s}$ for $h \geq -s$ and $\alpha_{h,0}$ for $h \geq 0$. This gives a method of constructing Markov processes with marginal distributions in the natural exponential family generated by

μ. We shall consider an example after the proof of the theorem. There are some restrictions on how $\alpha_{h+s,s}$ and $\alpha_{h,0}$ can be chosen. Since $\alpha_{h+i,i} \geq 0$ for $h \neq 0$ and $\alpha_{i,i} < 0$, it follows from (ii) that

$$\alpha_{ss} \leq \alpha_{00} \text{ and } \alpha_{h+s,s} \geq \alpha_{h,0} \text{ for all } h \in \mathbb{N}\backslash\{0\}. \tag{12.1.8}$$

Proof: We will not give the details of the proof of Theorem 12.1.2. The main idea is to equate two expressions for $dP_{\theta(t)}(i)/dt$. (That $P_{\theta(t)}$ and $\theta(t)$ are differentiable follows from the assumption that X is a Feller process.) First, we differentiate the expression for $P_{\theta(t)}(i)$ given by (12.1.1) and obtain

$$\frac{d}{dt}P_{\theta(t)}(i) = \frac{\theta'(t)P_{\theta(t)}(i)\sum_{h=0}^{\infty}(i-h)\mu(h)e^{h\theta(t)}}{\sum_{h=0}^{\infty}\mu(h)e^{h\theta(t)}}. \tag{12.1.9}$$

Then the Chapmann-Kolmogorov equations

$$\frac{d}{dt}P_{\theta(t)}(i) = \sum_{k=0}^{\infty} a_{ki}P_{\theta(t)}(k)$$

(cf. Karlin and Taylor, 1975) give

$$\frac{d}{dt}P_{\theta(t)}(i) = P_{\theta(t)}(i)\sum_{k=0}^{\infty}\alpha_{ki}e^{(k-i)\theta(t)} \tag{12.1.10}$$

for all $i \in S$. By equating (12.1.9) and (12.1.10) for $i, j \in S$ it follows that

$$\sum_{h=0}^{\infty}\sum_{k=0}^{\infty}(i-h)\mu(h)\alpha_{k,j}e^{(h+k-j)\theta(t)} \tag{12.1.11}$$
$$= \sum_{h=0}^{\infty}\sum_{k=0}^{\infty}(j-h)\mu(h)\alpha_{k,i}e^{(h+k-i)\theta(t)}$$

for all $i, j \in S$ and $t \in I$.

Now, the uniqueness of the Laplace transform (see Barndorff-Nielsen (1978, Theorem 7.2) is applied. Here it is important that the mapping $\theta(t)$ is open so that (12.1.11) holds for θ is an open interval. It follows that the sum of the coefficients of terms with $h+k-j=n$ on the lefthand side must equal the sum of the coefficients of terms with $h+k-i=n$ on the righthand side of (12.1.11). For instance, take $j=0$. Then it follows that the sum of the coefficients of terms on the righthand side with $h+k-i=-n$ for $n \in \mathbb{N}\backslash\{0\}$ must be zero. For $h \geq s$ these terms are all negative, so (i) follows. By clever manipulations of (12.1.11) it is possible to prove (ii) in a similar way.

Now insert (ii) in $\sum_{j=0}^{i+s} a_{ij} = 0$, which holds because A is an intensity matrix. Then we find that

$$s^{-1}\sum_{j=0}^{i+s}\mu(j)[j\alpha_{i-j+s,s} - (j-s)\alpha_{i-j,0}] = 0,$$

and by multiplying this expression by z^i and summing over i, we obtain (iii). We have now shown that $P_{\theta(t)}$ solves the Chapmann-Kolmogorov equation if and only if (i), (ii), and (iii) hold. Thus the claim at the end of the theorem has also been proved.

Finally, the differential equation (iv) is obtained by equating again (12.1.9) and (12.1.10), using (ii), identifying the generating functions g, g_s, and g_0, and finally, using (iii). □

Example 12.1.3 *Birth-and-death processes.* If X is a birth-and-death process, it is natural to take $s = 1$. Among the $\alpha_{h+1,1}$'s and the $\alpha_{h,0}$'s only five coefficients can be different from zero, viz., α_{00}, α_{10}, α_{01}, α_{11}, and α_{21}. Obviously, $\alpha_{00} < 0$, $\alpha_{11} < 0$, $\alpha_{01} > 0$, $\alpha_{10} \geq 0$, and $\alpha_{21} \geq 0$. Moreover, by (12.1.8) $\alpha_{00} \geq \alpha_{11}$ and $\alpha_{21} \geq \alpha_{10}$. Assume that five real numbers satisfying these conditions have been chosen. Then, by Theorem 12.1.2 the corresponding Markov process is stable with respect to the natural exponential family generated by a measure μ on $\mathbb{N}$ if and only if the generating function $g(z)$ of μ is a solution of the differential equation

$$\frac{g'(z)}{g(z)} = -\frac{\alpha_{00} + \alpha_{10}z}{\alpha_{01} + (\alpha_{11} - \alpha_{00})z + (\alpha_{21} - \alpha_{10})z^2}. \tag{12.1.12}$$

If (12.1.12) admits a solution g that is an integer series with positive coefficients and radius of convergence larger than one, then g determines a measure μ, and by means of (ii) in Theorem 12.1.2 the whole intensity matrix can be found. Thus a complete list of all birth-and-death processes that are stable with respect to a natural exponential family can be obtained by finding the solutions to all equations of the form (12.1.12) that are generating functions of a measure on $\mathbb{N}$.

It turns out that the possibilities are rather restricted. If it is assumed that μ is strictly positive on $\mathbb{N}$, then the generating function must be one of the following three types. The first type is

$$g(z) = C(a - z)^{-r}\exp(kz)$$

with $C > 0$, $a > 0$, $r \geq 0$, and $k \geq 0$. The corresponding natural exponential families consist of convolutions of a Poisson distribution and a negative binomial distribution. Particular cases are the class of Poisson processes and the class of negative binomial distributions. The corresponding birth-and-death processes are linear.

The second type is

$$g(z) = C(a - z)^{-r}\exp(k/(a - z))$$

with $C > 0$, $a > 0$, $r \geq 0$, and $k \geq 0$. The corresponding natural exponential families consist of convolutions of a negative binomial distribution and a compound of a geometric distribution by a Poisson distribution.

The third type is

$$g(z) = C(a-z)^{-r}(b-z)^{-s},$$

where $C > 0$, $b > a > 0$, $r \geq 0$, and $r+s \geq 0$. Here the natural exponential family includes the convolutions of two negative binomial distributions, obtained when r and s are both positive.

If, on the other hand, μ has a generating function of one of these three types, then there exists a birth-and-death process that is stable with respect to the natural exponential family generated by μ. The intensity matrix of the process is uniquely determined up to a multiplicative coefficient (which determines only the speed of the process).

If X is stable with respect to the class of negative binominal distributions with parameters r and p, where r is fixed, then for some $\lambda > 0$, $\mu > 0$, and $p_0 \in [0,1]$,

$$a_{i,i+1} = \lambda(i+r), \quad i = 0, 1, \cdots,$$

and

$$a_{i,i-1} = \mu i, \quad i = 1, 2 \cdots,$$

while the dynamic parameter function is

$$p(t) = \begin{cases} \frac{\lambda - (\lambda - \mu p_0)(1-p_0)^{-1}\exp[(\lambda-\mu)t]}{\mu - (\lambda - \mu p_0)(1-p_0)^{-1}\exp[(\lambda-\mu)t]} & \text{if } \lambda \neq \mu \\ 1 - (\lambda t + (1-p_0)^{-1})^{-1} & \text{if } \lambda = \mu. \end{cases}$$

If X is stable with respect to the class of Poisson distributions, then for some $\lambda > 0, \mu > 0$, and $\rho_0 > 0$,

$$\begin{aligned} a_{i,i+1} &= \lambda, \quad i = 0, 1, 2, \cdots, \\ a_{i,i-1} &= \mu i, \quad i = 0, 1, 2, \cdots, \end{aligned}$$

and the mean value parameter of the Poisson distribution at time $t \geq 0$ is

$$\rho(t) = \lambda\mu^{-1} + (\rho_0 - \lambda\mu^{-1})\exp(-\mu t).$$

A non-standard example is obtained if $\alpha_{00} = -3$, $\alpha_{10} = 1$, $\alpha_{01} = 2$, $\alpha_{11} = -6$, and $\alpha_{21} = 2$. Then X is a birth-and-death process with

$$a_{i,i+1} = 2(i+1)(i+3)/(i+2)$$

and

$$a_{i,i-1} = i(i+1)/(i+2).$$

The process is stable with respect to the exponential family given by

$$P_\theta(n) = (n+2)e^{n\theta}(1-e^\theta)^2/(2-e^\theta),$$

where $n \in \mathbb{N}$ and $\theta < 0$. More explicitly, if the distribution of X_0 is P_{θ_0}, then the distribution of X_t is $P_{\theta(t)}$, where

$$\theta(t) = \log[1 + (1 + Ke^t)^{-1}],$$

where $K = (e^{\theta_0} - 1)^{-1} - 1$. □

The following result is a corollary to Theorem 12.1.2.

Corollary 12.1.4 *Suppose the state space of X is finite, i.e., $E = \{x_0, x_1, \cdots, x_n\}$, where $x_0 < x_1 < \cdots < x_n$, and assume that X is stable with respect to a natural exponential family $\mathcal{P}$. Then*

(i) There exists an affine transformation g of the real axis such that

$$g(x_i) = i, \quad \textit{for } i = 0, 1, \cdots, n.$$

(ii) The exponential family $\mathcal{P}$ is the image by g^{-1} of the binomial distributions on $\{0, 1, \cdots, n\}$.

(iii) The intensity matrix of X is given by $a_{ij} = 0$ if $|i - j| \geq 2$ and

$$a_{i,i+1} = \lambda(n - i) \quad \textit{and} \quad a_{i,i-1} = i\mu \quad \textit{for} \quad i = 0, 1, \cdots, n,$$

where λ and μ are non-negative real numbers such that $\lambda\mu > 0$.

(iv) If $X_0 \sim b(n, p_0)$, then $X_t \sim b(n, p(t))$ for all $t \geq 0$, where

$$p(t) = \lambda(\lambda + \mu)^{-1} + [p_0 - \lambda(\lambda + \mu)^{-1}] \exp[-(\lambda + \mu)t].$$

It is clear that there are many fewer models for Markov processes that are stable with respect to a natural exponential family than there are models that are exponential families of Markov processes in the sense of the definition in Chapter 3. For instance, the class of all Markov processes on a finite set E is an exponential family of stochastic processes, whereas Corollary 12.1.4 asserts that the Markov processes on E that are stable with respect to a natural exponential family is a very limited class.

Of course, somewhat richer classes of processes are obtained by considering families of Markov processes that are stable with respect to general exponential families without the assumption that the canonical statistic is the identity mapping. We shall discuss such processes briefly.

Suppose a stochastic process X is stable with respect to an exponential family of distributions $\mathcal{P}_{f,\mu}$ with the densities

$$\exp[\theta f(x)] / \int_E \exp[\theta f(y)] \mu(dy) \tag{12.1.13}$$

with respect to a measure μ, and that the dynamic parameter function is $\theta(t)$. Here θ and f are one-dimensional. Then obviously the process $\{f(X_t)\}$

is stable with respect to the natural exponential family $\mathcal{P}$ generated by the image ν of μ by f with the same dynamic parameter function. However, if X is a Markov process, $\{f(X_t)\}$ need not be Markovian; but it is often possible to find a Markov process $\tilde{X}$ that is stable with respect to $\mathcal{P}_\nu$ with dynamic parameter function $\theta(t)$. We will just state the result with some necessary preliminaries. The proofs can be found in Ycart (1992b).

Let ν be the image of μ by f, and denote by S_μ and S_ν the supports of μ and ν respectively.

Proposition 12.1.5 *An operator $M : L_2(S_\mu, \mu) \mapsto L_2(S_\nu, \nu)$ is defined by*

$$\int_B M(\varphi) d\nu = \int_{f^{-1}(B)} \varphi d\mu$$

for all Borel subsets of S_ν, where $\varphi \in L_2(S_\mu, \mu)$.

Without loss of generality it can be assumed that μ and ν are probability measures such that any bounded continuous function on E is in $L_2(S_\mu, \mu)$.

Definition 12.1.6 *Let A be the generator of a Markov process on E with domain $\mathcal{D}(A)$, and let $\mathcal{D}(B)$ be the set of functions u from $\mathbb{R}$ into $\mathbb{R}$ such that $u \circ f \in \mathcal{D}(A)$. Then an operator B on $\mathcal{D}(B)$ is defined by*

$$B(u) = M(A(u \circ f)), \tag{12.1.14}$$

where $u \in \mathcal{D}(B)$.

Theorem 12.1.7 *Let A be the generator of a Markov process on S_μ that is stable with respect to $\mathcal{P}_{f,\mu}$ with dynamic parameter function $\theta(t)$. Assume that*

(i) $\varphi \in \mathcal{D}(A) \Rightarrow M(\varphi) \circ f \in \mathcal{D}(A)$.

(ii) $\mathcal{D}(B)$ is dense in the space of bounded continuous functions on S_ν.

Then the operator B given by (12.1.14) is the generator of a Markov process $\tilde{X}$ on $S_\nu = f(S_\mu)$ that is stable with respect to $\mathcal{P}_\nu$ with dynamic parameter function $\theta(t)$.

Theorem 12.1.8 *Let X be a Markov process on S_μ with generator A, and let $\tilde{X}$ be a Markov process on S_ν with generator G. Suppose $\mathcal{D}(G) = M(\mathcal{D}(A))$ and that $G(M(\varphi)) = M(A(\varphi))$ for all $\varphi \in \mathcal{D}(A)$. Then X is stable with respect to $\mathcal{P}_{f,\mu}$ if and only if $\tilde{X}$ is stable with respect to $\mathcal{P}_\nu$. Moreover, if the processes are stable, the dynamic parameter function will be the same for both.*

Example 12.1.9 *Spin systems.* Suppose $E = \{0,1\}^S$, where S is a finite set with $N+1$ elements. For $\eta \in E$ define

$$\eta_x(y) = \begin{cases} \eta(y) & \text{if } y \neq x \\ 1 - \eta(y) & \text{if } x = y. \end{cases}$$

The configuration η_x is obtained from η by a flip at site x. A spin system on S is a Markov process on E with generator A defined by

$$A\varphi(\eta) = \sum_{x \in S} c(x,\eta)[\varphi(\eta_x) - \varphi(\eta)],$$

where the flip rates $c(x,\eta)$ are non-negative real numbers.

Let $\mathcal{P}_{f,\mu}$ be an exponential family of distributions on E with densities of the form (12.1.13), where μ is a measure on E and f is the mapping from E into $\{0,\cdots,N\}$ given by $f(\mu) = \sum_{x\in S} \mu(x)$.

By combining Corollary 12.1.4 and Theorem 12.1.7 we see that a spin system on S is stable with respect to $\mathcal{P}_{f,\mu}$ if and only if the image of μ by f generates the family of binomial distributions on $\{0,1,\cdots,N\}$ and there exist two positive real numbers β and δ such that

$$\sum_{\{x:\eta(x)=1\}} c(x,\eta_x)\mu(\eta_x)/\mu(\eta) = \beta f(\eta)$$

and

$$\sum_{\{x:\eta(x)=0\}} c(x,\eta_x)\mu(\eta_x)/\mu(\eta) = \delta(N - f(\eta)).$$

An explicit example is obtained by choosing a measure μ satisfying

$$\sum_{\{\eta: f(\eta)=n\}} \mu(\eta) = \binom{N}{n}.$$

Then the measures in $\mathcal{P}_{f,\mu}$ are

$$P_\pi(\eta) = \mu(\eta)\pi^{f(\eta)}(1-\pi)^{N-f(\eta)},$$

where $\pi = (1+e^\theta)^{-1} \in (0,1)$. Let X be the spin system on E with flip rates

$$c(x,\eta) = \begin{cases} \beta\mu(\eta_x)/\mu(\eta) & \text{if } \eta(x)=0 \\ \delta\mu(\eta_x)/\mu(\eta) & \text{if } \eta(x)=1, \end{cases}$$

then by Theorem 12.1.8 and Corollary 12.1.4 X is stable with respect to $\mathcal{P}_{f,\mu}$ with dynamic parameter function

$$\pi(t) = \beta(\beta+\delta)^{-1} + [\pi_0 - \beta(\beta+\delta)^{-1}]\exp[-(\beta+\delta)t].$$

□

12.2 Families with a sufficient reduction

Let $(\Omega, \mathcal{F}, \{\mathcal{F}_t\}_{t\in[0,\infty)})$ be a stochastic basis, and let $\mathcal{P} = \{P_\theta : \theta \in \Theta\}$, $\Theta \subseteq \mathbb{R}^k$ be a class of probability measures on $(\Omega, \mathcal{F})$. We call an m-dimensional

$\{\mathcal{F}_t\}$-adapted process B a *sufficient process* for $\mathcal{P}$ if B_t is sufficient for $\{P_\theta^t : \theta \in \Theta\}$ with respect to $\mathcal{F}_t$ for all $t > 0$, i.e., if for every $t > 0$ and $A \in \mathcal{F}_t$ there exists a random variable $Y_t(A)$ that does not depend on θ such that $P_\theta(A|B_t) = Y_t(A)$ P_θ-almost surely for all $\theta \in \Theta$.

We have seen that for an exponential family on $(\Omega, \mathcal{F}, \{\mathcal{F}_t\})$ the canonical process is a sufficient process. When $\{\mathcal{F}_t\}$ is generated by a sequence of independent, identically distributed random variables, the assumption that a sufficient process exists is equivalent to assuming that $\mathcal{P}$ is an exponential family. When $\{\mathcal{F}_t\}$ is generated by a stochastic process, where the observations are dependent, the situation is much more complicated. A more general definition of an exponential family of stochastic processes than the one given in Chapter 3 would be to call $\mathcal{P}$ an exponential family when a sufficient process exists. This approach was taken in the special case where $\{\mathcal{F}_t\}$ is generated by a Markov process X and where $B_t = X_t$ in Küchler (1982a,b). In this section we give some results on existence of a sufficient process and relations to our definition of an exponential family of stochastic processes.

Let X be a stochastic process on $(\Omega, \mathcal{F})$ with state space $E \subseteq \mathbb{R}^d$ that is a conservative Markov process with respect to $\{\mathcal{F}_t\}$ under $\mathbb{P} = \{P_x : x \in E\}$ with transition function $P(t,x,C) = P_x(X_t \in C)$, $t \in [0,\infty)$, $x \in E$, $C \in \mathcal{B}(E)$. For definitions concerning Markov processes see Section 6.2. As usual, we assume that X is right-continuous with limits from the left and that $\{\mathcal{F}_t\}$ is the right-continuous filtration generated by X; see Section 3.1.

Define $\Xi_{\mathbb{P}}$ as the set of pairs (α, g), where α is a real number and g is a strictly positive measurable function on E such that

$$\int_E g(y)P(t,x,dy) = e^{\alpha t} g(x) \tag{12.2.1}$$

for all $t > 0$ and $x \in E$. Assume that $(\alpha, g) \in \Xi_{\mathbb{P}}$. Then the function $P_{\alpha,g}$ defined by

$$P_{\alpha,g}(t,x,dy) = e^{-\alpha t} P(t,x,dy)\frac{g(y)}{g(x)}, \quad t > 0, \ x,y \in E,$$

is obviously a transition function on $(E, \mathcal{B}(E))$ satisfying $P_{\alpha,g}(t,x,E) \equiv 1$. Hence there exists a class of probability measures $\{P_{\alpha,g,x} : x \in E\}$ on $(\Omega, \mathcal{F})$ such that X is a conservative Markov process with respect to $\{\mathcal{F}_t\}$ under $\{P_{\alpha,g,x} : x \in E\}$, and such that $P_{\alpha,g}(t,x,C) = P_{\alpha,g,x}(X_t \in C)$. Define for some fixed $x \in E$

$$\mathcal{Q}(P_x) = \{P_{\alpha,g,x} : (\alpha, g) \in \Xi_{\mathbb{P}}\}. \tag{12.2.2}$$

Now let X be a conservative Markov process with respect to $\{\mathcal{F}_t\}$ under $\mathbb{P}^\theta = \{P_{\theta,x} : x \in E\}$ for all $\theta \in \Theta$. The state space is $E \subseteq \mathbb{R}^d$. We denote the transition function by $P_\theta(t,x,B) = P_\theta(X_t \in B | X_0 = x)$, where $x \in E$, $B \in \mathcal{B}(E)$, and $\theta \in \Theta$. Define, as in Chapter 6, the set $\mathcal{P}_x = \{P_{\theta,x} : \theta \in \Theta\}$.

Condition 12.2.1 *There exists a σ-finite measure ν on $\mathcal{B}(E)$, with $\nu(U) > 0$ for every non-void open set $U \subseteq E$, such that $P_\theta(t,x,\cdot)$ and ν are equivalent measures for every $t > 0$ and $x \in E$ and such that the densities $P_\theta(t,x,dy)/\nu(dy)$ have a version that is continuous in (t,x,y).*

Then we have the following result.

Theorem 12.2.2 *Under Condition 12.2.1 the following two statements are equivalent:*
(a) *X is a sufficient process for $\mathcal{P}_x$ for every $x \in E$.*
(b) *There exists a $\theta_0 \in \Theta$ such that $\mathcal{P}_x \subseteq \mathcal{Q}(P_{\theta_0,x})$ for every $x \in E$.*

Proof: First assume (b). Fix a $t > 0$ and choose $A \in \mathcal{F}_t$. We can assume that A is of the form

$$A = \cap_{i=1}^n \{X_{s_i} \in B_i\}, \quad n \in \mathbb{N}, \ B_i \in \mathcal{B}(E), \ 0 < s_1 < \cdots < s_n \le t.$$

For simplicity we assume that $n = 2$. For $n > 2$ the proof is completely analogous. For every $\theta \in \Theta$ there exists a pair $(\alpha, g) \in \Xi_{P^{\theta_0}}$ such that $P_\theta(s,x,dy) = e^{-\alpha s} P_{\theta_0}(s,x,dy) g(y)/g(x)$ for all $s > 0$ and $x, y \in E$. Hence for every $B \in \mathcal{B}(E)$ and $x \in E$ we have

$$
\begin{aligned}
&P_{\theta,x}(A \cap \{X_t \in B\}) \\
&= \int_{B_1}\int_{B_2}\int_{B} P_\theta(t-s_2,z,dw)P_\theta(s_2-s_1,y,dz)P_\theta(s_1,x,dy) \\
&= e^{-\alpha t}\int_{B_1}\int_{B_2}\int_{B} \frac{g(w)}{g(x)} P_{\theta_0}(t-s_2,z,dw)P_{\theta_0}(s_2-s_1,y,dz)P_{\theta_0}(s_1,x,dy) \\
&= \int_B P_{\theta_0,x}(A|X_t = w) e^{-\alpha t}\frac{g(w)}{g(x)} P_{\theta_0}(t,x,dw) \\
&= \int_{\{X_t \in B\}} P_{\theta_0,x}(A|X_t) dP_{\theta,x}.
\end{aligned}
$$

Thus $P_{\theta,x}(A|X_t) = P_{\theta_0,x}(A|X_t)$ $P_{\theta,x}$-almost surely for every $\theta \in \Theta$, so X is a sufficient process for $\mathcal{P}_x$.

Now suppose (a) holds. If $\mathcal{P}$ contains only one element, the theorem is trivial, so we can assume that there exists an element $P_\theta \in \mathcal{P}$ such that $P_\theta \neq P_{\theta_0}$. Under Condition 12.2.1 the density function

$$h(t,x,y) = \frac{P_{\theta_0}(t,x,dy)}{P_\theta(t,x,dy)}$$

is finite, strictly positive, and continuous. Moreover, it satisfies the equation

$$h(t,x,z) = h(s,x,y)h(t-s,y,z), \quad 0 < s < t, \quad x,y,z \in E. \tag{12.2.3}$$

To see this, choose $x \in E$ and u, s, t such that $0 < u < s < t$. Because X is a sufficient process for $\mathcal{P}_x$, we have for all $B, C \in \mathcal{B}(E)$ that

$$\begin{aligned} P_{\theta_0,x}&(X_s \in B, X_t \in C) \\ &= \int_{\{X_t \in C\}} P_{\theta_0,x}(X_s \in B|X_t) dP_{\theta_0,x} \\ &= \int_{\{X_t \in C\}} P_{\theta,x}(X_s \in B|X_t) dP_{\theta_0,x} \\ &= \int_C h(t,x,z) P_{\theta,x}(X_s \in B|X_t = z) P_{\theta,x}(X_t \in dz) \\ &= \int_C h(t,x,z) P_{\theta,x}(X_s \in B, X_t \in dz). \end{aligned}$$

On the other hand, we have by the Markov property that for every $B, C \in \mathcal{B}(E)$,

$$\begin{aligned} P_{\theta_0,x}&(X_s \in B, X_t \in C) \\ &= \int_{\{X_s \in B\}} P_{\theta_0,X_s}(X_{t-s} \in C) dP_{\theta_0,x} \\ &= \int_B P_{\theta_0,y}(X_{t-s} \in C) P_{\theta_0,x}(X_s \in dy) \\ &= \int_B \int_C h(t-s,y,z) P_{\theta,y}(X_{t-s} \in dz) h(s,x,y) P_{\theta,x}(X_s \in dy) \\ &= \int_C \int_B h(s,x,y) h(t-s,y,z) P_{\theta,x}(X_s \in dy, X_t \in dz). \end{aligned}$$

By comparing these two expressions for $P_{\theta_0,x}(X_s \in B, X_t \in C)$, we see that (12.2.3) holds for every $x \in E$ and for all s, t satisfying $0 < s < t$ for $\nu \times \nu$-almost all y and z (ν is the measure in Condition 12.2.1). The equation is obtained as stated by using the continuity of h and that $\nu(U) > 0$ for every non-void open set $U \subseteq E$.

Now, by choosing $x = y = z$ in (12.2.3) we obtain

$$h(t,x,x) = h(s,x,x)h(t-s,x,x), \quad 0 < s < t,$$

so, since h is continuous,

$$h(t,x,x) = e^{\alpha t} l(x), \quad t > 0, \tag{12.2.4}$$

for some real numbers $\alpha = \alpha(x)$ and $l(x)$, which might depend on x. Moreover, it follows from (12.2.3) that

$$h(t,x,z) = h(t/2,x,x)h(t/2,x,z)$$

and

$$h(t,x,z) = h(t/2,x,z)h(t/2,z,z).$$

Using (12.2.4), this implies that

$$\alpha(x) = \alpha(z) = \alpha \text{ and } l(x) = l(z) = l$$

for all $x, z \in E$. From (12.2.3) (with $x = y = z$) and (12.2.4) it follows that $l^2 = l$, and since h is strictly positive, we see that $l = 1$. Next define

$$H(x,z) = e^{-\alpha t} h(t,x,z)$$

for $t > 0$ and $x, z \in E$. From the equations (12.2.3) and (12.2.4) we have for $0 < s < t$ that

$$e^{-\alpha t} h(t,x,z) = e^{-\alpha s} h(s,x,z) e^{-\alpha(t-s)} h(t-s,z,z) = e^{-\alpha s} h(s,x,z).$$

Hence the function $H(x,z)$ does not depend on t. We have thus proved that there exists a real number $\alpha \neq 0$ and a non-constant continuous function $H : E \times E \mapsto (0,\infty)$ satisfying $H(x,x) = 1$ such that

$$h(t,x,z) = e^{\alpha t} H(x,z).$$

Obviously, α and H are uniquely determined.

As the final step in the proof of Theorem 12.2.2 define a strictly positive, continuous function g on E by

$$g(x) = H(x, y_0)$$

for some arbitrary, but fixed, $y_0 \in E$. Note that g is not constant. By multiplying (12.2.3) by $e^{\alpha t}$ we obtain

$$H(x,z) = H(x,y_0) H(y_0,z). \tag{12.2.5}$$

In particular,

$$H(z,y_0) H(y_0,z) = H(z,z) = 1, \tag{12.2.6}$$

so

$$H(x,z) = \frac{H(x,y_0)}{H(z,y_0)} = \frac{g(x)}{g(z)}. \tag{12.2.7}$$

By the definition of h we have that $g(y) P_{\theta_0}(t,x,dy) = e^{\alpha t} g(x) P_\theta(t,x,dy)$, and by integrating over E we see that (12.2.1) is satisfied, so that $(\alpha, g) \in \Xi_{\mathbb{P}^{\theta_0}}$. This completes the proof of Theorem 12.2.2. □

Example 12.2.3 Let X be a *Wiener process* with drift zero and diffusion coefficient σ under $\{P_x : x \in \mathbb{R}\}$. The transition function is

$$P(t,x,dy) = (2\pi t)^{-\frac{1}{2}} \exp\left(-\frac{1}{2\sigma^2 t}(y-x)^2\right) dy. \tag{12.2.8}$$

It is thus equivalent to the Lebesgue measure. Let us determine the largest class of conservative time-homogeneous Markov processes with transition functions that are equivalent to the Lebesgue measure and for which X is a sufficient process, i.e., we will find $\mathcal{Q}(P_x)$ for all $x \in \mathbb{R}$. To do this we need to find the strictly positive continuous solutions to (12.2.1). By elementary calculations, (12.2.1) implies that

$$\tfrac{1}{2}\frac{d^2}{dx^2}g = \alpha g. \tag{12.2.9}$$

A strictly positive solution g to this equation exists if and only if $\alpha \geq 0$. In this case, the set of all strictly positive solutions g is given by $\{g_{\alpha,a} \in \Theta\}$, where $\Theta = \{(\alpha, a) : ((\alpha > 0) \wedge (a \in [0,\infty)) \vee ((\alpha \geq 0) \wedge (a = \infty))\}$ and

$$g_{\alpha,a}(x) = \begin{cases} a\exp(x\sqrt{2\alpha}) + \exp(-x\sqrt{2\alpha}) & \text{if } 0 < \alpha,\ 0 \leq a < \infty \\ \exp(x\sqrt{2\alpha}) & \text{if } 0 \leq \alpha,\ a = \infty. \end{cases}$$

We identify functions that differ by a constant factor only. An easy calculation shows that these functions are also solutions of (12.2.1). We can therefore define $P_{(\alpha,a),x} = P_{\alpha,g_{\alpha,a},x}$.

Under $\mathbb{P}^{\alpha,a}$ the process X is Markovian with transition function

$$P_{\alpha,a}(t,x,dy) = e^{-\alpha t}\frac{g_{\alpha,a}(y)}{g_{\alpha,a}(x)}P(t,x,dy),$$

where P is given by (12.2.8). By applying Ito's formula to the process $\log(dP^t_{(\alpha,a),x}/dP^t_x) = \log[e^{-\alpha t}g_{\alpha,a}(X_t)/g_{\alpha,a}(x)]$ and using (12.2.9), it follows (cf. Exercise 12.6) that X under $\mathbb{P}^{\alpha,a}$ is a diffusion process on $\mathbb{R}$ with diffusion coefficient σ and drift

$$b_{\alpha,a}(x) = \frac{g'_{\alpha,a}(x)}{g_{\alpha,a}(x)} = \begin{cases} \sqrt{2\alpha}\frac{a\exp(x\sqrt{2\alpha}) - \exp(-x\sqrt{2\alpha})}{a\exp(x\sqrt{2\alpha}) + \exp(-x\sqrt{2\alpha})} & \text{if } 0 \leq a < \infty \\ \sqrt{2\alpha} & \text{if } a = \infty. \end{cases}$$

Note that by restricting to the cases $a = 0$ and $a = \infty$ we obtain the class of Wiener processes with diffusion coefficient σ and with a drift that varies freely in $\mathbb{R}$, i.e., the natural exponential family generated by the Wiener process with drift zero and diffusion coefficient σ. □

Example 12.2.4 (Watanabe, 1975) Consider the *Bessel process* given as the solution of

$$dX_t = \theta X_t^{-1}dt + dW_t, \quad X_0 = x > 0,$$

where W is a standard Wiener process and $\theta \in \Theta = [\frac{1}{2}, \infty)$; see Example 6.2.2. The state space is $(0,\infty)$. Denote by $P_\theta(t,x,B)$ the transition function of the Bessel process given by (6.2.5). Then for $\alpha \geq 0$ we can find a positive function $g_\alpha(x)$ satisfying (12.2.1) or, equivalently,

$$\tfrac{1}{2}g''_\alpha(x) + \theta x^{-1}g'_\alpha(x) = \alpha g_\alpha(x), \tag{12.2.10}$$

with the initial conditions $g_\alpha(0) = 1,\ g'_\alpha(0) = 0$. The unique solution of (12.2.10) with these initial conditions is given by

$$g_\alpha(x) = 2^{\theta-\frac{1}{2}}\Gamma(\theta+\tfrac{1}{2})(\sqrt{2\alpha})^{\frac{1}{2}-\theta}I_{\theta-\frac{1}{2}}(x\sqrt{2\alpha}),\quad x \geq 0.$$

Here Γ denotes the gamma function, and I_γ denotes the modified Bessel function of the first kind of order γ. Thus, for fixed θ we can define a class of probability measure $\mathcal{Q}(P_{\theta,x}) = \{P_{\theta,\alpha,x} : (\alpha, g_\alpha) \in \Xi_{\boldsymbol{P}}\}$ of the type (12.2.2). Under $P_{\theta,\alpha,x}$ the process X is a solution of

$$dX_t = \theta X_t^{-1}dt + \frac{g'_\alpha(X_t)}{g_\alpha(X_t)}dt + dW_t,\quad X_0 = x > 0,$$

and has transition function

$$P_\theta(t,x,dy)\frac{g_\alpha(y)}{g_\alpha(x)}e^{-\alpha t}$$
$$= e^{-\alpha t}t^{-1}g_\alpha(y)g_\alpha(x)^{-1}\exp\left[-\frac{x^2+y^2}{2t}\right](xy)^{\frac{1}{2}-\theta}y^{2\theta}I_{\theta-\frac{1}{2}}\left(\frac{xy}{t}\right)dy.$$ □

Example 12.2.5 Consider a *birth-and-death process* X with birth intensities λ_i, $i \geq 0$, and death intensities μ_i, $i \geq 1$. It is known (see, e.g., Karlin and McGregor (1957)) that X is conservative (non-exploding) if and only if

$$\sum_{i=0}^{\infty}\frac{1}{\lambda_i\pi_i}\sum_{k=0}^{i}\pi_k = \infty, \tag{12.2.11}$$

where $\pi_0 = 1$ and

$$\pi_i = \frac{\lambda_0\lambda_1\cdots\lambda_{i-1}}{\mu_1\mu_2\cdots\mu_i},\quad i \geq 1.$$

Suppose that (12.2.11) and

$$\sum_{i=0}^{\infty}\pi_i\sum_{k=0}^{i-1}\frac{1}{\lambda_k\pi_k} = \infty \tag{12.2.12}$$

hold. Then $\mathcal{Q}(P_x)$ consists of the birth-and-death processes with intensities

$$\lambda_i^{(\alpha)} = \lambda_i\frac{Q_{i+1}(\alpha)}{Q_i(\alpha)},\quad i \geq 0,\quad \text{and}\quad \mu_i^{(\alpha)} = \mu_i\frac{Q_{i-1}(\alpha)}{Q_i(\alpha)}\quad i \geq 1,$$

where $\alpha \in [\alpha_0, \infty)$. Here $Q_i(\alpha)$ is the polynomial

$$Q_i(\alpha) = \sum_{k=0}^{i}u_k(i)\alpha^i,\quad i \geq 0,$$

where $u_0(i) = 1$, $i \geq 0$, $u_k(i) = 0$, $k > i$, and

$$u_k(i) = \sum_{j=k-1}^{i-1} \frac{1}{\lambda_j \pi_j} \sum_{l=k-1}^{j} \pi_l u_{k-1}(l), \quad 1 \leq k \leq i.$$

These polynomials were studied thoroughly in Karlin and McGregor (1957). The number α_0 is defined by

$$\alpha_0 = \lim_{i \to \infty} \xi_i,$$

where ξ_i is the largest zero point of the polynomial Q_i. It can be shown that the limit exists and is non-positive.

A proof of the result in this example can be found in Küchler (1982b). Here we just note that $g_\alpha(i) = Q_i(\alpha)$ solves (12.2.1). When (12.2.11) and (12.2.12) are both satisfied, ∞ is a so-called natural boundary. For a concrete example, see Exercise 12.4. □

Assume that $X = \{X_t : t \geq 0\}$ is (for simplicity) a real-valued Lévy process satisfying $X_0 = 0$ under P_0, and let $\mathcal{P}$ be the natural exponential family generated by X. The following results can be generalized to the case when X takes values in $\mathbb{R}^d$ or even in a more general semi-group. The process X is sufficient for $\mathcal{P}$, and (under weak regularity conditions) $\mathcal{P}$ is the largest set of probability measures under which X has independent stationary increments and is a sufficient process. Denote by $\mathcal{M}(P_0)$ the largest family of probability measures containing P_0 for which X is a sufficient process. In general, the class $\mathcal{M}(P_0)$ is not an exponential family in the sense of this book. We shall study its structure and its connections with the natural exponential family generated by P_0.

It can be shown that for every $P \in \mathcal{M}(P_0)$ the process X is Markovian; indeed, it has interchangeable increments (see Freedman, 1963). In general it is not time-homogeneous. We have not earlier defined time-inhomogeneous Markov processes. Here it is enough to say that they satisfy an obvious generalization of the time-homogeneous Markov property (6.2.1). The finite-dimensional distributions of X under P are determined by the transition function $P(s, x; t, dy) = P(X_t \in dy | X_s = x)$. Specifically,

$$P(X_{t_1} \in dy_1, \cdots, X_{t_n} \in dy_n) = \prod_{i=1}^{n} P(t_{i-1}, y_{i-1}; t_i, dy_i),$$

for $0 < t_1 < \cdots < t_n$, where $t_0 = y_0 = 0$ since we have assumed that $X_0 = 0$.

Under mild regularity conditions (see Küchler and Lauritzen, 1989), every $P \in \mathcal{M}(P_0)$ has a transition function given by

$$P(s, x; t, dy) = P_0(x + X_{t-s} \in dy) \frac{h(t, y)}{h(s, x)} \qquad (12.2.13)$$

for $(s,x) \in \{(u,z) : h(u,z) > 0\}$, where h is a uniquely determined non-negative function on $(0,\infty) \times \mathbb{R}$ satisfying

$$\int h(t,y) P_0(X_t \in dy) = 1, \quad t > 0,$$

and

$$h(s,x) = \int_{\mathbb{R}} h(t,y) P_0(x + X_{t-s} \in dy), \quad 0 < s < t, \quad x \in \mathbb{R}. \tag{12.2.14}$$

Such functions are called *space-time invariant functions.* Moreover, under the mentioned regularity conditions, the mapping $P \mapsto h$ is one-to-one from $\mathcal{M}(P_0)$ onto the set $\mathcal{H}$ of all space-time invariant functions. It is possible to describe the structure of the set $\mathcal{H}$ and therefore of the set $\mathcal{M}(P_0)$. This is done as follows. First note that $\mathcal{M}(P_0)$ and $\mathcal{H}$ are convex sets. The idea is to find the sets of extreme points of $\mathcal{M}(P_0)$ and $\mathcal{H}$, and to represent every element of these sets as a mixture of their extreme points.

By the mapping $P \mapsto h$ the natural exponential family $\mathcal{P}$ generated by P_0 corresponds to the functions

$$h_\theta(s,x) = \exp[\theta x - s\psi(\theta)], \quad s > 0, \ x \in \mathbb{R},$$

where $\theta \in \Theta = \{u \in \mathbb{R} : \exp[\psi(u)] = \int e^{ux} P_0(X_1 \in dx) < \infty\}$. We will consider three cases: (i) the support of $P_0(X_t \in \cdot)$ is $\mathbb{N}$ and $P_0(X_t = 0) > 0$, $t > 0$ (example: a Poisson process), (ii) the support of $P_0(X_t \in \cdot)$ is $\mathbb{R}$ and $P_0(X_t = 0) > 0$, $t > 0$ (example: a compound Poisson process with absolutely continuous jump distribution), or (iii) the support of $P_0(X_t \in \cdot)$ is $\mathbb{R}$ and $P_0(X_t = 0) = 0$, $t > 0$ (example: a Wiener process). Note that $P_0(X_t = 0) > 0$ implies that $P_0(X_t = 0) = e^{-at}$ for some $a > 0$. Define $E(P_0) = \{h_\theta : \theta \in \Theta\}$. In case (iii) the set of extreme points of the convex set $\mathcal{H}$ is $E(P_0)$. In the cases (i) and (ii) we have to extend $E(P_0)$ by the function $h_{-\infty}(s,x) = 1_{\{0\}}(x) \cdot e^{as}$, where a equals the jump intensity in both cases, to get the set of extreme points. This corresponds formally, at least in case (i), to $\theta = -\infty$, which explains the notation. In the cases (i) and (ii) we include the element $-\infty$ in Θ.

Theorem 12.2.6 *Every element h of $\mathcal{H}$ can be represented uniquely by*

$$h(s,x) = \int_\Theta h_\theta(s,x) \mu_h(d\theta),$$

where μ_h is a probability measure on Θ depending on h. In particular, every element of $\mathcal{M}(P_0)$ has a unique representation

$$P_h(A) = \int_\Theta P_\theta(A) \mu_h(d\theta), \quad A \in \mathcal{F}, \tag{12.2.15}$$

where P_h is the probability measure corresponding to h, cf. (12.2.13), and P_θ denotes the probability measure corresponding to h_θ, $\theta \in \Theta$ (in the extended sense).

A proof can be found in Küchler and Lauritzen (1989), where the reader will also find a treatment of more cases than (i), (ii), and (iii). The proof is based on a result by Lauritzen (1984) on representation of elements of convex sets by the extreme points of the set.

The idea of constructing or motivating statistical models by finding the extreme points of the convex set of probability measures under which a given (possibly multi-dimensional) process is sufficient has been pursued in more general settings by Lauritzen (1984, 1988). A model obtained in this way is called an *extreme point model.* Thus Theorem 12.2.6 shows that $\mathcal{M}(P_0)$ is an extreme point model. It is not, in general, the case that exponential families of stochastic processes are extreme point models. Indeed, non-ergodic models in the sense explained in Chapter 8 are not extreme point models; see Lauritzen (1984). For instance, the extreme point model corresponding to the exponential family of pure birth processes does not contain this exponential family, but consists of inhomogeneous Poisson processes, see Jensen and Johansson (1990).

Note that by mixing measures $\{P_\theta\}$ under each of which a process X is Markovian one does not, in general, obtain a probability measure under which X is Markovian. The following is a simple example.

Example 12.2.7 (Salminen, 1983) Let X be a standard Wiener process under P_0 and an Ornstein-Uhlenbeck process under P_1. Then X is not a Markov process under $P = \frac{1}{2}(P_0 + P_1)$. □

Example 12.2.3 (continued). Assume that $X = \{X_t : t \geq 0\}$ is a real-valued Wiener process on $(\Omega, \mathcal{F}, P_0)$, and choose a probability measure U on $\mathbb{R}$ such that

$$h(s,x) = \int_{-\infty}^{\infty} e^{\theta x - \theta^2 s/2} U(d\theta) \tag{12.2.16}$$

is finite for all $x \in \mathbb{R}$ and $s \geq 0$. Then h is a space-time invariant function for X under P_0, and under the corresponding probability measure $P^U \in \mathcal{M}(P_0)$, defined by (12.2.13), the process X solves the stochastic differential equation

$$dX_t = a(t, X_t)dt + dW_t, \quad t \geq 0, \quad X_0 = 0,$$

where

$$a(t,x) = \frac{\partial}{\partial x} \log h(t,x), \quad x \in \mathbb{R}, \quad t \geq 0,$$

and where W denotes a standard Wiener process; see Exercise 12.6.

If $U = \delta_a$ (the Dirac measure at a) for some $a \in \mathbb{R}$, we get $a(s,x) = a$; i.e., X is a Wiener process with drift a under P_a, and $E_a(W_t) = at$. In the general case, X is a drift-mixture of Wiener processes under P^U. □

Related work has been done by Grigelionis (1996). We shall briefly review his main results, which are interesting first steps towards a theory for infinite-dimensional exponential families of stochastic processes.

Grigelionis starts with a process X with values X_t in a separable Hilbert space H that has independent increments under P_0. In a way similar to the finite-dimensional case, he constructs the natural exponential family $\mathcal{P} = \{P_\theta : \theta \in \Theta\}$ generated by P_0 by means of the functions

$$f_\theta(t,h) = \exp[(\theta,h) - \psi(t,\theta)], \quad h \in H, \quad \theta \in \Theta,$$

where $(\cdot,\cdot)$ denotes the inner product in H. The set $\Theta \subseteq H$ is the largest set of parameters θ for which f_θ is integrable with respect to P_0, while the function $\psi(t,\theta)$ is determined such that the integral of f_θ with respect to P_0 is one.

Then he constructs space-time invariant functions $q^U(t,h)$, $t \geq 0, h \in H$, by mixing the functions $f_\theta(t,h)$ with a mixing probability U on Θ, and corresponding probability measures P^U by mixing the measures P_θ with the same probability U. Under P^U the process X is a (time-inhomogeneous) Markov process with transition function

$$P^U(s,h;t,\Gamma) = \int_\Gamma \frac{q^U(t,k)}{q^U(s,h)} P_0(h + X_t - X_s \in dk).$$

Thus, a considerable part of Section 12.2 is likely to go through also for Hilbert-space-valued processes and exponential families of such processes with infinite-dimensional parameter.

Let us mention another variation of the theme exponential families of processes. Assume (X,Y) is a Markov additive process with respect to $\{\mathcal{F}_t\}$. This means, roughly speaking, that Y is Markovian with respect to $\{\mathcal{F}_t\}$ with state space E and X is an R^d-valued process with increments $X_t - X_s$ that conditionally on Y_s are independent of $\mathcal{F}_s$. Another way of saying this is that X is a process with independent increments in a random environment. By combining the definition of natural exponential families of Lévy process (Chapter 2) and the families $\mathcal{Q}$ constructed for Markov processes in this section, one can define exponential families of a certain mixed type. This was done by Grigelionis (1994), who called the elements of a thus defined exponential family the Doob-Esscher-transforms of the underlying Markov additive process and studied the existence and the properties of the so-called Lundberg exponent. These results extend the classical theory of Lundberg exponents for risk processes in insurance mathematics; see, e.g., Grandell (1992).

It is of interest to discuss what may happen if we relax Condition 12.2.1. We shall do this for one-dimensional diffusion-type processes. Consider a stochastic differential equation of the type

$$dX_t = b_t(X)dt + c_t(X)dW_t, \quad X_0 = x, \tag{12.2.17}$$

where X is one-dimensional and W is a standard one-dimensional Wiener process, while b_t and c_t are functionals depending on $\{X_s : s \leq t\}$ only. If a probability measure exists on the canonical space $C([0,\infty))$ of continuous functions from $[0,\infty)$ into $\mathbb{R}$ such that under this measure the canonical process, $X_t(\omega) = \omega(t)$, $\omega \in C([0,\infty))$, is a solution to (12.2.17), then we call this measure a weak solution to the equation (12.2.17). We will consider classes of weak solutions given by stochastic differential equations where $c > 0$ is a fixed functional, while the drift functional b is parametrized by a set $\mathcal{G}$ of functions. We denote the measure corresponding to the function $g \in \mathcal{G}$ by P_g and define $\mathcal{P}(\mathcal{G}) = \{P_g : g \in \mathcal{G}\}$. The class of drift functionals is denoted by $\mathcal{A}(\mathcal{G}) = \{A_g : g \in \mathcal{G}\}$.

First we discuss the Markovian case, where $b_t(X) = h(t, X_t)$ and $c_t(X) = \sigma(t, X_t)$ for some functions h and σ (drift and diffusion coefficient). In the Markovian case a sufficient condition that a unique weak solution exist is that h and σ be continuous functions and $\sigma > 0$; see Jacod (1979). In the following, $\sigma > 0$ is a fixed continuous function. Let a be a given continuous drift function. The stochastic differential equation with drift a and diffusion coefficient σ has a weak solution P_a. In the following, Condition 12.2.1 is replaced by a condition ensuring that $P_g^t \ll P_a^t$ for all $g \in \mathcal{G}$ and all $t > 0$.

Theorem 12.2.8 *Let $\mathcal{G}$ be the class of functions $g : [0,\infty) \times \mathbb{R} \to \mathbb{R}$ satisfying*

(1) *g solves the differential equation*

$$\frac{\partial g}{\partial t} + \tfrac{1}{2}\frac{\partial}{\partial x}\left[\sigma(t,x)^2\left(g(t,x)^2 + \frac{\partial g}{\partial x}\right)\right] = 0. \tag{12.2.18}$$

(2) $P_g\left(\int_0^t \sigma(s,X_s)^2\left[g(s,X_s) - \sigma(s,X_s)^{-2}a(s,X_s)\right]^2 ds < \infty\right) = 1$ *for all* $t > 0$.

Then X is a sufficient process for $\mathcal{P}(\mathcal{G})$, where $\mathcal{A}(\mathcal{G}) = \{\sigma^2 g : g \in \mathcal{G}\}$.

Proof: Condition 2 in the theorem ensures that $P_g^t \ll P_a^t$ for all $g \in \mathcal{G}$ and all $t > 0$; see Jacod and Shiryaev (1987, Theorem III-5-34). The likelihood function $L_t(g) = dP_g^t/dP_a^t$ for $\mathcal{P}(\mathcal{G})$ based on observation of X in $[0,t]$ has the form

$$\log L_t(g) = \int_0^t g(s,X_s)dX_s - \tfrac{1}{2}\int_0^t [\sigma(s,X_s)g(s,X_s)]^2 ds + H_t,$$

where H_t does not depend on g; see Appendix A. Since g satisfies the equation (12.2.18), there exists a function $\Lambda : [0,\infty) \times \mathbb{R} \to \mathbb{R}$ such that $\partial\Lambda/\partial x = g$ and

$$-\tfrac{1}{2}[\sigma(t,x)g(t,x)]^2 = \tfrac{1}{2}\sigma^2(t,x)\frac{\partial^2\Lambda}{\partial x^2} + \frac{\partial\Lambda}{\partial t}.$$

Hence

$$\log L_t(g) \quad = \quad H_t + \int_0^t \frac{\partial\Lambda}{\partial x}(s,X_s)dX_s$$

$$
\begin{aligned}
&+\int_0^t \frac{\partial \Lambda}{\partial s}(s,X_s)ds + \tfrac{1}{2}\int_0^t \frac{\partial^2 \Lambda}{\partial x^2}(s,X_s)\sigma^2(s,X_s)ds \\
= \ & \Lambda(t,X_t) - \Lambda(0,x) + H_t,
\end{aligned}
$$

where we have used Ito's formula; see Section A.6. We see that the factor of the likelihood function dependent on g depends on the data through X_t only, so by the factorization criterion (see, e.g., Lehmann, 1983, p. 39) X is a sufficient process for $\mathcal{P}(\mathcal{G})$. □

The next corollary follows immediately from the theorem.

Corollary 12.2.9 *Suppose σ is a function of x only, i.e., $\sigma(t,x) = \sigma(x)$. Then X is a sufficient process for $\mathcal{P}(\mathcal{G})$, where $\mathcal{A}(\mathcal{G}) = \{\sigma^2 g : g \in \mathcal{G}\}$, provided that $\mathcal{G}$ is the class of functions $g : \mathbb{R} \to \mathbb{R}$ satisfying condition (2) of Theorem 12.2.8 and the differential equation*

$$
g'(x) + g(x)^2 + \alpha\sigma(x)^{-2} = 0
$$

for some $\alpha \in \mathbb{R}$.

The following example is an application of Corollary 12.2.9.

Example 12.2.10 Consider the case where σ is constant. For simplicity we assume $\sigma(\cdot) \equiv 1$. The solutions to the Riccati equation

$$
g' + g^2 + \alpha = 0 \tag{12.2.19}
$$

are

$$
g_{\alpha,\kappa}(x) = \begin{cases}
\sqrt{-\alpha}\,\frac{\kappa - \exp(-2x\sqrt{-\alpha})}{\kappa + \exp(-2x\sqrt{-\alpha})} & \alpha < 0 \\
(\kappa + x)^{-1} & \alpha = 0 \\
\sqrt{\alpha}\,\frac{1+\kappa\cot(x\sqrt{\alpha})}{\kappa - \cot(x\sqrt{\alpha})} & \alpha > 0,
\end{cases}
$$

where $\kappa \in \mathbb{R} \cup \{\infty\}$ in all cases. To each of these solutions corresponds a diffusion with drift $g_{\alpha,\kappa}$ for which Condition 2 in Theorem 12.2.8 is satisfied; see Sørensen (1986b). For $\alpha = 0$ and $\kappa = \infty$ we obtain the standard Wiener process. We have thus constructed a class of diffusions containing the Wiener process for which X is a sufficient process. Because we have not imposed Condition 12.2.1, we have obtained a larger class than the one found in Example 12.2.3, which corresponds to the case $\alpha < 0$ and $\kappa \in [0,\infty]$ or $(\alpha,\kappa) = (0,\infty)$. When $\alpha < 0$ and $\kappa < 0$, the diffusion has state space $(-\infty, -\frac{1}{2}\log(-\kappa)/\sqrt{-\alpha})$ or $(-\frac{1}{2}\log(-\kappa)/\sqrt{-\alpha}, \infty)$. When $\alpha = 0$ the state space is $(-\infty,-\kappa)$ or $(-\kappa,\infty)$, while finally, for $\alpha > 0$ the state space is one of the intervals $(\alpha^{-\frac{1}{2}}\mathrm{Arccot}(\kappa)+n\pi, \alpha^{-\frac{1}{2}}\mathrm{Arccot}(\kappa)+(n+1)\pi)$, where $n \in \mathbb{Z}$. In all cases the finite boundaries are entrance boundaries. □

We have seen that a diffusion process can be embedded in a class of diffusions for which a sufficient process exists. It is natural to ask whether this is possible for the more general process that solves the stochastic differential equation (12.2.17).

It can certainly be embedded in the exponential family given by the weak solutions to

$$dX_t = \theta b_t(X)dt + c_t(X)dW_t, \quad X_0 = x,$$

where $\theta \in \Theta \subseteq \mathbb{R}$, that provided a weak solution exists for all $\theta \in \Theta$ and that condition (3.5.3) is satisfied. The sufficient process for this exponential family is the canonical process (A, B) defined by (3.5.2) and (3.5.5). We have the following result, the proof of which is omitted as it is rather similar to the proof of Theorem 12.2.8; see Sørensen (1986b).

Theorem 12.2.11 *Let $\mathcal{G}$ be the class of functions $g : [0,\infty) \times \mathbb{R} \mapsto \mathbb{R}$ satisfying the differential equation*

$$\frac{\partial g}{\partial t} + g\frac{\partial g}{\partial x} + \tfrac{1}{2}\frac{\partial^2 g}{\partial x^2} = 0, \tag{12.2.20}$$

for which the stochastic differential equation with drift $g(B_t, A_t)b_t(X)$ and diffusion coefficient $c_t(X)$ has a unique weak solution, and

$$P_g\left(\int_0^t c_s(X)^{-2} b_s(X)^2 \left[g(B_s, A_s) - 1\right]^2 ds < \infty\right) = 1$$

for all $t > 0$. Then (A, B) is a sufficient process for for $\mathcal{P}(\mathcal{G})$, where $\mathcal{A}(\mathcal{G}) = \{gb : g \in \mathcal{G}\}$.

Note that a constant function g is a solution of (12.2.20), so the exponential family mentioned above is included in $\mathcal{P}(\mathcal{G})$. In this case the last condition in Theorem 12.2.11 equals condition (3.5.3).

Note also that for functions g of the form $g(t,x) = g(x)$, equation (12.2.20) reduces to the Riccati equation (12.2.19) for some $\alpha \in \mathbb{R}$. The solutions of equation (12.2.19) were given in Example 12.2.10.

12.3 Exercises

12.1 Let X be a birth-and-death process that is stable with respect to the natural exponential family $\mathcal{P}_\mu$ generated by a strictly positive measure μ on $\{0, 1, \cdots\}$. Assume that $\alpha_{00} = -3$, $\alpha_{10} = 2$, $\alpha_{01} = 2$, $\alpha_{11} = -6$, and $\alpha_{21} = 3$, where α_{ij} is given by (12.1.2). Show that the birth rate is given by

$$2(n+1)(1 - 2^{-(n+2)})/(1 - 2^{-(n+1)}), \quad n = 0, 1, 2, \ldots,$$

and the death rate by

$$(n+1)(1-2^{-n})/(1-2^{-(n+1)}), \quad n=1,2,\ldots.$$

Show also that

$$\mu(n)=1-2^{-(n+1)}, \quad n=0,1,2,\ldots,$$

and that $\mathcal{P}_\mu=\{P_\theta : \theta<0\}$, where

$$P_\theta(n)=(1-e^\theta)(2-e^\theta)(1-2^{-(n+1)})e^{n\theta}, \quad n=0,1,2,\ldots.$$

12.2 Let X be a standard Wiener process on $\mathbb{R}^n$ (in particular, $X_0=0$). Show that X is stable with respect to the exponential familiy of n-dimensional normal distributions with zero mean and covariance matrix $\sigma^2 I_n$, $\sigma^2>0$, where I_n denotes the n-dimensional identity matrix. Then show that $Y_t=\|X_t\|^2$ is a Markov process on $[0,\infty)$ that is stable with respect to the exponential family of $\sigma^2\chi^2(n)$-distributions, $\sigma^2>0$.

12.3 Let X be a Brownian motion on $\mathbb{R}^n$ ($n\geq 2$) with drift $(1,0,\cdots,0)^T$ and diffusion matrix I_n. Suppose $X_0=0$. Show that X is stable with respect to the exponential family generated by μ and f, where $\mu(dx)=\exp(x_1)dx$ and $f(x)=x_1^2+\cdots+x_n^2$, with $x=(x_1,\cdots,x_n)^T$ and with dx denoting the Lebesque-measure on $\mathbb{R}^n$. Then show that the image of μ by f has a density g with respect to the Lebesgue measure on $\mathbb{R}_+$ given by

$$g(y)=\tfrac{1}{2}V_{n-1}\int_{-1}^{1}[t\sqrt{y}+n]y^{n/2-1}(1-t^2)^{(n-1)/2}e^{t\sqrt{y}}dt,$$

where V_n denotes the volume of the unit ball in $\mathbb{R}^n$. Finally show that the diffusion process Y on $[0,\infty)$ that solves the stochastic differential equation

$$dY_t=[n+2h(Y_t)]dt+\sqrt{2Y_t}dW_t,$$

where

$$h(y)=\tfrac{1}{2}V_{n-1}g(y)^{-1}\int_{-1}^{1}t(t\sqrt{y}+n+1)[y(1-t^2)]^{(n-1)/2}e^{t\sqrt{y}}dt,$$

is stable with respect to the natural exponential family generated by ν (the image of μ by f) with dynamic parameter function $\theta(t)=-\tfrac{1}{2}t^{-1}$.

12.4 Consider a birth-and-death process with birth intensities $\lambda_i=i+1$, $i\geq 0$, and death intensities $\mu_i=i$, $i\geq 1$. Show that in this case the polynomials $Q_i(\alpha)$ in Example 12.2.5 are the Laguerre polynomials

$$Q_i(\alpha)=\sum_{k=0}^{i}\binom{i}{k}\frac{\alpha^k}{k!}, \quad i=0,1,\cdots, \quad \alpha\in\mathbb{R}.$$

It can be shown that $\alpha_0=0$.

12.5 Assume that $X = \{X_t : t \geq 0\}$ is a real-valued process with stationary increments under P_0 and with $X_0 = 0$, P_0-a.s.; let h be a strictly positive space-time invariant function (cf. (12.2.14)); and define

$$P_h(s, x; t, dy) = P_0(x + X_{t-s} \in dy)\frac{h(t, y)}{h(s, x)},$$

where $x, y \in \mathbb{R}$, $0 < s < t$. Moreover, let P_h^t be the measure on $\mathcal{F}_t^X$ (the σ-algebra generated by X) corresponding to $P_h(s, x; t, dy)$ and given by

$$\begin{aligned} &P_h^t(X_{t_1} \in dy_1, \cdots, X_{t_n} \in dy_n) \\ &\quad = P_h(0, 0; t_1, dy_1) \cdots P_h(t_{n-1}, y_{n-1}; t_n, dy_n). \end{aligned}$$

Deduce that P_h^t is absolutely continuous with respect to the restriction of P_0 to $\mathcal{F}_t^X$, and show that

$$\frac{dP_h^t}{dP_0^t} = h(t, X_t), \;\; P_0\text{-almost surely}, \;\; t > 0.$$

12.6 Consider Example 12.2.3 (continued). Assume that $h(s, x)$ is a function of the form (12.2.16), and let P_h denote the probability measure defined by P_0 and h; cf. (12.2.13). Then the likelihood function

$$L_t(h) = \frac{dP_h^t}{dP_0^t} = h(t, X_t)$$

is a strictly positive continuous P_0-martingale. Apply Ito's formula (Appendix A) to $\log L_t$ and use Girsanov's theorem (Appendix A) to show that under P_h the process X satisfies

$$dX_t = \left[\frac{\partial}{\partial x} \log h(t, X_t)\right] dt + dW_t, \quad X_0 = 0,$$

for some standard Wiener process W.

The process X will obviously be time-homogeneous under P_h if and only if $h(s, x) = h_0(s)g_0(x)$ for some functions h_0 and g_0. Prove that in this case there exists an $\alpha \in [0, \infty)$ such that $h_0(s) = \exp[\alpha s]$ and $g(x) = a_1 e^{x\sqrt{2\alpha}} + a_2 e^{-x\sqrt{2\alpha}}$. Compare this with (12.2.1) and Example 12.2.3.

12.7 Let $N = \{N_t : t \geq 0\}$ be a Poisson process with parameter $\lambda = 1$. Then every space-time invariant function $g(s, j)$ has the form

$$g(s, j) = \int_{[-\infty, \infty)} g_\alpha(s, j)\mu(d\alpha)$$

for some probability measure μ on $[-\infty,\infty)$ depending on g, and

$$\begin{aligned} g_\alpha(s,j) &= \exp[\alpha j - (e^\alpha - 1)s], \quad \alpha \in (-\infty,\infty), \\ g_{-\infty}(s,j) &= 1_{\{0\}}(j)\cdot e^{-s}; \end{aligned}$$

see Küchler and Lauritzen (1989) and compare with Example 2.1.1. Reparametrization by $\theta = e^\alpha$ yields for $h_\theta = g_\alpha$

$$h_\theta(s,j) = \exp[j\log(\theta) - (\theta-1)s], \quad \theta \in (0,\infty),$$

$$h_0(s,j) = 1_{\{0\}}(j)\cdot e^{-s}, \qquad \theta = 0.$$

Choose

$$\mu(d\theta) = \beta^\gamma \theta^{\gamma-1} e^{-\beta\theta}\Gamma(\gamma)^{-1} d\theta, \quad \theta > 0,$$

and show that under P_μ constructed as in (12.2.15) the process N has the marginal distributions

$$P^U(N_t = j) = \binom{\gamma+j-1}{j}\left(\frac{\beta}{\beta+t}\right)^\gamma \left(\frac{t}{\beta+t}\right)^j,$$

for $t \geq 0$ and $j = 0,1,2,\cdots$. This process is called the Pólya process. For a multi-dimensional version of this exercise, see Grigelionis (1996).

12.8 Consider Example 12.2.4. Show that $(X_t, \int_0^t X_s^{-2}ds)$ is a sufficient statistic for $\{P_{\theta,\alpha,x} : (\theta,\alpha) \in [\frac{1}{2},\infty)\times[0,\infty)\}$ when X has been observed continuously up to time t. This sufficient process is also minimal sufficient for the exponential family of Bessel processes, which is the subfamily obtained for $\alpha = 0$. Compare this result to Theorem 12.2.11.

12.4 Bibliographic Notes

The results presented in Section 12.1 are taken from Ycard (1988, 1989, 1992a,b), while the results in the first part of Section 12.2 are mainly from Küchler (1982a,b) and Küchler and Lauritzen (1989). Related work has been done by Grigelionis (1996). Mixed Poisson processes and their martingale characterizations were studied in Grigelionis (1997). Exponential families of stochastic processes with values in m-dimensional differentiable manifolds containing penetrable boundaries can be constructed by means of results in Grigelionis (1991). More material on extreme point models and related subjects can be found in Lauritzen (1984, 1988), Jensen and Johansson (1990), Diaconis and Freedman (1980), and Björk and Johansson (1995). The last part of Section 12.2 is based on work by Sørensen (1986b).

Definitions of exponential families of stochastic processes other than those presented in Chapter 12 have been proposed. Feigin (1981) called a class of discrete time Markov processes an exponential family if the transition densities formed an exponential family of distributions. Feigin's exponential families are only exponential families of stochastic processes in our sense when the transition densities have the special exponential structure (3.2.1); see also Chapter 6. Feigin gave a representation theorem for the canonical process in this special case (for γ and h one-dimensional). A generalization of this result is given in Section 5.5. Franz and Winkler (1987) used the term exponential family to denote a model with likelihood function of the form (3.1.2) where, however, $\phi_t(\theta)$ was allowed to be random. These models are not generally exponential families of stochastic processes in our sense, but could be seen as a generalization of Feigin's exponential families. Recently, Dinwoodie (1995) has proposed yet another definition of an exponential family of stochastic processes, which, however, seems to have rather limited applicability.

Appendix A

A Toolbox from Stochastic Calculus

In this appendix we give without proof a number of results from stochastic calculus that are used in this book. The results are generally useful within the field of statistical inference for stochastic processes. Some statements hold only up to a null-set; we will, however, ignore this complication when it is of no consequence for statistical applications. A number of concepts and results that are explained in detail in the main chapters of the book are not presented here again.

Much of the material in this appendix is based on Jacod and Shiryaev (1987) and Rogers and Williams (1987, 1994), so if the reader wishes to study stochastic calculus more thoroughly, it is a good idea to take a look at these books.

A.1 Stochastic basis

A *stochastic basis* is a probability space $(\Omega, \mathcal{F}, P)$ equipped with a *right-continuous filtration*, i.e., a family $\{\mathcal{F}_t\}_{t \in \mathbb{R}_+}$ of σ-algebras on Ω satisfying

$$\mathcal{F}_s \subseteq \mathcal{F}_t \subseteq \mathcal{F}, \quad s \leq t,$$
$$\mathcal{F}_t = \bigcap_{s>t} \mathcal{F}_s.$$

Sometimes we will refer to a stochastic basis as a *filtered probability space*. If no probability measure P is specified, we call $(\Omega, \mathcal{F}, \{\mathcal{F}_t\})$ a *filtered space*. Note that we always assume that $\{\mathcal{F}_t\}$ is right-continuous. A stochastic process $X = \{X_t\}_{t \in \mathbb{R}_+}$ is called *adapted* to a filtration $\{\mathcal{F}_t\}$ if X_t is $\mathcal{F}_t$-measurable for all $t \geq 0$.

This continuous time setting also covers *discrete time processes.* Let $\{X_n\}_{n \in \mathbb{N}}$ be a discrete time process and $\{\mathcal{F}_n\}_{n \in \mathbb{N}}$ a discrete time filtration such that X_n is $\mathcal{F}_n$-measurable for all $n \in \mathbb{N}$. Then we can consider X a continuous time process by the definition

$$Y_t = X_{[t]}, \quad t \in \mathbb{R}_+,$$

where $[t]$ denotes the integer part of t. The process Y is adapted to the (right-continuous) continuous time filtration $\{\mathcal{G}_t\}$ defined by

$$\mathcal{G}_t = \mathcal{F}_{[t]}, \quad t \in \mathbb{R}_+.$$

Given a stochastic basis $(\Omega, \mathcal{F}, \{\mathcal{F}_t\}, P)$, we can define the *predictable σ-algebra* $\mathcal{P}$ as the smallest σ-algebra on $\Omega \times \mathbb{R}_+$ with respect to which the mapping $(\omega, t) \mapsto X_t(\omega)$ is measurable for every left-continuous stochastic process X that is adapted to $\{\mathcal{F}_t\}$. The predictable σ-algebra does not depend on P.

A stochastic process X is called *predictable* if the mapping $(\omega, t) \mapsto X_t(\omega)$ is $\mathcal{P}$-measurable. We will sometimes more precisely say that X is predictable with respect to $\{\mathcal{F}_t\}$. Obviously, every left-continuous or continuous adapted process is predictable. If X is right-continuous with limits from the left, then the process $\{X_{t-}\}$ is predictable.

A mapping $\tau : \Omega \mapsto \mathbb{R}_+ \cup \{\infty\}$ such that $\{\tau \leq t\} \in \mathcal{F}_t$ for all $t \in \mathbb{R}_+$ is called a *stopping time* on $(\Omega, \mathcal{F}, \{\mathcal{F}_t\}, P)$ or a stopping time with respect to $\{\mathcal{F}_t\}$.

The σ-algebra $\mathcal{F}_\tau$ of events up to a stopping time τ is defined by $A \in \mathcal{F}_\tau$ if and only if $A \cap \{\tau \leq t\} \in \mathcal{F}_t$ for all $t \geq 0$ and $A \in \sigma\{\mathcal{F}_t : t \geq 0\}$. The σ-algebra $\mathcal{F}_{\tau-}$ is the σ-algebra generated by $\mathcal{F}_0$ and all sets of the form $A \cap \{t < \tau\}$, where $t \geq 0$ and $A \in \mathcal{F}_t$.

A stochastic basis is called *complete* if

$$M \subset N \in \mathcal{F},\ P(N) = 0 \Rightarrow M \in \mathcal{F}$$

and

$$N \in \mathcal{F},\ P(N) = 0 \Rightarrow N \in \mathcal{F}_0.$$

We will usually not assume that the stochastic basis is complete. This is because in statistical considerations we are typically interested in a class of probability measures on $(\Omega, \mathcal{F})$ that are not equivalent. The following lemmas can be used to "translate" classical results for a complete basis to results for a basis $(\Omega, \mathcal{F}, \{\mathcal{F}_t\}, P)$ that is not assumed complete. For a more thorough discussion of problems in this connection, see Jacod and Shiryaev (1987, Chapter I). Define a σ-algebra $\mathcal{F}^P$ by requiring that $A \in \mathcal{F}^P$ if there exist $A_1, A_2 \in \mathcal{F}$ such that $A_1 \subseteq A \subseteq A_2$ and $P(A_1) = P(A_2)$. Moreover, let $\mathcal{N}^P$ be the class of sets with P-measure zero in $\mathcal{F}^P$, and let $\mathcal{F}_t^P$ be the

σ-algebra generated by $\mathcal{F}_t$ and $\mathcal{N}^P$. Then $\left(\Omega, \mathcal{F}^P, \left\{\mathcal{F}_t^P\right\}, P\right)$ is a complete stochastic basis.

Lemma A.1.1 *Every stopping time on the complete stochastic basis* $(\Omega, \mathcal{F}^P, \left\{\mathcal{F}_t^P\right\}, P)$ *is* P*-almost surely equal to a stopping time on the original stochastic basis* $\left(\Omega, \mathcal{F}^P, \{\mathcal{F}_t\}, P\right)$.

Proof: Jacod and Shiryaev (1987, p. 5). □

Lemma A.1.2 *To every stochastic process* X *that is predictable with respect to* $\left\{\mathcal{F}_t^P\right\}$ *there exists a process* $\tilde{X}$ *that is predictable with respect to* $\{\mathcal{F}_t\}$ *such that* $P(X_t = \tilde{X}_t,\ t \geq 0) = 1$.

Proof: Jacod and Shiryaev (1987, p. 19). □

A.2 Local martingales and increasing processes

A *martingale* on a filtered probability space $(\Omega, \mathcal{F}, \{\mathcal{F}_t\}, P)$ is a stochastic process X adapted to $\{\mathcal{F}_t\}$ with the following properties:

(1) X is right-continuous with limits from the left.

(2) $E|X_t| < \infty, \quad t \geq 0$.

(3) $E\left(X_t|\mathcal{F}_s\right) = X_s$ P-almost surely, $\quad s \leq t$.

If the equality in (3) is replaced by $\geq$ or $\leq$, the process X is called a *submartingale* or a *supermartingale*, respectively. A martingale X is called *square integrable* if $E(X_t^2) < \infty,\ t \geq 0$. Some results about martingales are usually proved for a stochastic basis where the filtration is complete. Many results hold as well for filtrations that are not complete; see Jacod and Shiryaev (1987, p. 10). Concerning construction of an adapted process $\tilde{X}$ satisfying (1), (2), and (3) from an adapted process satisfying only (2) and (3), combine Rogers and Williams (1994, II.65–68) and Jacod and Shiryaev (1987, p. 10).

Example A.2.1 Every Lévy process X with $E|X_t| < \infty$ and $E(X_t) = 0$ for all $t \geq 0$ is a martingale. In particular, a standard Wiener process is a martingale. □

Let $\mathcal{V}$ denote the class of $\{\mathcal{F}_t\}$-adapted right-continuous processes A with limits from the left, $A_0 = 0$, and finite variation in finite intervals, i.e., processes for which

$$\operatorname{Var}(A)_t = \lim_{n\to\infty} \sum_{k=1}^{n} |A_{tk/n} - A_{t(k-1)/n}|$$

is finite for all $t \in \mathbb{R}_+$. By $\mathcal{V}^+$ we denote the class of adapted, non-decreasing, right-continuous processes with $A_0 = 0$ (a non-decreasing process always has limits from the left). For short we call a process in $\mathcal{V}$ a *process with finite variation* and a process in $\mathcal{V}^+$ an *increasing process.* Obviously, $\mathcal{V}^+ \subseteq \mathcal{V}$, and $\text{Var}(A) \in \mathcal{V}^+$ for $A \in \mathcal{V}$. Moreover, every element in $\mathcal{V}$ can be uniquely written as the difference between two elements of $\mathcal{V}^+$, viz., $(\text{Var}(A)+A)/2$ and $(\text{Var}(A)-A)/2$ (Jacod and Shiryaev, 1987, I.3.3).

A process $A \in \mathcal{V}^+$ is called *integrable* if $E(A_\infty) < \infty$, and the class of integrable processes is denoted by $\mathcal{A}^+(P)$. Accordingly, $\mathcal{A}(P)$ denotes the class of *processes with integrable variation*, i.e., the elements of $\mathcal{V}$ for which $\text{Var}(A) \in \mathcal{A}^+(P)$.

Finally, let $\mathcal{K}$ be a class of stochastic processes. Then we say that X belongs to $\mathcal{K}$ locally and write $X \in \mathcal{K}_{\text{loc}}$ if there exists an increasing sequence of stopping times $\{\tau_n\}$ such that

(1) $\tau_n \uparrow \infty$ as $n \to \infty$ P-almost surely,

(2) $X^{\tau_n} \in \mathcal{K}$ for every $n \in \mathbb{N}$, where

$$X_t^{\tau_n} = X_{t\wedge\tau_n} = \begin{cases} X_t & \text{if } t \le \tau_n \\ X_{\tau_n} & \text{if } t > \tau_n. \end{cases}$$

In particular, X is called a *local martingale* if there exists a sequence of stopping times $\{\tau_n\}$ satisfying (1) above such that X^{τ_n} is a martingale for every $n \in \mathbb{N}$. If the martingale X^{τ_n} is square integrable, X is called a *locally square integrable martingale.* Note that trivially a martingale is a local martingale (put $\tau_n = n$) and that for a locally square integrable martingale X there exists a sequence of stopping times $\{\sigma_n\}$ satisfying (1) such that $\sup_{t>0} E[(X_t^{\sigma_n})^2] < \infty$ for every n (e.g., $\sigma_n = \tau_n \wedge n$, where τ_n is the original sequence of stopping times for which X^{τ_n} is a square integrable martingale).

Example A.2.2 Every continuous local martingale X is a locally square integrable martingale (put $\tau_n = \inf\{t \ge 0 : |X_t| \ge n\}$). □

An increasing process A is said to be *locally integrable* if there exists a sequence of stopping times $\{\tau_n\}$ satisfying (1) above such that $A^{\tau_n} \in \mathcal{A}^+(P)$ for all $n \in \mathbb{N}$. The class of locally integrable increasing processes is denoted by $\mathcal{A}^+_{\text{loc}}(P)$. By $\mathcal{A}_{\text{loc}}(P)$ we denote the class of processes with locally integrable variation, i.e., the processes in $\mathcal{V}$ for which $\text{Var}(A) \in \mathcal{A}^+_{\text{loc}}(P)$.

Example A.2.3 If $A \in \mathcal{V}^+$ and $E(A_t) < \infty$ for all $t \in \mathbb{R}_+$, then obviously $A \in \mathcal{A}^+_{\text{loc}}(P)$ (put $\tau_n = n$). □

Example A.2.4 If $A \in \mathcal{V}^+$, and if A is continuous, then $A \in \mathcal{A}^+_{\text{loc}}(P)$ (put $\tau_n = \inf\{t \ge 0 : A_t \ge n\}$). □

It can, in fact, more generally be shown that if $A \in \mathcal{V}$, and if A is predictable, then $A \in \mathcal{A}_{\text{loc}}(P)$ (Jacod and Shiryaev, 1987, p. 29).

The technique of proof where one first replaces a process X by X^{τ_n} and then lets n tend to infinity is called *localization*.

A class $\mathcal{R}$ of real-valued stochastic variables is called *uniformly integrable* if given $\epsilon > 0$, there exists $K \in [0, \infty)$ such that $E(1_{\{|X|>K\}}|X|) < \epsilon$ for all $X \in \mathcal{R}$; i.e., if $\sup_{X \in \mathcal{R}} E(1_{\{|X|>K\}}|X|) \to 0$ as $K \to \infty$. In particular, a stochastic process X is called uniformly integrable if the class $\mathcal{R} = \{X_t : t \geq 0\}$ is uniformly integrable. If there exists a $p > 0$ such that $\sup_{X \in \mathcal{R}} E(|X|^{1+p}) < \infty$, then $\mathcal{R}$ is uniformly integrable. Another sufficient condition that $\mathcal{R}$ be uniformly integrable is that there exist a non-negative random variable Y such that $E(Y) < \infty$ and $|X| < Y$ for all $X \in \mathcal{R}$. The sufficiency of both conditions is easy to prove; see, e.g., Rogers and Williams (1994, p.115).

We can now state (a version of) Doob's optional stopping theorem; see, e.g., Rogers and Williams (1994, p. 189) or Jacod and Shiryaev (1987, p. 10).

Theorem A.2.5 *Suppose that X is a uniformly integrable martingale, and let τ and σ be two stopping times with $\sigma \leq \tau$. Then X_σ and X_τ are integrable and*

$$E(X_\tau | \mathcal{F}_\sigma) = X_\sigma, \quad P\text{-almost surely.}$$

A.3 Doob-Meyer decomposition

A stochastic process X is said to be of class (D) if the set of random variables $\{X_\tau : \tau \text{ is a finite stopping time }\}$ is uniformly integrable. With this definition we can state the celebrated Doob-Meyer decomposition theorem; see Jacod and Shiryaev (1987, p. 32) or Rogers and Williams (1987, VI.29).

Theorem A.3.1 *Let X be a submartingale of class (D). Then there exists a unique predictable process $A \in \mathcal{A}^+(P)$ such that $M = X - A$ is a uniformly integrable martingale.*

This theorem states that any submartingale X of class (D) can be uniquely written as the sum of a martingale M and a predictable increasing process A. There is also a local version of Theorem A.3.1; see Rogers and Williams (1987, p. 375).

Theorem A.3.2 *Let X be a local submartingale. Then there exists a unique predictable process $A \in \mathcal{V}^+$ such that $M = X - A$ is a local martingale.*

The predictable increasing process is called the *compensator* of the submartingale. In general, the compensator depends on P and the filtration as well as on the submartingale. The compensator of a submartingale X can often be constructed as follows. Define

$$A_t^{(n)} = \sum_{j=1}^{2^n} E\left(X_{jt2^{-n}} - X_{(j-1)t2^{-n}} | \mathcal{F}_{(j-1)t2^{-n}}\right). \tag{A.3.1}$$

Then for every $t > 0$, $A_t^{(n)} \to A_t$ weakly in $L_1(P)$ for $n \to \infty$; i.e., $E(A_t^{(n)} 1_B) \to E(A_t 1_B)$ for every $B \in \mathcal{F}$, as proved by Rao (1969). In practice it is preferable to have convergence in probability, and when this is the case the compensator is called *calculable.* It can be shown that a continuous compensator is calculable, and that a broad class of continuous-time submartingales have a continuous compensator. A discussion of these problems can be found, for instance, in Helland (1982, pp. 85–86).

Corollary A.3.3 *Let $M \in \mathcal{V}$ be a predictable local martingale. Then*

$$M \equiv 0.$$

The following results follow from Theorem A.3.1 by using, inter alia, localization arguments.

Theorem A.3.4 *Suppose $A \in \mathcal{A}_{loc}^{+}(P)$. Then there exists a unique predictable process $A^p \in \mathcal{A}_{loc}^{+}(P)$ such that $A - A^p$ is a local martingale.*

Here A^p is also called the compensator of A. When A is quasi-left-continuous, then A^p is continuous, and it can be calculated as the limit in probability of $A_t^{(n)}$ given by (A.3.1). An adapted process X is called *quasi-left-continuous* if $X_{\tau_n} \to X_\tau$ almost surely on $\{\tau < \infty\}$ for any increasing sequence $\{\tau_n\}$ of stopping times with limit τ. This essentially means that the jumps of X cannot be predicted. For instance, the Poisson process is quasi-left-continuous.

A particularly important example is when $A \in \mathcal{A}_{\mathrm{loc}}^{+}(P)$ is a *counting process*, i.e., a process that starts at zero and only changes by upward jumps of size one. For such a process, A^p is called the *integrated intensity.* In many cases it is absolutely continuous with respect to t with a predictable derivative called the *intensity* of A.

Example A.3.5 Let N be a Poisson process with parameter λ. Then $N_t - \lambda t$ is a martingale, so the compensator of N is λt. It is easy to see that the limit of (A.3.1) is λt. A Poisson process is a counting process with intensity λ. □

Theorem A.3.6 *Let M be a locally square integrable martingale. Then there exists a unique predictable process $\langle M \rangle \in \mathcal{A}^{+}_{loc}(P)$ such that $M_t^2 - \langle M \rangle_t$ is a local martingale.*

Here it has been used that by Jensen's inequality the square of a martingale is a submartingale. The process $\langle M \rangle$ is called the *quadratic characteristic* of M or the predictable quadratic variation of M. Obviously,

$$\langle cM \rangle = c^2 \langle M \rangle \tag{A.3.2}$$

for any constant $c \in \mathbb{R}$.

Example A.3.7 Let W be a standard Wiener process. Then $W_t^2 - t$ is a martingale, so $\langle W \rangle = t$. It can be shown that a continuous local martingale (which is always locally square integrable; see Example A.2.2) is a standard Wiener process if and only if $\langle M \rangle = t$; see, e.g., Rogers and Williams (1987, p.63). □

The quadratic characteristic $\langle M \rangle$ of a locally square integrable martingale M is deterministic (non-random) if and only if M has independent increments. It is continuous if and only if M is quasi-left-continuous.

Suppose X and Y are two locally square integrable martingales. Then their *quadratic cocharacteristic*, or predictable quadratic covariation, is defined by

$$\langle X, Y \rangle_t = \tfrac{1}{4} \left(\langle X + Y \rangle_t - \langle X - Y \rangle_t \right). \tag{A.3.3}$$

It is easy to check that $XY - \langle X, Y \rangle$ is a local martingale. In fact, $\langle X, Y \rangle$ is the only predictable process in $\mathcal{V}$ for which $XY - \langle X, Y \rangle$ is a local martingale. If X and Y are proper square integrable martingales, $XY - \langle X, Y \rangle$ is a proper martingale (Jacod and Shiryaev, 1987, p. 38). Note that $\langle X, X \rangle = \langle X \rangle$ and that $\langle \cdot, \cdot \rangle$ is bilinear.

From the result about (A.3.1) it follows that

$$B_t^{(n)} = \sum_{j=1}^{2^n} E\left(\left[X_{jt2^{-n}} - X_{(j-1)t2^{-n}} \right] \left[Y_{jt2^{-n}} - Y_{(j-1)t2^{-n}} \right] | \mathcal{F}_{(j-1)t2^{-n}} \right) \tag{A.3.4}$$

for every $t > 0$ converges weakly in $L_1(P)$ to $\langle X, Y \rangle_t$ for $n \to \infty$. Often there is convergence in probability, in which case $\langle X, Y \rangle$ is called calculable. This happens, for instance, when $\langle X, Y \rangle_t$ is continuous, which it is if X and Y are quasi-left-continuous (Jacod and Shiryaev, 1987, p. 38).

For a vector of martingales $\underline{X} = \left(X^{(1)}, \cdots, X^{(k)} \right)$ we can define the $k \times k$ *cocharacteristic matrix* $\langle \underline{X} \rangle$ by $\langle \underline{X} \rangle_{ij} = \langle X^{(i)}, X^{(j)} \rangle$. The matrix $\langle \underline{X} \rangle_t$ is symmetric and positive semi-definite (because of the bilinearity of $\langle \cdot, \cdot \rangle$). The process $\langle \underline{X} \rangle$ is increasing in the sense that $\langle \underline{X} \rangle_t - \langle \underline{X} \rangle_s$ is positive semi-definite for $s < t$.

For a discrete time martingale M, the quadratic characteristic is given by

$$\langle M \rangle_n = \sum_{i=1}^{n} E\left((M_i - M_{i-1})^2 | \mathcal{F}_{i-1} \right). \tag{A.3.5}$$

A.4 Semimartingales and stochastic integration

A stochastic process X is called a *semimartingale* if it has a decomposition of the form

$$X_t = X_0 + M_t + A_t, \tag{A.4.1}$$

where $A \in \mathcal{V}$ and M is a local martingale with $M_0 = 0$. The decomposition (A.4.1) is typically not unique. If A can be chosen predictable, X is called a *special semimartingale.* A semimartingale with bounded jumps is an example of a special semimartingale (Jacod and Shiryaev, 1987, p. 44). A decomposition where A is predictable is unique. A vector process is called a semimartingale if every coordinate has a decomposition of the form (A.4.1).

Example A.4.1 A Poisson process N with parameter λ can be decomposed as $N_t = (N_t - \lambda t) + \lambda t$ or $N_t = 0 + N_t$, where $N_t - \lambda t$ and 0 are martingales, while λt and N are in $\mathcal{V}$. Thus a Poisson process is a special semimartingale. In fact, any Lévy process is a semimartingale, and any integrable Lévy process is a special semimartingale, cf. Example A.2.1. □

Theorem A.3.2 shows that a submartingale is a special semimartingale, and by Theorem A.3.4 a counting process is a special semimartingale. We shall later see several other examples of semimartingales.

Let us now consider the construction of the stochastic integral of a one-dimensional process with respect to a one-dimensional semimartingale. If a semimartingale X belongs to $\mathcal{V}$, and if H is a bounded, predictable process, the integral $H \cdot X_t$ of H with respect to X over $[0, t]$ is defined pathwise as the Stieltjes integral

$$H \cdot X_t(\omega) = \int_0^t H_s(\omega)\, dX_s(\omega), \quad t > 0. \tag{A.4.2}$$

For general semimartingales a pathwise construction is not possible because M in (A.4.1) might have infinite variation in $[0, t]$, so that dX does not define a signed measure on $\mathbb{R}$. An exception is when H is of the simple form

$$H_t = Y 1_{\{0\}}(t) \text{ or } H_t = Y 1_{(r,s]}(t), \quad r < s,$$

where Y is bounded and in the first case $\mathcal{F}_0$-measurable, in the second case $\mathcal{F}_r$-measurable. Let $\mathcal{E}$ denote the class of processes of this elementary form. The stochastic integral of $H \in \mathcal{E}$ with respect to X over $[0, t]$ can be defined as

$$H \cdot X_t = \int_0^t H_s dX_s = Y\left(X_{s\wedge t} - X_{r\wedge t}\right) \tag{A.4.3}$$

if $H = Y1_{(r,s]}$, and as $H \cdot X_t = 0$ if $H = Y1_{\{0\}}$. As a function of time $H \cdot X_t$ is a stochastic process, in fact a semimartingale. Hence $H \mapsto H \cdot X$ is a map from $\mathcal{E}$ into the class of semimartingales.

Theorem A.4.2 *Let X be a semimartingale. Then the map $H \mapsto H \cdot X$ defined on $\mathcal{E}$ has a unique extension to the space of* locally bounded predictable processes *with the following properties (H denotes a locally bounded predictable process):*

(a) $H \cdot X$ is a semimartingale.

(b) $H \mapsto H \cdot X$ is linear.

(c) Let $\{H^{(n)}\}$ be a sequence of predictable processes that converges pointwise to the limit process H and satisfies $|H^{(n)}| \le K$, where K is a locally bounded predictable process. Then

$$\sup_{s\le t} |H^{(n)} \cdot X_s - H \cdot X_s| \to 0 \quad as\ n \to \infty$$

in probability for every $t > 0$.

(See Jacod and Shiryaev, 1987, pp. 46–47).

The formula (A.4.3) makes sense for every process X, but the extension in Theorem A.4.2 is only possible when X is a semimartingale. This fundamental result by Bichteler, Dellacherie, and Mokobodzki indicates why the class of semimartingales is a natural and important class of stochastic processes.

The process $H \cdot X$ is called the *stochastic integral* of H with respect to X. Sometimes we will also use the following notation for the stochastic integral,

$$H \cdot X_t = \int_0^t H_s dX_s,$$

even when dX does not define a signed measure. In general, $H \cdot X_t$ cannot be constructed pathwise like the Stieltjes integral. When H is left-continuous, it can, however, be approximated by *Riemann sums.*

Theorem A.4.3 *Suppose that X is a semimartingale and that H is an adapted left-continuous process. Consider a sequence of subdivisions of $\mathbb{R}_+$ given by*

$$0 = t_0^{(n)} < t_1^{(n)} < \cdots < t_m^{(n)} < \cdots < \infty,$$

where

$$\sup_m \left(t_{m+1}^{(n)} - t_m^{(n)}\right) \to 0 \quad as\ n \to \infty.$$

Then

$$\sup_{s\le t} \left| \sum_{m=0}^{\infty} H_{t_m^{(n)}} \left(X_{s\wedge t_{m+1}^{(n)}} - X_{s\wedge t_m^{(n)}}\right) - H \cdot X_s \right| \to 0$$

in probability as $n \to \infty$ (see Jacod and Shiryaev, 1987, p. 51).

Theorem A.4.4 *The stochastic integral of a locally bounded predictable process H with respect to a semimartingale X has the following properties:*

(a) $X \mapsto H \cdot X$ *is linear.*

(b) If X *is a local martingale, then so is* $H \cdot X$.

(c) If $X \in \mathcal{V}$, *then* $H \cdot X \in \mathcal{V}$, *and the stochastic integral coincides with the pathwise Stieltjes integral mentioned above.*

(d) $H \cdot X_0 = 0$, $H \cdot X = H \cdot (X - X_0)$.

(e) $\Delta(H \cdot X)_t = H_t \Delta X_t$.

(f) $K \cdot (H \cdot X) = (KH) \cdot X$, *where* H *and* K *are both locally bounded predictable processes.*

(See Jacod and Shiryaev, 1987, p. 47.)

In (e) we have used the notation $\Delta Y_t = Y_t - Y_{t-}$ for a semimartingale Y. Here $Y_{t-} = \lim_{u \uparrow t} Y_u$. This limit is well-defined since a semimartingale has limits from the left. Note that it follows from (e) that if X is continuous, then so is $H \cdot X$.

When X is a locally square integrable martingale, the stochastic integral can be defined for more general integrands than the locally bounded predictable processes. Let $L^2(X)$ denote the class of all predictable processes H for which the process $H^2 \cdot \langle X \rangle \in \mathcal{A}^+$, and $L^2_{\text{loc}}(X)$ the predictable processes H for which $H^2 \cdot \langle X \rangle \in \mathcal{A}^+_{\text{loc}}$. Note that the class of locally bounded predictable processes belongs to $L^2_{\text{loc}}(X)$ because $\langle X \rangle \in \mathcal{A}^+_{\text{loc}}$.

Theorem A.4.5 *Suppose* X *is a locally square integrable martingale. Then the map* $H \mapsto H \cdot X$ *has a further unique extension to the set* $L^2_{\text{loc}}(X)$ *that satisfies (a) and (b) of Theorem A.4.2 and (for* $H \in L^2_{\text{loc}}(X)$*)*

(1) Let $\{H^{(n)}\}$ *be a sequence of predictable processes that converges pointwise to a limit process* H *and satisfies* $|H^{(n)}| \le K$, *where* $K \in L^2_{\text{loc}}(X)$. *Then*

$$\sup_{s \le t} |H^{(n)} \cdot X_s - H \cdot X_s| \to 0 \quad \text{as } n \to \infty$$

in probability for every $t > 0$.

(2) $H \cdot X$ *is a locally square integrable martingale.*

(3) $H \cdot X$ *is a proper square integrable martingale if and only if* $H \in L^2(X)$.

(3) $K \cdot (H \cdot X) = (KH) \cdot X$ *for* $H \in L^2_{\text{loc}}(X)$ *and* $K \in L^2_{\text{loc}}(H \cdot X)$.

(4) $\langle H\cdot X, K\cdot Y\rangle = (HK)\cdot\langle X,Y\rangle$ *if* X *and* Y *are locally square integrable martingales,* $H \in L^2_{\text{loc}}(X)$ *and* $K \in L^2_{\text{loc}}(Y)$.

(See Jacod and Shiryaev, 1987, p. 48.)

For a d-dimensional semimartingale $\underline{X} = (X_1, \cdots, X_d)$ and a d-dimensional process $\underline{H} = (H_1, \cdots, H_d)$, where H_i is a locally bounded predictable process (or $H_i \in L^2_{\text{loc}}(X_i)$, when X_i is a locally square integrable martingale) for every $i = 1, \cdots, d$, the stochastic integral $\underline{H}\cdot\underline{X}$ of $\underline{H}$ with respect to $\underline{X}$ is simply defined by

$$\underline{H}\cdot\underline{X}_t = \sum_{i=1}^{d}(H_i\cdot X_i)_t.$$

For a d-dimensional continuous local martingale $\underline{X} = (X_1, \cdots, X_d)$, the class of possible integrands is larger than the class of d-dimensional predictable processes $\underline{H} = (H_1, \cdots, H_d)$ for which $H_i \in L^2_{\text{loc}}(X_i)$, $i = 1, \ldots, d$. Indeed, suppose $\langle\underline{X}\rangle = c\cdot A$, where c is a symmetric positive-semidefinite matrix of predictable processes and where A is increasing and predictable. Then define $L^2_{\text{loc}}(\underline{X})$ as the class of d-dimensional predictable processes $\underline{H}$ for which $(\underline{H}^T c\underline{H})\cdot A \in \mathcal{A}^+_{\text{loc}}(P)$. Note that a continuous local martingale is locally square integrable, and that we can always find a representation of $\langle\underline{X}\rangle$ of the form $\langle\underline{X}\rangle = c\cdot A$. With this definition the following theorem holds.

Theorem A.4.6 *Let* $\underline{X} = (X_1, \ldots, X_d)$ *be a continuous local martingale, and suppose* $\underline{H} \in L^2_{\text{loc}}(\underline{X})$. *Then* $\underline{H}^{(n)}\cdot\underline{X}$, *where* $\underline{H}^{(n)} = \underline{H}1_{\{|\underline{H}|\leq n\}}$, *converges in probability to a limit* $\underline{H}\cdot\underline{X}$. *The convergence is uniform on compact intervals. The stochastic integral* $\underline{H}\cdot\underline{X}$ *is a continuous local martingale with*

$$\langle\underline{H}\cdot\underline{X}, \underline{K}\cdot\underline{X}\rangle = (\underline{H}^T c\underline{K})\cdot A,$$

for $\underline{H}, \underline{K} \in L^2_{\text{loc}}(X)$. *The mapping* $\underline{H} \mapsto \underline{H}\cdot\underline{X}$ *is linear, and* $\underline{H}\cdot\underline{X}$ *is square integrable if and only if* $(\underline{H}^T c\underline{H})\cdot A \in \mathcal{A}^+(P)$ (see Jacod and Shiryaev, 1987, p. 167).

Let X and Y be two semimartingales, and define for every $n \in \mathbb{N}$

$$V_t^{(n)}(X,Y) = \sum_{j=1}^{2^n}\left(X_{jt2^{-n}} - X_{(j-1)t2^{-n}}\right)\left(Y_{jt2^{-n}} - Y_{(j-1)t2^{-n}}\right). \quad (A.4.4)$$

Then there exists a process $[X,Y]$ that is adapted and right-continuous with limits from the left such that for every $t > 0$

$$\sup_{s\leq t}|V_s^{(n)}(X,Y) - [X,Y]_s| \to 0$$

in probability as $n \to \infty$, see Jacod and Shiryaev (1987, pp. 51–52). The process $[X, Y]$ is called the *quadratic covariation* of X and Y. Its discontinuities are given by

$$\Delta [X,Y]_t = (\Delta X_t)(\Delta Y_t). \tag{A.4.5}$$

The *quadratic variation* of X is defined by

$$[X] = [X, X].$$

It can be proved that

$$[X] \in \mathcal{V}^+ \text{ and } [X,Y] \in \mathcal{V}.$$

Obviously, $[\cdot,\cdot]$ is bilinear and

$$[X,Y] = \tfrac{1}{4}([X+Y] - [X-Y]).$$

If Y is continuous and $Y \in \mathcal{V}$, then

$$[X,Y] = 0 \tag{A.4.6}$$

for any semimartingale X.

For a k-dimensional semimartingale $\underline{X} = (X^{(1)}, \cdots, X^{(k)})$ we can define the $k \times k$ *covariation matrix* $[\underline{X}]$ by $[\underline{X}]_{ij} = [X^{(i)}, X^{(j)}]$. The matrix $[\underline{X}]_t$ is symmetric and positive semi-definite (because $[\cdot,\cdot]$ is bilinear). The process $[\underline{X}]$ is increasing in the sense that $[\underline{X}]_t - [\underline{X}]_s$ is positive semi-definite for $s < t$.

For *discrete time* semimartingales X and Y,

$$[X,Y]_n = \sum_{i=1}^{n} (X_i - X_{i-1})(Y_i - Y_{i-1}). \tag{A.4.7}$$

If X and Y are (local) martingales, the process

$$X_t Y_t - [X,Y]_t$$

is a (local) martingale (Jacod and Shiryaev, 1987, p. 53). If X and Y are, moreover, locally square integrable, then $[X,Y] \in \mathcal{A}_{\text{loc}}(P)$ and

$$[X,Y] - \langle X,Y \rangle$$

is a local martingale; i.e., $\langle X,Y \rangle$ is the compensator of $[X,Y]$.

We conclude this section by a decomposition property of local martingales. First we need two definitions.

Two local martingales M and N are called *orthogonal* if their product MN is a local martingale. Note that if M and N are locally square integrable, then they are orthogonal if and only if $\langle M, N \rangle = 0$.

A local martingale X is called a *purely discontinuous local martingale* if $X_0 = 0$ and if it is orthogonal to all continuous local martingales. A purely discontinuous local martingale is not really discontinuous. Thus it is not equal to the sum of its jumps, a sum that in fact often diverges. An example of a purely discontinuous martingale is $N_t - \lambda t$, where N is a Poisson process with parameter λ. This example is typical. A purely discontinuous local martingale is generally the sum of its compensated jumps.

Theorem A.4.7 *Any local martingale M has a unique decomposition*

$$M_t = M_0 + M_t^c + M_t^d,$$

where M^c is a continuous local martingale with $M_0^c = 0$, and M^d is a purely discontinuous local martingale (see Jacod and Shiryaev, 1987, pp. 42–43).

The process M^c is called the *continuous part* of M, while M^d is called the *purely discontinuous part* of M.

Theorem A.4.8 *Let X be a semimartingale. Then there exists a unique continuous local martingale X^c with $X_0^c = 0$ such that for any decomposition of the form (A.4.1) we have $M^c = X^c$* (see Jacod and Shiryaev, 1987, p. 45).

The process X^c is called the *continuous martingale part* of X.

For two semimartingales X and Y we have the following expression for their quadratic covariation (Jacod and Shiryaev, 1987, p. 55):

$$[X,Y]_t = \langle X^c, Y^c \rangle_t + \sum_{s \le t} (\Delta X_s)(\Delta Y_s). \tag{A.4.8}$$

The (possibly infinite) sum on the righthand side always converges. Using this formula it can be proved that for two semimartingales X and Y and for two locally bounded predictable processes H and K we have

$$[H \cdot X, K \cdot Y] = (HK) \cdot [X,Y]. \tag{A.4.9}$$

A.5 Stochastic differential equations

Let $\underline{Y}$ be a k-dimensional semimartingale. By a solution to the stochastic differential equation

$$dX_t = a^{(1)}(t, X_t) dY_t^{(1)} + \cdots + a^{(k)}(t, X_t) dY_t^{(k)}, \; X_0 = U, \tag{A.5.1}$$

where U is F_0-measurable, we mean a one-dimensional stochastic process X satisfying

$$X_t = U + \int_0^t a^{(1)}(s, X_s)\, dY_s^{(1)} + \cdots + \int_0^t a^{(k)}(s, X_s)\, dY_s^{(k)}.$$

This has meaning, of course, only when the integrals exist.

We first consider stochastic differential equations of the type

$$dX_t = a(t, X_t)\, dt + b(t, X_t)\, dW_t, \quad X_0 = U, \tag{A.5.2}$$

where W is a standard Wiener process. A solution to (A.5.2) is called a (one-dimensional) *diffusion process*. The functions a and b are called, respectively, the *drift coefficient* and the *diffusion coefficient*. Note that a diffusion process is a semimartingale.

Theorem A.5.1 *Suppose the functions a and b satisfy*

$$a^2(t, x) + b^2(t, x) \leq K\left(1 + x^2\right) \tag{A.5.3}$$

and

$$|a(t, x) - a(t, y)| + |b(t, x) - b(t, y)| \leq K|x - y| \tag{A.5.4}$$

for all x, y, and t, where K is a constant. Assume, moreover, that $E(U^2) < \infty$. Then the stochastic differential equation (A.5.2) has a unique solution that is a continuous Markov process.

If the conditions (A.5.3) and (A.5.4) are satisfied only for x and y in a finite interval (and possibly only up to a stopping time), then there is a unique solution on that interval (up to the stopping time). If a and b are continuously differentiable, the conditions are satisfied on any bounded interval.

Another important type of stochastic differential equation is

$$dX_t = a(t, X_{t-})\, dt + b(t, X_{t-})\, dW_t + c(t, X_{t-})\, dN_t, \quad X_0 = U, \tag{A.5.5}$$

where W is a standard Wiener process and N is a counting process. The solution of the stochastic differential equation (A.5.5) is a particular example of a *diffusion with jumps*. A result similar to Theorem A.5.1 about existence of a unique solution can be given for equations of the type (A.5.5); see Doléans-Dade (1976).

A different type of generalization of (A.5.2) is

$$dX_t = A_t dt + B_t dW_t, \; X_0 = U, \tag{A.5.6}$$

where A and B are adapted right-continuous processes with limits from the left and U is $\mathcal{F}_0$-measurable. A solution of (A.5.6) is called an *Ito process* and is in general not a Markov process. An important case is when $A_t(\omega) = \alpha_t(X(\omega))$ and $B_t(\omega) = \beta_t(X(\omega))$, where the functionals α_t and β_t depend only on X through $\{X_s : s \leq t\}$. An Ito process of this particular type is called a *diffusion-type process*. For coefficients of this form a unique solution of (A.5.6) exists if there exist constants L_1 and L_2 and a non-decreasing right-continuous function K such that

$$\alpha_t(x)^2 + \beta_t(x)^2 \leq L_1 \int_0^t \left(1 + x(s)^2\right) dK(s) + L_2\left(1 + x(t)^2\right)$$

and

$$|\alpha_t(x) - \alpha_t(y)|^2 + |\beta_t(x) - \beta_t(y)|^2 \le L_1 \int_0^t |x(s) - y(s)|^2 dK(s) + L_2 |x(t) - y(t)|^2$$

for all $t > 0$ and all continuous real functions x and y (Liptser and Shiryaev, 1977, Theorem 4.6).

Example A.5.2 An example of a process of the diffusion type is the velocity process for the Duffing-van der Pol oscillator, which is a model for an oscillator that moves in a certain force field. Let V_t denote the speed of the oscillator at time t. Then its movement is determined by the stochastic differential equation

$$dV_t = \left[\alpha X_t - \left(V_t + X_t^3\right)\right] dt + \sigma X_t dW_t,$$

where

$$X_t = X_0 + \int_0^t V_s ds$$

is the position of the oscillator at time t. The process V_t is a diffusion type process with coefficients that depend on the past back to time zero. The two-dimensional process (X_t, V_t), however, is a (two-dimensional) ordinary diffusion process. □

Let us finally consider *weak solutions* to stochastic differential equations. The existence of a weak solution is usually enough in statistical applications. Consider the canonical space $D([0,\infty))$ of functions ω from $[0,\infty)$ into $\mathbb{R}$ that are right-continuous with limits from the left with the σ-algebra generated by the coordinate functions $\{\omega(t) : t \ge 0, \ \omega \in D([0,\infty))\}$. A probability measure on the canonical space is called a weak solution to the stochastic differential equation (A.5.1) if the canonical process $X_t(\omega) = \omega(t), \ \omega \in D([0,\infty))$ is a solution to (A.5.1). The generalization to multivariate processes is obvious. Only rather weak conditions are needed to ensure the existence of a weak solution. Consider, for instance, a stochastic differential equation of the type (A.5.5), where we assume that N is a Lévy process the Lévy measure of which is absolutely continuous with respect to the Lebesgue measure and has finite first moment. This equation has a weak solution provided only that a is bounded on $[0,n] \times [-n,n]$ for all $n \in \mathbb{N}$, that $b > 0$, and that b and c are continuous on $[0,\infty) \times \mathbb{R}$. This follows from results in Jacod (1979, Chapter 13).

A.6 Ito's formula

The following transformation result is the central tool of stochastic calculus. Note how it differs from classical calculus.

Theorem A.6.1 *Let $\underline{X}_t = (X_{1,t}, \cdots, X_{k,t})$ be a k-dimensional semimartingale, and let $f : \mathbb{R}^k \mapsto \mathbb{R}$ be a twice continuously differentiable function. Then $f(\underline{X}_t)$ is a semimartingale, and*

$$\begin{aligned} f(\underline{X}_t) &= f(\underline{X}_0) + \sum_{i=1}^{k} \int_0^t \dot{f}_i(\underline{X}_{s-})\, dX_{i,s} \\ &\quad + \tfrac{1}{2} \sum_{i,j=1}^{k} \int_0^t \ddot{f}_{ij}(\underline{X}_{s-})\, d\langle X_i^c, X_j^c \rangle_s \\ &\quad + \sum_{s \le t} \left[f(\underline{X}_s) - f(\underline{X}_{s-}) - \sum_{i=1}^{k} \dot{f}_i(\underline{X}_{s-})\, \Delta X_{i,s} \right], \end{aligned} \tag{A.6.1}$$

where $\dot{f}_i = \partial f / \partial x_i$ and $\ddot{f}_{ij} = \partial^2 f / \partial x_i \partial x_j$.

The formula (A.6.1) is called *Ito's formula.* When $\underline{X}$ is continuous, the last term on the righthand side is zero. For a proof see, e.g., Jacod and Shiryaev (1987, p. 57). A more general version of Ito's formula was given by Föllmer, Protter, and Shiryaev (1995). Next we give a few typical applications.

Example A.6.2 Suppose X is a continuous one-dimensional semimartingale, and let $f : \mathbb{R} \mapsto \mathbb{R}$ be a real function with primitive F, i.e., $F' = f$. Then by Ito's formula

$$\int_0^t f(X_s)\, dX_s = F(X_t) - F(X_0) - \tfrac{1}{2} \int_0^t f'(X_s)\, d\langle X^c \rangle_s.$$

If W is a standard Wiener process, then $\langle W \rangle_t = t$ (cf. Example A.3.7) and

$$\int_0^t W_s dW_s = \tfrac{1}{2}\left(W_t^2 - t\right).$$

Let X be the solution of

$$dX_t = X_t dt + X_t dW_t, \quad X_0 = 1. \tag{A.6.2}$$

Then

$$X_t^c = \int_0^t X_s dW_s,$$

and by Theorem A.4.5 (5),

$$\langle X^c \rangle_t = \int_0^t X_s^2 ds.$$

Hence

$$\int_0^t X_s^{-1} dX_s = \log(X_t) + \tfrac{1}{2} t.$$

In Example A.6.4 we shall see that the solution of (A.6.2) is always positive. □

Example A.6.3 By applying Ito's formula to the function $f(x,y) = xy$, we see that for two semimartingales X and Y we have

$$d(X_tY_t) = Y_{t-}dX_t + X_{t-}dY_t + d[X,Y]_t. \qquad \text{(A.6.3)}$$

If X or Y is continuous, it follows from (A.4.8) that

$$d(X_tY_t) = Y_{t-}dX_t + X_{t-}dY_t + d\langle X^c, Y^c\rangle_t. \qquad \text{(A.6.4)}$$

□

Example A.6.4 Suppose $F(x,t)$ is a real function that is twice continuously differentiable with respect to x and continuously differentiable with respect to t. Assume further that for each $t \geq 0$ the function $x \mapsto F(x,t)$ is invertible with inverse function $F_t^{-1}(y)$, and consider the stochastic differential equation

$$\begin{aligned} dX_t &= \left[F_t'\left(F_t^{-1}(X_t),t\right) + \tfrac{1}{2}F_{xx}''\left(F_t^{-1}(X_t),t\right)\right]dt \qquad \text{(A.6.5)} \\ &\quad + F_x'\left(F_t^{-1}(X_t),t\right)dW_t, \quad X_0 = F(0,0). \end{aligned}$$

By Ito's formula it follows that

$$X_t = F(W_t,t)$$

is a solution.

A particular example is the equation (A.6.2), where $F(x,t) = \exp\left(x + \frac{1}{2}t\right)$. The (unique) solution is

$$X_t = \exp\left(W_t + \tfrac{1}{2}t\right),$$

which is positive for all $t > 0$. In fact, $X_t \to \infty$ almost surely as $t \to \infty$.

Another example of an equation of the form (A.6.5) is

$$dX_t = -\tfrac{1}{2}\exp(-2X_t)\,dt + \exp(-X_t)\,dW_t, \quad X_0 = 0. \qquad \text{(A.6.6)}$$

Here $F(x,t) = \log(x+1)$, and

$$X_t = \log(W_t + 1)$$

solves (A.6.6) in the random time interval $[0,\tau)$, where τ is the stopping time $\tau = \inf\{t : W_t = -1\}$. □

Example A.6.5 The stochastic differential equation

$$dX_t = X_{t-}dt + X_{t-}dW_t + X_{t-}dN_t, \quad X_0 = x_0, \qquad \text{(A.6.7)}$$

where N is a Poisson process that is independent of the standard Wiener process W, is solved by

$$X_t = x_0 \exp\left(W_t + \tfrac{1}{2}t\right)2^{N_t}.$$

This follows easily from Ito's formula. □

We conclude this section by an important application of Ito's formula.

Theorem A.6.6 *Let X be a one-dimensional semimartingale. Then the equation*

$$dY_t = Y_{t-} dX_t, \quad Y_0 = 1, \tag{A.6.8}$$

has one and only one adapted solution that is right-continuous with limits from the left. The solution $\mathcal{E}(X)$ is a semimartingale and is given by

$$\mathcal{E}(X)_t = \exp\left[X_t - X_0 - \tfrac{1}{2}\langle X^c\rangle_t\right] \prod_{s\le t} \left[(1+\Delta X_s)\exp(-\Delta X_s)\right], \tag{A.6.9}$$

where the (possibly infinite) product is absolutely convergent. Furthermore:

(a) If $X \in \mathcal{V}$, then $\mathcal{E}(X) \in \mathcal{V}$.

(b) If X is a local martingale, then so is $\mathcal{E}(X)$.

(c) Define $\tau = \inf\{t : \Delta X_t = -1\}$. Then $\mathcal{E}(X)_t \neq 0$ for $t < \tau$, and $\mathcal{E}(X)_t = 0$ for $t \geq \tau$.

For a proof see, e.g., Jacod and Shiryaev (1987, p. 59). The process $\mathcal{E}(X)$ is called the *Doléan-Dade exponential* of X. If $X \in \mathcal{V}$, $\mathcal{E}(X)$ has the simpler form

$$\mathcal{E}(X)_t = \exp(X_t - X_0) \prod_{s\le t} \left[(1+\Delta X_s)\exp(-\Delta X_s)\right]. \tag{A.6.10}$$

Note also that if X is a continuous local martingale, then

$$\mathcal{E}(X)_t = \exp\left(X_t - X_0 - \tfrac{1}{2}\langle X\rangle_t\right) \tag{A.6.11}$$

is a continuous local martingale.

A.7 Martingale limit theorems

In this section we have collected some useful limit results for martingales. We begin with two classical *martingale convergence theorems.*

Theorem A.7.1 *Let X be a supermartingale for which $\sup_{t\geq 0} E\left(X_t^-\right) < \infty$. Then X_t converges almost surely to a limit random variable X_∞ satisfying $E|X_\infty| < \infty$. If, moreover, $\{X_t : t \geq 0\}$ is uniformly integrable, then X_t also converges in L_1 to X_∞.*

By $X_t^- = -(X_t \wedge 0)$ we denote, as usual, the negative part of X_t. Note that for a supermartingale that is bounded from below the first condition is always satisfied.

Theorem A.7.2 *Let X be a martingale, and assume that $\{X_t : t \geq 0\}$ is uniformly integrable. Then X_t converges almost surely to a limit random variable X_∞ satisfying $E|X_\infty| < \infty$. Moreover, $E(|X_t - X_\infty|) \to 0$ as $t \to \infty$, and $X_t = E(X_\infty|\mathcal{F}_t)$ for all $t \geq 0$.*

Proofs of these results can be found in Elliott (1982, Chapter 4) or (in the discrete time case) in Hoffmann-Jørgensen (1994, Chapter 7). Next we give two *laws of large numbers* for martingales.

Theorem A.7.3 (Liptser, 1980) *Let M be a locally square integrable martingale with $M_0 = 0$, and suppose $A \in \mathcal{V}^+$ is predictable. Further, define the increasing process*

$$B_t = \int_0^t (1 + A_s)^{-2} \, d\langle M \rangle_s.$$

Then

$$\frac{M_t}{A_t} \to 0 \quad as \quad t \to \infty$$

almost surely on $\{A_\infty = \infty\} \cap \{B_\infty < \infty\}$.

Liptser (1980) also proved results for local martingales that are not locally square integrable. The following result, which was first proved by Lepingle (1978), follows easily from Theorem A.7.3.

Theorem A.7.4 *Suppose M is a locally square integrable martingale, and let $f : \mathbb{R}_+ \mapsto \mathbb{R}_+$ be an increasing function satisfying*

$$\int_0^\infty (1 + f(t))^{-2} \, dt < \infty.$$

Then

$$\frac{M_t}{f(\langle M \rangle_t)} \to 0 \quad as \quad t \to \infty$$

almost surely on $\{\langle M \rangle_\infty = \infty\}$.

By choosing $f(t) = t$ the next result follows.

Corollary A.7.5 *Let M be a locally square integrable martingale. Then*

$$\frac{M_t}{\langle M \rangle_t} \to 0 \quad as \quad t \to \infty$$

almost surely on $\{\langle M \rangle_\infty = \infty\}$.

A law of large numbers for a multivariate martingale is, of course, easily obtained by applying one of the above results to each of the coordinates of the martingale. However, in many statistical applications such a result is not sufficient (see, e.g., Example 8.3.6), and a result of the following more general type is needed (see also the weak law of large numbers in Corollary A.7.8 below). For a positive semi-definite $n \times n$-matrix A, we denote the smallest and the largest eigenvalues by $\lambda_1(A)$ and $\lambda_n(A)$, respectively.

Theorem A.7.6 (Le Breton and Musiela, 1989) *Let M be an n-dimensional locally square integrable martingale, and let Γ be a positive semi-definite $n \times n$-matrix of increasing right-continuous predictable processes such that Γ_0 is non-singular. Suppose there exists a positive random variable ξ such that the process $\xi\Gamma - \langle M \rangle$ is almost surely positive semi-definite, and let $g : [0, \infty) \mapsto (0, \infty)$ be an increasing continuous function satisfying*

$$\int_0^\infty 1/g(u)du < \infty.$$

Then

$$\Gamma_t^{-1} M_t \to 0 \quad as \quad t \to \infty$$

almost surely on $\{\lim_{t\to\infty} \lambda_1(\Gamma_t) = \infty, \; \sup_{t\geq 0} \lambda_1^{-1}(\Gamma_t) g(\log[1 + \lambda_n(\Gamma_t)]) < \infty\}$.

Other strong laws of large numbers for multivariate martingales have been given by Melnikov (1986) and Kaufmann (1987). Moreover, Le Breton and Musiela (1987) and Dzhaparidze and Spreij (1993) have proved laws of large numbers for multivariate martingales that are Gaussian or that have deterministic quadratic variation, respectively.

We conclude this section with a *central limit theorem* for multivariate martingales. The result can be formulated more elegantly by introducing the concepts of stable and mixing convergence. We have, however, chosen to avoid this complication, which is not necessary for our purposes.

For a positive semi-definite matrix A we denote by $A^{\frac{1}{2}}$ and $\det(A)$ its positive semi-definite square root and its determinant, respectively. By $\mathrm{diag}(x_1, \cdots, x_n)$ we mean the $n \times n$ diagonal matrix with $x_1, \cdots, x_n$ on the diagonal, and $I_n = \mathrm{diag}(1, \cdots, 1)$.

Theorem A.7.7 *Suppose $M = (M_1, \cdots, M_n)^T$ is an n-dimensional square integrable martingale with mean value zero and quadratic covariation matrix $[M]$. Let H_t denote the covariance matrix of M_t, i.e., $H_t = E(M_t M_t^T)$. Assume, furthermore, that there exists a family of invertible non-random $n \times n$-matrices $\{K_t : t > 0\}$ such that as $t \to \infty$ we have $K_t \to 0$ and*

(i)

$$\bar{K}_{it} E\left(\sup_{s\leq t} |\Delta M_{is}|\right) \to 0, \quad i = 1, \ldots, n, \tag{A.7.1}$$

where $\bar{K}_{it} = \sum_{j=1}^n |K_{jit}|$,

(ii)

$$K_t \left[M\right]_t K^T \to W \qquad \text{(A.7.2)}$$

in probability, where W is a random positive semi-definite matrix satisfying $P(\det(W) > 0) > 0$, and

(iii)

$$K_t H_t K_t^T \to \Sigma, \qquad \text{(A.7.3)}$$

where Σ is a positive definite (deterministic) matrix.

Then

$$\left(K_t M_t, K_t \left[M\right]_t K_t^T\right) \to \left(W^{1/2} Z, W\right) \qquad \text{(A.7.4)}$$

and, conditionally on $\{\det(W) > 0\}$,

$$W^{-\frac{1}{2}} K_t M_t \to Z \qquad \text{(A.7.5)}$$

in distribution as $t \to \infty$, where Z is an n-dimensional standard normal distributed random vector independent of W. The convergence result (A.7.4) also holds conditionally on $\{\det(W) > 0\}$.

A proof of the theorem can be found in Küchler and Sørensen (1996b). The proof is based on results for $n = 1$ in Feigin (1985). Theorem A.7.7 generalizes an earlier result by Sørensen (1991). Note that the distribution of $W^{1/2}Z$ is the normal variance-mixture with characteristic function $u \mapsto E\left(\exp\left[-\frac{1}{2}u^T W u\right]\right)$, $u = (u_1, \cdots, u_n)^T$. Note also that under the conditions of the theorem,

$$M_t^T \left[M\right]_t^{-1} M_t \to \chi^2(n) \quad \text{as} \quad t \to \infty \qquad \text{(A.7.6)}$$

in distribution conditionally on $\{\det(W) > 0\}$.

In applications, a main problem is to find the family of matrices $\{K_t : t > 0\}$. It is not enough to study the rate of increase of the entries of H_t, because they may all increase at the same rate even when M grows at different rates in directions that are not parallel to the coordinate axes. Often, one must search for a family $\{K_t : t > 0\}$ of the form $D_t C$ where D_t is a diagonal matrix, while C changes the coordinate axes appropriately; see Example 8.3.6. Since H_t is positive semi-definite, there exists an orthogonal matrix O_t and a diagonal matrix D_t with non-negative diagonal elements such that $H_t = O_t^T D_t O_t$. If we can use $K_t = D_t^{-\frac{1}{2}} O_t$ or $K_t = O_t^T D_t^{-\frac{1}{2}} O_t$, then condition (iii) in Theorem A.7.7 is automatically satisfied.

There are several versions of the central limit theorem for martingales in the literature. For instance, (i) is sometimes replaced by a kind of Lindeberg condition, or the result is given in terms of $\langle M \rangle$ rather than $[M]$. For a discussion of these matters, see, e.g., Hall and Heyde (1980).

From Theorem A.7.7 follows a weak law of large numbers for multivariate martingales under considerably weaker conditions than the strong law in

Theorem A.7.6. This can be important in some applications; see Example 8.3.6.

Corollary A.7.8 *Assume the conditions of Theorem A.7.7. Then $[M]_t$ is invertible on $\{\det(W) > 0\}$ for t sufficiently large, and*

$$[M]_t^{-1} M_t \to 0 \tag{A.7.7}$$

in probability on $\{\det(W) > 0\}$ as $t \to \infty$.

A.8 Stochastic integration with respect to random measures

In the following, $E = \mathbb{R}^d \setminus \{0\}$, and $\mathcal{E}$ denotes the Borel sets of E. By a *random measure* on $[0, \infty) \times E$ we mean a kernel $\rho(\omega; dt, dx)$ such that $\rho(\omega; \cdot, \cdot)$ is a measure on $[0, \infty) \times E$ for every $\omega \in \Omega$ satisfying $\rho(\omega; \{0\} \times E) = 0$ and $\omega \mapsto \rho(\omega; B)$ is $\mathcal{F}$-measurable for every Borel subset B of $[0, \infty) \times E$. For every $\mathcal{F} \times \mathcal{B}([0, \infty)) \times \mathcal{E}$-measurable function $Y : \Omega \times [0, \infty) \times E \mapsto \mathbb{R}$ we can define the *integral process* $Y * \rho$ by

$$Y * \rho_t(\omega) =$$

$$\begin{cases} \int_{[0,t]\times E} Y(\omega, s, x)\, \rho(\omega; ds, dx) & \text{if } \int_{[0,t]\times E} |Y(\omega, s, x)|\, \rho(\omega; ds, dx) < \infty \\ \infty & \text{otherwise.} \end{cases}$$

We call a function $Y : \Omega \times [0, \infty) \times E \mapsto \mathbb{R}$ that is $\mathcal{P} \times \mathcal{E}$-measurable a *predictable random function* from $\mathbb{R}_+ \times E$ into $\mathbb{R}$. As in Section A.1, $\mathcal{P}$ denotes the predictable σ-algebra. A random measure ρ is called *predictable* if $Y * \rho$ is a predictable process for every predictable random function Y. Let ρ be an arbitrary random measure. If there exists a predictable random measure ρ^p such that the predictable process $Y * \rho^p$ is the compensator of the process $Y * \rho$ for every positive predictable random function Y for which $Y * \rho \in \mathcal{A}^+_{\text{loc}}(P)$, then ρ^p is called the *compensator* of ρ. Not every random measure has a compensator.

Now consider the *random measure associated with the jumps* of an adapted right-continuous stochastic process X with limits from the left:

$$\mu(\omega; dt, dx) = \sum_{s \geq 0} 1_{\{\Delta X_s(\omega) \neq 0\}} \epsilon_{(s, \Delta X_s(\omega))}(dt, dx), \tag{A.8.1}$$

where ϵ_a denotes the Dirac measure at a; see Jacod and Shiryaev (1987, p. 69). It can be proved that μ has a compensator ν that satisfies $\nu(\omega; \{t\} \times E) \leq 1,\ t \geq 0$. The process X is quasi-left-continuous (see Section A.3) if and only if the set $\{\omega : \nu(\omega; \{t\} \times E) > 0 \text{ for some } t \in \mathbb{R}_+\}$ is a P-null set. Proofs of these results can be found in Jacod and Shiryaev (1987, pp. 69–70).

We can now discuss the stochastic integral with respect to the compensated measure $\mu - \nu$. If Y is a predictable random function such that $|Y| * \mu \in \mathcal{A}^+_{\text{loc}}(P)$, the situation is simple. Then $|Y| * \nu$ is the compensator of $|Y| * \mu$, i.e., $|Y| * \nu \in \mathcal{A}^+_{\text{loc}}(P)$, and we can define the stochastic integral (process) of Y with respect to $\mu - \nu$ by

$$Y * (\mu - \nu) = Y * \mu - Y * \nu, \tag{A.8.2}$$

which is a local martingale. The stochastic integral can, however, be defined for much more general integrands.

For a predictable random function Y define

$$\hat{Y}_t(\omega) = \begin{cases} \int_E Y(\omega, t, x)\, \nu(\omega; \{t\} \times dx) & \text{if } \int_E |Y(\omega, t, x)|\, \nu(\omega; \{t\} \times dx) < \infty \\ \infty & \text{otherwise,} \end{cases}$$

and let $\mathcal{G}_{\text{loc}}(\mu, P)$ denote the set of all predictable random functions Y for which $Y^* \in \mathcal{A}^+_{\text{loc}}(P)$, where

$$Y^*_t(\omega) = \left\{ \sum_{s \le t} \left(Y(\omega, s, \Delta X_s) 1_{\{\Delta X_s \ne 0\}} - \hat{Y}_s(\omega) \right)^2 \right\}^{1/2}.$$

For $Y \in \mathcal{G}_{\text{loc}}(\mu, P)$ there exists a unique purely discontinuous local martingale M such that $\Delta M_t = Y(t, \Delta X_t) 1_{\{\Delta X_t \ne 0\}} - \hat{Y}_t$; see Jacod and Shiryaev (1987, p. 72). We call this local martingale the *stochastic integral of Y with respect to $\mu - \nu$* and denote it by $Y * (\mu - \nu)$. It is identical to the integral defined by (A.8.2) when $|Y| * \mu \in \mathcal{A}^+_{\text{loc}}(P)$. This and the following results about the stochastic integral with respect to $\mu - \nu$ are proved in Jacod and Shiryaev (1987, pp. 72–74). The space $\mathcal{G}_{\text{loc}}(\mu, P)$ is linear, and the mapping $Y \mapsto Y * (\mu - \nu)$ is linear on $\mathcal{G}_{\text{loc}}(\mu, P)$.

The integral with respect to $\mu - \nu$ fits the integral with respect to a semimartingale in the following sense. If H is a locally bounded predictable process and $Y \in \mathcal{G}_{\text{loc}}(\mu, P)$, then $HY \in \mathcal{G}_{\text{loc}}(\mu, P)$ and

$$H \cdot \{Y * (\mu - \nu)\} = (HY) * (\mu - \nu).$$

For a predictable random process Y define the increasing predictable processes

$$C(Y)_t = \left(Y - \hat{Y}\right)^2 * \nu_t + \sum_{s \le t} (1 - a_s) \hat{Y}_s^2$$

and

$$\bar{C}(Y)_t = |Y - \hat{Y}| * \nu_t + \sum_{s \le t} (1 - a_s) |\hat{Y}_s|,$$

where

$$a_s = \nu(\omega; \{s\} \times E). \tag{A.8.3}$$

Theorem A.8.1 *Let Y be a predictable random function from $[0, \infty) \times E$ into $\mathbb{R}$.*

(a) $C(Y) \in \mathcal{A}^+(P)$ $(\mathcal{A}^+_{loc}(P))$ if and only if $Y \in \mathcal{G}_{loc}(\mu, P)$ and $Y * (\mu - \nu)$ is a (locally) square integrable martingale. In this case,

$$\langle Y * (\mu - \nu) \rangle = C(Y).$$

(b) $\bar{C}(Y) \in \mathcal{A}^+(P)$ $(\mathcal{A}^+_{loc}(P))$ if and only if $Y \in \mathcal{G}_{loc}(\mu, P)$ and $Y * (\mu - \nu) \in \mathcal{A}(P)$ $(\mathcal{A}_{loc}(P))$.

When X is quasi-left-continuous, $\hat{Y} = 0$, and the results are simpler. This is also the case if a_s given by (A.8.3) is either one or zero. Since a_s is the probability (given the past) that a jump occurs at time s, the condition $a_s \in \{0, 1\}$ for all $s \ge 0$ is, for instance, satisfied for discrete time processes.

Theorem A.8.2 *Suppose X is quasi-left-continuous, and let Y be a predictable random function from $[0, \infty) \times E$ into $\mathbb{R}$.*

(a) $Y \in \mathcal{G}_{loc}(\mu, P)$ and $Y * (\mu - \nu)$ is a (locally) square integrable martingale if and only if $Y^2 * \nu \in \mathcal{A}^+(P)$ $(\mathcal{A}^+_{loc}(P))$. In this case,

$$\langle Y * (\mu - \nu) \rangle = Y^2 * \nu.$$

(b) $Y \in \mathcal{G}_{loc}(\mu, P)$ and $Y * (\mu - \nu) \in \mathcal{A}(P)$ $(\mathcal{A}_{loc}(P))$ if and only if $|Y| * \nu \in \mathcal{A}^+(P)$ $(\mathcal{A}^+_{loc}(P))$.

(c) $Y \in \mathcal{G}_{loc}(\mu, P)$ if and only if

$$\left\{ |Y| 1_{\{|Y| > 1\}} + Y^2 1_{\{|Y| \le 1\}} \right\} * \nu \in \mathcal{A}^+_{loc}(P).$$

(d) Suppose $Y \ge -1$. Then $Y \in \mathcal{G}_{loc}(\mu, P)$ if and only if

$$\left(1 - \sqrt{1 + Y}\right)^2 * \nu \in \mathcal{A}^+_{loc}(P).$$

A.9 Local characteristics of a semimartingale

The local behaviour of a d-dimensional semimartingale $\underline{X} = (X_1, \cdots, X_d)$ is described by its local characteristics (α, β, ν), which consist of a d-dimensional predictable process α, a $d \times d$-matrix of predictable processes β, and a predictable random measure ν on $\mathbb{R}_+ \times E$, where $E = \mathbb{R}^d \setminus \{0\}$.

The matrix β is defined by

$$\beta_{ij} = \langle X_i^c, X_j^c \rangle,$$

where X_i^c is the continuous martingale part of X_i; cf. Theorem A.4.7. The predictable random measure ν is the compensator of the random measure associated with the jumps of X; cf. (A.8.1). In order to define α, consider the process

$$\tilde{X}_t = X_0 + \sum_{s \leq t} \Delta X_s 1_{E \setminus [-1,1]^d} (\Delta X_s).$$

There are only finitely many jumps of a size outside $[-1,1]^d$, so $\tilde{X}$ is well-defined. Now, the process $X - \tilde{X}$ has bounded jumps, so it is a special semimartingale. Hence a unique decomposition of $X - \tilde{X}$ of the form (A.4.1) exists, where $A \in \mathcal{V}$ is predictable (and hence $A \in \mathcal{A}^+_{\text{loc}}(P)$). The last local characteristic is simply given by $\alpha = A$. Thus (α, β, ν) are uniquely determined by X and P. We shall later give an example that the converse is not true (Example A.9.1).

The following representation of X shows how the local characteristics determine the local behaviour of X:

$$X = \tilde{X} + X^c + U * (\mu - \nu) + \alpha, \tag{A.9.1}$$

where

$$U(\omega, t, x) = 1_{[-1,1]^d}(x). \tag{A.9.2}$$

It can be shown that $U \in \mathcal{G}_{\text{loc}}(\mu, P)$. The process α is an integrated drift for the process $X - \tilde{X}$, β generalizes the integrated diffusion coefficient, and ν can be thought of as a generalized jump intensity.

The processes in β are always continuous (because X^c is continuous), and if X is quasi-left-continuous, α is continuous too. More generally,

$$\Delta\alpha_t = \int_{[-1,1]^d} x\nu(\{t\} \times dx), \quad t > 0. \tag{A.9.3}$$

The random measure ν has the property

$$\left(|x|^2 \wedge 1\right) * \nu \in \mathcal{A}^+_{\text{loc}}(P). \tag{A.9.4}$$

These results can be found in Jacod and Shiryaev (1987, p. 77). When X is a *special semimartingale*, ν satisfies (Jacod and Shiryaev, 1987, p. 82)

$$\left(|x|^2 \wedge |x|\right) * \nu \in \mathcal{A}^+_{\text{loc}}(P),$$

so the identity mapping belongs to $\mathcal{G}_{\text{loc}}(\mu, P)$. By combining this with the fact that a special semimartingale has a unique decomposition (A.4.1) where $A \in \mathcal{V}$ is predictable, it follows that for special semimartingales,

$$X_t = X_0 + X_t^c + x * (\mu - \nu)_t + A_t.$$

It is always possible to find a version of the local characteristics of the form

$$\alpha_i = a_i \cdot A, \quad i = 1, \cdots, d, \tag{A.9.5}$$

$$\beta_{ij} = c_{ij} \cdot A, \quad i, j = 1, \cdots, d, \tag{A.9.6}$$

$$\nu(\omega; dt, dx) = K_{\omega,t}(dx)\, dA_t(\omega), \tag{A.9.7}$$

where $A \in \mathcal{A}_{\text{loc}}^{+}(P)$ is predictable, a is a d-dimensional predictable process, c is a $d \times d$-matrix of predictable processes that at all times is symmetric and positive semi-definite, and $K_{\omega,t}(dx)$ is a transition kernel from $(\Omega \times [0, \infty), \mathcal{P})$ into $(E, \mathcal{E})$; see Jacod and Shiryaev (1987, p. 77).

Example A.9.1 A *Lévy process* is a d-dimensional semimartingale with deterministic local characteristics of the form $(t\alpha, t\Sigma, \nu(dx)dt)$, where α is a d-dimensional vector; Σ is a symmetric, positive semi-definite matrix; and ν is a measure on E. Traditionally, a Lévy process is characterized by the Lévy characteristics (α, Σ, ν), where ν is called a Lévy measure. Usually, the decomposition (A.9.1) is written in the form (2.1.11) (for a one-dimensional Lévy process). Particular examples are the Brownian motion with drift θ, which has local characteristics $(\theta t, t, 0)$; the Poisson process with intensity λ, which has characteristics $(\lambda t, 0, \lambda \epsilon_1(dx)dt)$; and a compound Poisson process with jump intensity λ, where the jump size has distribution function F and expectation μ. In the last case, the local characteristics are $(\lambda \mu t, 0, \lambda F(dx)dt)$. □

Example A.9.2 We have seen earlier that a counting process N with compensator Λ is a semimartingale with the decomposition

$$N = (N - \Lambda) + \Lambda.$$

The local characteristics are $(\Lambda_t, 0, \epsilon_1(dx)d\Lambda_t)$. □

Example A.9.3 A d-dimensional diffusion-type process that solves the stochastic differential equation (3.5.1) is obviously a semimartingale with local characteristics

$$\left(\int_0^t a_s(X)\, ds + \int_0^t b_s(X)\, ds\theta, \int_0^t c_s(X)\, c_s(X)^T\, ds, 0 \right).$$

The diffusion with jumps which solves

$$dX_t = X_{t-}dt + X_{t-}dW_t + dN_t, \quad X_0 = 1,$$

where W is a standard Wiener process independent of the Poisson process N with intensity λ, is a semimartingale with local characteristics

$$\left(\int_0^t X_s ds + \lambda t, \int_0^t X_s^2 ds, \lambda \epsilon_1 (dx)\, dt\right).$$

□

We conclude this section with an example that the local characteristics do not determine X and P.

Example A.9.4 Let $(\Omega, \mathcal{F}, P)$ be a probability space with a filtration $\{\mathcal{F}_t\}$, and let W be a standard real-valued Wiener process with respect to $\{\mathcal{F}_t\}$. Define the function

$$a(x) = \begin{cases} a_1 > 0 & \text{if } x > 0 \\ \frac{1}{2}(a_1 + a_2) & \text{if } x = 0 \\ a_2 > 0 & \text{if } x < 0, \end{cases}$$

where $a_1 \neq a_2$. Then the equation

$$dX_t = a(X_t) dW_t$$

has a unique strong solution X adapted to $\{\mathcal{F}_t\}$ that is a locally square integrable martingale with

$$\langle X \rangle_t = \int_0^t a(X_s)^2 \, ds,$$

and that is strongly Markovian; see Harrison and Shepp (1981). Now define $\tau_u = \inf\{t : \langle X \rangle_t \geq u\}$. Then τ_u is an $\{\mathcal{F}_t\}$-stopping time for every $u \geq 0$, and $u \mapsto \tau_u$ is continuous and strongly increasing. Hence $\tilde{W}$, defined by

$$\tilde{W}_u = X_{\tau_u},$$

is a Wiener process with respect to the filtration $\{\mathcal{F}_{\tau_u}\}$; see Kazamaki (1972).

Extending $(\Omega, \mathcal{F}, \{\mathcal{F}_{\tau_u}\}, P)$ if necessary, we can assume that it carries a second Wiener process W^* that is independent of $\tilde{W}$ and adapted to $\{\mathcal{F}_{\tau_u}\}$. Because $\langle X \rangle_t$ is an $\{\mathcal{F}_{\tau_u}\}$-stopping time for every $t > 0$, the processes $X = \tilde{W}_{\langle X \rangle}$ and $W^*_{\langle X \rangle}$ are continuous local $\{\mathcal{F}_t\}$-martingales with the same local characteristics $(0, 0, \langle X \rangle)$. But these two processes are different in law. The former is strongly Markovian, the latter is not. To prove the second statement, assume the contrary. Then $\langle X \rangle$ must be the inverse of an additive functional with respect to $\{\mathcal{F}_{\tau_u}^{W^*}\}$; see Engelbert and Schmidt (1988). This is, however, impossible because of the independence of W^* and X. □

A.10 A Girsanov-type theorem for semimartingales

Let P and $\tilde{P}$ be two probability measures on the filtered space $(\Omega, \mathcal{F}, \{\mathcal{F}_t\})$. The measure $\tilde{P}$ is said to be *locally absolutely continuous* with respect to P if $\tilde{P}^t \ll P^t$ for all $t \geq 0$. Here P^t denotes the restriction of P to $\mathcal{F}_t$. The likelihood ratio between $\tilde{P}$ and P corresponding to observation of $\mathcal{F}_t$ is the Radon-Nikodym derivative $d\tilde{P}^t/dP^t$. In order to calculate likelihood functions for semimartingale models, we thus need results that give Radon-Nikodym derivatives in terms of local characteristics. We also need results that tell how the local characteristics are changed under locally absolutely continuous changes of probability measures. In this section we shall summarize such results and conditions in terms of the local characteristics ensuring local absolute continuity of the probability measures. The results are taken from Jacod and Mémin (1976) and Jacod and Shiryaev (1987, Chapter III), see also the discussion in Küchler and Sørensen (1989).

We shall need the following somewhat technical definition. Define a measure M_μ^P on $\tilde{\mathcal{F}} = \mathcal{F} \times \mathcal{B}(\mathbb{R}_+) \times \mathcal{E}$ by $M_\mu^P(Y) = E(Y * \mu_\infty)$ for every $\tilde{\mathcal{F}}$-measurable function Y from $\tilde{\Omega} = \Omega \times \mathbb{R}_+ \times E$ into $\mathbb{R}_+$. As in the previous sections, μ is defined by (A.8.1). Let Z be a local martingale. We shall define the conditional expectation $M_\mu^P(\Delta Z|\tilde{\mathcal{P}})$ of the increments $\Delta Z = \{\Delta Z_t\}$ of Z with respect to the σ-algebra $\tilde{\mathcal{P}} = \mathcal{P} \times \mathcal{E}$ on $\tilde{\Omega}$. First, we define another measure on $\tilde{\mathcal{F}}$ by $(\Delta Z) \cdot M_\mu^P$. Then $M_\mu^P(\Delta Z|\tilde{\mathcal{P}})$ can be defined as the Radon-Nikodym derivative of the restriction of this second measure to $\tilde{\mathcal{P}}$ with respect to the restriction of M_μ^P to $\tilde{\mathcal{P}}$. That this Radon-Nikodym derivative exists was shown in Jacod (1976).

Let X be a d-dimensional semimartingale defined on the filtered probability space $(\Omega, \mathcal{F}, \{\mathcal{F}_t\}, P)$ with local characteristics $\mathcal{L} = (\alpha, \beta, \nu)$. We assume that $X_0 = x_0$ is non-random and use that β has the form (A.9.6). The process X will be fixed in this section.

Theorem A.10.1 *Let $\tilde{P}$ be a second probability measure on $(\Omega, \mathcal{F})$ satisfying $\tilde{P}^t \ll P^t$ for all $t > 0$, and define a positive martingale L by $L_t = d\tilde{P}^t/dP^t$. Let U be given by (A.9.2). Then X is a semimartingale under $\tilde{P}$, and there exists a d-dimensional predictable process z and a non-negative predictable random function Y from $\mathbb{R}_+ \times E^d$ into $\mathbb{R}_+$ satisfying*

$$[U(Y-1)] * \nu \in \mathcal{V}, \tag{A.10.1}$$

$$(cz)_i \cdot A \in \mathcal{V}, \quad i = 1, \cdots, d, \tag{A.10.2}$$

such that the local characteristics of X under $\tilde{P}$ are given by

$$\tilde{\alpha} = \alpha + (cz) \cdot A + [U(Y-1)] * \nu, \tag{A.10.3}$$

$$\tilde{\beta} = \beta, \tag{A.10.4}$$

$$\tilde{\nu} = Y\nu. \tag{A.10.5}$$

The random function Y and the process z are determined by the equations

$$M_\mu^P\left(\Delta L|\tilde{\mathcal{P}}\right) = L_-(Y-1) \tag{A.10.6}$$

and

$$\langle L^c, X^c\rangle = (czL_-)\cdot A. \tag{A.10.7}$$

The continuous martingale part of L and the bracket in (A.10.7) are relative to P. If X is quasi-left-continuous under P, then it is so too under $\tilde{P}$.

Theorem A.10.2 *Let z be a d-dimensional predictable process satisfying $(z^T cz)\cdot A \in \mathcal{A}_{loc}(P)$, and Y a non-negative predictable random function from $\mathbb{R}_+ \times E^d$ into $\mathbb{R}_+$. Suppose $W \in \mathcal{G}_{loc}(\mu, P)$, where*

$$W(s,x) = Y(s,x) - 1 + \frac{1_{\{\nu(\{s\}\times E^d)\neq 1\}}}{1-\nu(\{s\}\times E^d)}\int_{E^d}(Y(s,x)-1)\,\nu(\{s\},dx), \tag{A.10.8}$$

and define Z by

$$Z_t = \exp\left(N_t - \tfrac{1}{2}\left(z^T cz\right)\cdot A_t\right)\prod_{s\leq t}\left[(1+\Delta N_s)\exp(-\Delta N_s)\right],\quad t\geq 0, \tag{A.10.9}$$

with

$$N = z\cdot X^c + W*(\mu-\nu). \tag{A.10.10}$$

Suppose, moreover, that

$$\int_{E^d} Y(s,x)\,\nu(\{s\},dx) = 1 \tag{A.10.11}$$

P-almost surely for all s such that $\nu(\{s\}\times E^d) = 1$, and that $E(Z_t) = 1$ for all $t\geq 0$. Then there exists a probability measure $\tilde{P}$ on $\sigma(\mathcal{F}_t : t\geq 0)$ such that $d\tilde{P}^t/dP^t = Z_t$ and such that under $\tilde{P}$ the process X is a semimartingale with local characteristics $(\tilde{\alpha},\tilde{\beta},\tilde{\nu})$ given by (A.10.3)–(A.10.5).

For the existence of $\tilde{P}$ given $\{\tilde{P}^t\}_{t\geq 0}$, where $d\tilde{P}^t = Z_t dP^t$, we need to assume that $(\Omega, \sigma(\mathcal{F}_t : t\geq 0))$ is standard measurable, which is not a restriction in our case; see Ikeda and Watanabe (1981, p. 176).

Note that if X is *quasi-left-continuous,* (A.10.11) is not needed, $W = Y-1$, and $W \in \mathcal{G}_{\text{loc}}(\mu, P)$ is satisfied if and only if

$$y = \left(1-\sqrt{Y}\right)^2 * \nu \in \mathcal{A}^+_{\text{loc}}(P); \tag{A.10.12}$$

cf. Theorem A.8.2.

If $\tilde{P}$ is a probability measure on $(\Omega,\mathcal{F})$ such that under $\tilde{P}$ the process X is a semimartingale with local characteristics $\tilde{\mathcal{L}} = (\tilde{\alpha},\tilde{\beta},\tilde{\nu})$ given by (A.10.3)–(A.10.5), then $\tilde{P}$ is not necessarily locally absolutely continuous with respect to P. This is because the local characteristics of a semimartingale do not in general determine the probability measure uniquely as we saw in Example A.9.4.

Definition A.10.3 *We say that the problem $(X, \mathcal{L}, x_0)$ has a solution P if P is a probability measure on (Ω, F), satisfying $P(X_0 = x_0) = 1$, under which X is a semimartingale with local characteristics $\mathcal{L}$. If P is the only solution of the problem $(X, \mathcal{L}, x_0)$, we say that $(X, \mathcal{L}, x_0)$ has a unique solution.*

If T is an $(\mathcal{F}_t)$-stopping time, then $(X, \mathcal{L}, x_0)$ is said to have a T-unique solution if every pair of solutions of $\left(X^T, \mathcal{L}^T, x_0\right)$ coincide on $\mathcal{F}_{T-}$. Here X^T denotes the stopped process defined by $X_t^T = X_{t \wedge T}$, and $\mathcal{L}^T = (\alpha^T, \beta^T, \nu^T)$ is defined analogously.

It is well-known that for processes with independent increments, for diffusion processes, and for some counting processes the corresponding $(X, \mathcal{L}, x_0)$-problems have a unique solution provided that Ω is the space of right-continuous functions with limits from the left, that $\{\mathcal{F}_t\}$ is the right-continuous filtration generated by X, and that $\mathcal{F} = \sigma(\mathcal{F}_t : t \geq 0)$. See Jacod and Mémin (1976) for more details. In general, a unique solution of $(X, \mathcal{L}, x_0)$ does not imply T-uniqueness. Conditions under which this conclusion holds for every stopping time T relative to the filtration given by $\mathcal{F}_t^0 = \sigma(X_s : s \leq t)$ are given in Jacod and Mémin (1976, Section 7). These conditions are satisfied for diffusion processes and processes with independent increments. Kabanov, Liptser, and Shiryaev (1980) gave conditions on the coefficients of stochastic differential equations driven by a semimartingale under which the solution is a process that is the unique solution of the corresponding $(X, \mathcal{L}, x_0)$ problem.

In the last theorem we assume that $\mathcal{F} = \sigma(\mathcal{F}_t : t \geq 0)$.

Theorem A.10.4 *Suppose X is quasi-left-continuous, and let $\tilde{P}$ be a second probability measure on $(\Omega, \mathcal{F})$ under which X is a semimartingale with local characteristics $\tilde{\mathcal{L}} = (\tilde{\alpha}, \tilde{\beta}, \tilde{\nu})$ given by (A.10.3)–(A.10.5). Assume that the process y defined by (A.10.12) belongs to $\mathcal{A}_{loc}^+(P)$, that*

$$\tilde{P}\left(\left(z^T cz\right) \cdot A_t + y_t < \infty\right) = 1, \quad t > 0, \tag{A.10.13}$$

and that the problem $(X, \tilde{\mathcal{L}}, x_0)$ has a $\sigma_n \wedge t$-unique solution for every $n \geq n_0(t)$ and for all $t > 0$, where the stopping time σ_n is defined by

$$\sigma_n = \inf\left\{t > 0 : \left(z^T cz\right) \cdot A_t + y_t \geq n\right\}.$$

Then $\tilde{P}^t \ll P^t$ for all $t > 0$, and

$$\frac{d\tilde{P}^t}{dP^t} = Z_t, \qquad t > 0,$$

where Z is defined by (A.10.9) with $W = Y - 1$.

In case the uniqueness condition is not satisfied, Z still has an interpretation as a quasi-likelihood; see Sørensen(1990). See also Jacod (1987).

Example A.10.5 Suppose the stochastic process X under P_θ, $\theta \in \Theta \subseteq \mathbb{R}^k$, is a diffusion-type process that solves the equation

$$dX_t = a_t(X;\theta)\,dt + \sigma_t(X)\,dW_t, \quad X_0 = x_0,$$

where $c_t = \sigma_t(X)\sigma_t(X)^T$ is almost surely invertible under P_θ. For $P = P_{\theta_0}$ and $\tilde{P} = P_\theta$ consider (A.10.3)–(A.10.5). Here $\nu = \tilde{\nu}$ (i.e., $Y = 0$) because X has no jumps,

$$\tilde{\alpha}_t - \alpha_t = \int_0^t [a_s(X;\theta) - a_s(X;\theta_0)]\,ds,$$

and

$$\beta_t = \int_0^t \sigma_s(X)\,\sigma_s(X)^T\,ds.$$

Hence $A_t = t$,

$$\begin{aligned} z_t &= c_t^{-1}[a_t(X;\theta) - a_t(X;\theta_0)] \\ &\quad -\tfrac{1}{2}\int_0^t [a_s(X;\theta) - a_s(X;\theta_0)]^T c_t^{-1}[a_s(X;\theta) - a_s(X;\theta_0)]\,ds \\ &= \int_0^t [a_s(X;\theta) - a_s(X;\theta_0)]^T c_s^{-1} dX_s ds. \\ &\quad -\tfrac{1}{2}\int_0^t \left\{a_s(X;\theta)^T c_s^{-1} a_s(X;\theta) - a_s(X;\theta_0)^T c_s^{-1} a_s(X;\theta_0)\right\}, \end{aligned}$$

and, provided that the conditions of Theorem A.10.4 are satisfied,

$$\begin{aligned} \log\frac{dP_\theta^t}{dP_{\theta_0}^t} &= \int_0^t [a_s(X;\theta) - a_s(X;\theta_0)]^T c_s^{-1} dX_s \qquad \text{(A.10.14)} \\ &\quad -\int_0^t [a_s(X;\theta) - a_s(X;\theta_0)]^T c_s^{-1} a_s(X;\theta_0)\,ds \\ &\quad -\tfrac{1}{2}\int_0^t [a_s(X;\theta) - a_s(X;\theta_0)]^T c_t^{-1}[a_s(X;\theta) - a_s(X;\theta_0)]\,ds \\ &= \int_0^t [a_s(X;\theta) - a_s(X;\theta_0)]^T c_s^{-1} dX_s \\ &\quad -\tfrac{1}{2}\int_0^t \{a_s(X;\theta)^T c_s^{-1} a_s(X;\theta) - a_s(X;\theta_0)^T c_s^{-1} a_s(X;\theta_0)\}ds. \end{aligned}$$

Here we have used that under P_{θ_0},

$$X_t^c = X_t - \int_0^t a_s(X;\theta_0)\,ds.$$

□

Example A.10.6 Let X be a stochastic process that under P_θ, $\theta \in \Theta \subseteq \mathbb{R}^k$, is a counting process with intensity $\lambda(\theta)$, where $\lambda_t(\theta) > 0$ for all $t > 0$. Consider (A.10.3)–(A.10.5) for $P = P_{\theta_0}$ and $\tilde{P} = P_\theta$. Here $\beta = \tilde{\beta} = 0$, $\nu(dx, dt) = \epsilon_1(dx)\lambda_t(\theta_0)dt$, and $\tilde{\nu}(dx, dt) = \epsilon_1(dx)\lambda_t(\theta)dt$, so

$$Y(x,t) = Y(t) = \lambda_t(\theta)/\lambda_t(\theta_0).$$

Finally,

$$\tilde{\alpha}_t - \alpha_t = \int_0^t [\lambda_s(\theta) - \lambda_s(\theta_0)]\, ds = [U(Y-1)] * \nu_t,$$

so $z = 0$. Thus, provided that the conditions of Theorem A.10.4 hold,

$$\frac{dP_\theta^t}{dP_{\theta_0}^t} = \exp\left[(Y-1) * (\mu - \nu)_t\right] \prod_{\{i:\tau_i \le t\}} Y(\tau_i) \exp\left[-\left(Y(\tau_i) - 1\right)\right], \tag{A.10.15}$$

where $\tau_1, \tau_2, \cdots$ are the jump times of X. Here we have used that $\Delta\,(Y-1)*(\mu-\nu)_{\tau_i} = Y(\tau_i) - 1$. If we assume, moreover, that the counting process X has only finitely many jumps in every bounded interval, then

$$\frac{dP_\theta^t}{dP_{\theta_0}^t} = \exp\left\{\sum_{i=1}^{N_t} \left[\log\left(\lambda_{\tau_i}(\theta)\right) - \log\left(\lambda_{\tau_i}(\theta_0)\right)\right] - \int_0^t [\lambda_s(\theta) - \lambda_s(\theta_0)]\, ds\right\}. \tag{A.10.16}$$

For processes with independent increments that are semimartingales, it is, in general, not possible to simplify the structure of dP_θ^t/dP^t given by (A.10.9). However, in this case the situation is simplified by the fact that the triple of local characteristics is non-random (deterministic).

Appendix B

Miscellaneous Results

B.1 The fundamental identity of sequential analysis

It has for many years been statistical folklore that "the likelihood function is independent of the stopping rule." Several proofs of this result, which is known as the fundamental identity of sequential analysis or Wald's identity, have appeared; see, e.g., Sudakov (1969), Döhler (1981), Küchler (1985), or Jacod and Shiryaev (1987, Theorem III.3.4.(ii)). For completeness we include a proof of the result in the generality necessary for the present book.

Consider a filtered space $(\Omega, \mathcal{F}, \{\mathcal{F}_t\}, P_1, P_2)$ with two probability measures P_1 and P_2, which are typically non-equivalent or even singular on $\mathcal{F}$. This is why the result is not a trivial consequence of the optional stopping theorem. Let P_i^t denote the restriction of P_i to $\mathcal{F}_t$. We assume that P_2 is locally dominated by P_1, i.e.,

$$P_2^t \ll P_1^t, \quad t \geq 0, \tag{B.1.1}$$

and denote the Radon-Nikodym derivatives by

$$L_t = \frac{dP_2^t}{dP_1^t}, \quad t \geq 0. \tag{B.1.2}$$

For a stopping time τ with respect to $\{\mathcal{F}_t\}$, the σ-algebra of events happening before τ is given by

$$A \in \mathcal{F}_\tau \Leftrightarrow (A \cap \{\tau \leq t\} \in \mathcal{F}_t \text{ for all } t \in [0, \infty]) \, . \tag{B.1.3}$$

We denote the restriction of P_i to $\mathcal{F}_\tau$ by P_i^τ. Now we can formulate the main result.

Theorem B.1.1 *Let τ be a stopping time. Then*

$$P_2(A \cap \{\tau < \infty\}) = \int_{A\cap\{\tau<\infty\}} L_\tau dP_1 \tag{B.1.4}$$

for all $A \in \mathcal{F}_\tau$.

Proof: Since $\mathcal{F}_{\tau\wedge n} \subseteq \mathcal{F}_n$, we see that $P_2^{\tau\wedge n} \ll P_1^{\tau\wedge n}$, and as $\tau \wedge n$ is a bounded stopping time, it follows by the optional stopping theorem that

$$\frac{dP_2^{\tau\wedge n}}{dP_1^{\tau\wedge n}} = E_1(L_n|\mathcal{F}_{\tau\wedge n}) = L_{\tau\wedge n}. \tag{B.1.5}$$

Here E_i denotes expectation with respect to P_i, and $\tau \wedge n$ denotes the minimum of τ and n.

To show (B.1.4), it is enough to show this identity for $A \in \mathcal{F}_{\tau\wedge k}$ for all $k \in \mathbb{N}$ because

$$\{\tau < \infty\} \cap \mathcal{F}_\tau = \{\tau < \infty\} \cap \sigma(\cup_{k=1}^{\infty} \mathcal{F}_{\tau\wedge k}).$$

Therefore, fix k and choose $A \in \mathcal{F}_{\tau\wedge k}$. Since $A \cap \{\tau \le m\} \in \mathcal{F}_{\tau\wedge m}$ for all $m \ge k$, we have by (B.1.5) that

$$P_2(A \cap \{\tau \le m\}) = \int_{A\cap\{\tau\le m\}} L_{\tau\wedge m} dP_1 = \int_{A\cap\{\tau\le m\}} L_\tau dP_1.$$

Now for $m \to \infty$ we obtain, by the monotone convergence theorem, that

$$P_2(A \cap \{\tau < \infty\}) = \int_{A\cap\{\tau<\infty\}} L_\tau dP_1.$$

□

The following, in practice more useful, corollary follows immediately from Theorem B.1.1.

Corollary B.1.2 *Let τ be a stopping time satisfying*

$$P_1(\tau < \infty) = P_2(\tau < \infty) = 1. \tag{B.1.6}$$

Then

$$P_2^\tau \ll P_1^\tau \quad \textit{and} \quad \frac{dP_2^\tau}{dP_1^\tau} = L_\tau. \tag{B.1.7}$$

B.2 A conditional Radon-Nikodym derivative

Let the situation be as in the previous section, and let X be a stochastic process adapted to the filtration $\{\mathcal{F}_t\}$. Define for $s \leq t$ the σ-algebra

$$\mathcal{G}_{s,t} = \sigma(X_u : s \leq u \leq t). \tag{B.2.1}$$

Then we have the following lemma.

Lemma B.2.1 *For every $\mathcal{G}_{s,t}$-measurable random variable Z,*

$$E_2(Z|\mathcal{F}_s) = E_1(ZY_{s,t}|\mathcal{F}_s), \tag{B.2.2}$$

where

$$Y_{s,t} = L_t/L_s. \tag{B.2.3}$$

Remark: The lemma shows that we can interprete $Y_{s,t}$ as a conditional Radon-Nikodym derivative given $\mathcal{F}_s$.

Proof: Let A be an $\mathcal{F}_s$-measurable random variable. Then $E_2(AZ) = E_1(AZL_t) = E_1[E_1(AZL_sY_{s,t}|\mathcal{F}_s)] = E_1[AE_1(ZY_{s,t}|\mathcal{F}_s)L_s]$
$= E_2[AE_1(ZY_{s,t}|\mathcal{F}_s)]$. □

B.3 Three lemmas

In this section we state, for easy reference, some well-known results that are used in the book. The proofs are easy exercises. First the Toeplitz lemma.

Lemma B.3.1 *Let $a_k > 0$ and x_k $(k = 1, 2, \cdots)$ be sequences of real numbers such that $b_n = \sum_{k=1}^n a_k \uparrow \infty$ and $x_n \to x$ as $n \to \infty$. Then $b_n^{-1} \sum_{k=1}^n a_k x_k \to x$.*

Next follows the integral version of the Toeplitz lemma.

Lemma B.3.2 *Let $a(t) > 0$ and $x(t)$ $(t > 0)$ be real functions such that the following integrals exist and such that $b(t) = \int_0^t a(s)ds \uparrow \infty$ and $x(t) \to x$ as $t \to \infty$. Then $b(t)^{-1} \int_0^t a(s)x(s)ds \to x$.*

Finally, we give Groenwall's lemma.

Lemma B.3.3 *Let ρ be a continuous function on $[0, T]$ such that*

$$\rho(t) \leq c + K \int_0^t \rho(s)ds$$

for some positive constants c and K. Then $\rho(t) \leq ce^{Kt}$.

Appendix C

References

Abramowitz, M. and Stegun, I.A. (1970): *Handbook of Mathematical Functions.* Dover Publications, New York.

Aitchison, J. and Silvey, S.D. (1958): Maximum-likelihood estimation of parameters subject to restraints. *Ann. Math. Statist.* **29**, 813–828.

Aldous, D.J. (1978): Weak convergence of randomly indexed sequences of random variables. *Math. Proc. Camb. Phil. Soc.* **83**, 117–126.

Aldous, D.J. and Eagleson, G.K. (1978): On mixing and stability of limit theorems. *Ann. Probability* **6**, 325–331.

Amari, S.-I. (1985): *Differential-Geometrical Methods in Statistics.* Lecture Notes in Statistics 28. Springer, Heidelberg.

Anscombe, F.J. (1952): Large-sample theory of sequential estimation. *Proc. Camb. Philos. Soc.* **48**, 600–607.

Asmussen, S. (1987): *Applied Probability and Queues.* Wiley, Chichester.

Bai, D.S. (1975): Efficient estimation of transition probabilities in a Markov chain. *Ann. Statist.* **3**, 1305–1317.

Bar-Lev, S.K. and Enis, P. (1986): Reproducibility and natural exponential families with power variance functions. *Ann. Statist.* **14**, 1507–1522.

Barndorff-Nielsen, O.E. (1978): *Information and Exponential Families.* Wiley, Chichester.

Barndorff-Nielsen, O.E. and Cox, D.R. (1984): The effect of sampling rules on likelihood statistics. *Internat. Statist. Rev.* **52**, 309–326.

Barndorff-Nielsen, O.E. and Cox, D.R. (1989): *Asymptotic Techniques for Use in Statistics.* Chapman and Hall, London.

Barndorff-Nielsen, O.E. and Cox, D.R. (1994): *Inference and Asymptotics.* Chapman and Hall, London.

Barndorff-Nielsen, O.E. and Sørensen, M. (1994): A review of some aspects of asymptotic likelihood theory for stochastsic processes. *Internat. Statist. Rev.* **62**, 133–165.

Basawa, I.V. (1981): Efficiency of conditional maximum likelihood estimators and confidence limits for mixtures of exponential families. *Biometrika* **68**, 515–523.

Basawa, I.V. and Brockwell, P.J. (1978): Inference for gamma and stable processes. *Biometrika* **65**, 129–133.

Basawa, I.V. and Brockwell, P.J. (1984): Asymptotic conditional inference for regular nonergodic models with an application to autoregressive processes. *Ann. Statist.* **12**, 161–171.

Basawa, I.V. and Becker, N. (1983): Remarks on optimal inference for Markov branching processes: A sequential approach. *Aust. J. Statist.* **25**, 35–46.

Basawa, I.V. and Prakasa Rao, B.L.S. (1980): *Statistical Inference for Stochastic Processes.* Academic Press, London.

Basawa, I.V. and Scott, D.J. (1983): *Asymptotic Optimal Inference for Nonergodic Models.* Lecture Notes in Statistics 17. Springer, New York.

Bell, D.R. and Mohammed, S.-E. A. (1995): Smooth densities for degenerate stochastic delay equations with hereditary drift. *Ann. Probab.* **23**, 1875–1894.

Bellmann, R. and Cooke, K.L. (1963): *Differential-Difference Equations.* Academic Press, New York.

Bertoin, J. (1996): *Lévy Processes.* Cambridge University Press.

Bhat, B.R. (1988a): On exponential and curved exponential families in stochastic processes. *Math. Scientist* **13**, 121–134.

Bhat, B.R. (1988b): Optimal properties of SPRT for some stochastic processes. *Contemporary Mathematics* **80**, 285–299.

Billingsley, P. (1968): *Convergence of Probability Mesures.* Wiley, New York.

Björk, T. and Johansson, B. (1995): Parameter estimation and reverse martingales, Working Paper 79, Stockholm School of Economics.

Blumenthal, R.M. and Getoor, R.K. (1968): *Markov Processes and Potential Theory.* Academic Press, New York and London.

Bremaud, P. (1981): *Point Processes and Queues: Martingale Dynamics.* Springer, New York.

Brown, L.D. (1986): *Fundamentals of Statistical Exponential Families.* Institute of Mathematical Statistics, Hayward.

Brown, B.M. and Hewitt, J.T. (1975a): Asymptotic likelihood theory for diffusion processes. *J. Appl. Prob.* **12**, 228–238.

Brown, B.M. and Hewitt, J.T. (1975b): Inference for the diffusion branching process. *J. Appl. Prob.* **12**, 588–594.

Casalis, M. and Letac, G. (1994): Characterization of the Jorgensen set in generalized linear models. *TEST*, 145–162.

Chaleyat-Maurel, M. and Elie L. (1981): Diffusion Gaussiennes. *Astérisque* **84–85**, 255–279.

Chentsov, N.N. (1982): *Statistical Decision Rules and Optimal Inference.* American Mathematical Society, Rhode Island.

Çinlar, E., Jacod, J., Protter, P., and Sharpe, M.J. (1980): Semimartingales and Markov processes. *Z. Wahrscheinlichkeitstheorie verw. Gebiete* **54**, 161–219.

Chung, K.L. (1982): *Lectures from Markov Processes to Brownian Motion.* Springer, New York.

Courrège, P. and Priouret, P. (1965): Temps d'arrêt d'une fonction aléatoire. *Publ. Inst. Statist. Univ. Paris* **14**, 245–274.

Csörgö, M. and Fischler, R. (1973): Some examples and results in the theory of mixing and random-sum central limit theorems. *Period. Math. Hungar.* **3**, 41–57.

Curtain, R.F. and Pritchard, A.J. (1978): *Infinite Dimensional Linear Systems Theory.* Lecture Notes in Control and Information Science 8, Springer, Berlin.

Davis, M.H.A. (1977): *Linear Estimation and Stochastic Control.* Chapman and Hall, London.

Diaconis, P. and Freedman, D. (1980): de Finetti's theorem for Markov chains. *Ann. Prob.* **8**, 115–130.

Diaconis, P. and Ylvisaker, D. (1979): Conjugate priors for exponential families. *Ann. Statist.* **7**, 269–281.

Dietz, H.M. (1992): A non-Markovian relative of the Ornstein-Uhlenbeck process and some of its local statistical properties. *Scand. J. Statist.* **19**, 363–379.

Dinwoodie, I.H. (1995): Stationary exponential families. *Ann. Statist.* **23**, 327–337.

Döhler, R. (1981): Dominierbarkeit und Suffiziens in der Sequentialanalyse. *Math. Oper. Statist., Ser. Statist.* **12**, 101–134.

Doléans-Dade, C. (1976): On the existence and unicity of solutions of stochastic integral equations. *Z. Wahrscheinlichkeitstheorie verw. Gebiete* **36**, 93–101.

Dynkin, E.B. (1965): *Markov Processes, Vol. I.* Springer Verlag, Berlin, Göttingen, Heidelberg.

Dynkin, E.B. and Juschkevič A.A. (1969): *Markov Processes, Theorems and Problems.* Plenum Press, New York.

Dzhaparidze, K. and Spreij, P. (1993): The strong law of large numbers for martingales with deterministic quadratic variation. *Stochastics and Stochastic Reports* **42**, 53–65.

Eaton, M.L., Morris, C. and Rubin, H. (1971): On extreme stable laws and some applications. *J. Appl. Prob.* **8**, 794–801.

Efron, B. (1975): Defining the curvature of a statistical problem (with applications to second order efficiency). *Ann. Statist.* **3**, 1189–1242.

Efron, B. and Hinkley, D.V. (1978): Assessing the accuracy of the maximum likelihood estimator: Observed versus expected Fisher information. *Biometrika* **65**, 457–487.

Elliott, R.J. (1982): *Stochastic Calculus and Applications.* Springer, New York.

Engelbert, H.J. and Schmidt, W. (1989): Strong Markov continuous local martingales and solutions of one-dimensional stochastic differential equations (Part II). *Math. Nachr.* **144**, 241–281.

Feigin, P.D. (1976): Maximum likelihood estimation for continuous-time stochastic processes. *Adv. Appl. Prob.* **8**, 712–736.

Feigin, P.D. (1979): Some comments concerning a curious singularity. *J. Appl. Prob.* **16**, 440–444.

Feigin, P.D. (1981): Conditional exponential families and a representation theorem for asymptotic inference. *Ann. Statist.* **9**, 597–603.

Feigin, P.D. (1985): Stable convergence of semimartingales. *Stoch. Proc. Appl.* **19**, 125–134.

Feigin, P.D. and Reiser, B. (1979): On asymptotic ancillarity and inference for Yule and regular nonergodic processes. *Biometrika* **66**, 279–283.

Feller, W. (1966): *An Introduction to Probability Theory and its Applications, Vol. II.* Wiley, New York.

Föllmer, H., Protter, P., and Shiryaev, A.N. (1995): Quadratic covariation and an extension of Ito's formula. *Bernoulli* **1**, 149–169.

Franz, J. (1977): Niveaudurchgangszeiten zur Charakterisierung sequentieller Schätzverfahren. *Math. Oper. Statist., ser Statist.* **8**, 499–510.

Franz, J. (1982): Sequential estimation and asymptotic properties in birth-and-death processes. *Math. Oper. Statist., ser Statist.* **13**, 231–244.

Franz, J. (1985): Special sequential problems in Markov processes. *Banach Center Publ.* **16**, 95–114.

Franz, J. (1986): Beiträge zur sequentiellen Schätzung in Markovschen Prozessen der Exponentialklasse. Dissertation B, TU Dresden, Fakultät Mathematik/Naturwissenschaften.

Franz, J. and Magiera, R. (1978): On sequential plans for the exponential class of processes. *Zastosowania Matematyki* **16**, 153–165.

Franz, J. and Magiera, R. (1990): Admissible estimation in sequential plans for exponential-type processes. *Scand. J. Statist.* **17**, 275–285.

Franz, J. and Winkler, W. (1976): Über Stoppzeiten bei statistischen Problemen für homogene Prozesse mit unabhängigen Zuwächsen. *Math. Nachr.* **70**, 37–53.

Franz, J. and Winkler, W. (1987): Efficient sequential estimation for an exponential class of processes. In Sendler, W. (ed.): *Contributions to Stochastics*, 123–130. Physica-Verlag, Heidelberg.

Freedman, D.A. (1963): Invariants under mixing, which generalize de Finetti's theorem: continuous time parameter. *Ann. Math. Statist.* **34**, 1194–1216.

Genon-Catalot, V. and Picard, D. (1993): *Eléments de Statistique Asymptotique.* Mathématique et Applications 11. Springer, Paris.

Gerber, H.U. and Shiu, E.S.W. (1994): Option pricing by Esscher transforms. *Transactions of the Society of Actuaries*, XLVI, 99–191.

Gihman, I.I. and Skorohod, A.V. (1975): *The Theory of Stochastic Processes, Vol. 2.* Springer, Berlin.

Grambsch, P. (1983): Sequential sampling based on the observed Fisher information to guarantee the accuracy of the maximum likelihood estimator. *Ann. Statist.* **11**, 68–77.

Grandell, J. (1992): *Aspects of Risk Theory.* Springer, New York.

Grigelionis, B. (1991): On statistical inference for stochastic processes with boundaries. In M. Dozzi, H.J. Engelbert and D. Nualart (eds.): *Stochastic Processes and Related Topics, Math. Research*, **61**. Akademie-Verlag, Berlin.

Grigelionis, B. (1994): Conditionally exponential families and Lundberg exponents of Markov additive processes. Grigelionis et.al. (eds.): *Prob. Theory and Math. Stat.*, 337–350. VSPITEV.

Grigelionis,B. (1996): Mixed exponential processes. Preprint 96-7, Institute of Mathematics and Informatics, University of Vilnius.

Grigelionis, B. (1997): On mixed Poisson processes and martingales. To appear in *Scand. Actuarial J.*

Gushchin, A.A. and Küchler, U. (1996): Asymptotic properties of maximum-likelihood-estimators for a class of linear stochastic differential equations with time delay. Discussion Paper 29, SFB 373, Humboldt University, Berlin.

Hale, J and Lunel, S.M.V. (1993): *Introduction to Functional Differential Equations.* Springer.

Hall, P. and Heyde, C.C. (1980): *Martingale Limit Theory and Its Application.* Academic Press, New York.

Harrison, J.M. and Shepp, L.A. (1981): On skew Brownian motion. *Ann. Prob.* **9**, 309–313.

Helland, I.S. (1982): Central limit theorems for martingales with discrete and continuous time. *Scand. J. Statist.* **9**, 79–94.

Heyde, C.C. and Feigin, P.D. (1975): On efficiency and exponential families in stochastic process estimation. In *Statistical Distributions in Scientific Work, Vol. I*, 227–240 (G.P. Patil, S. Kotz & J.K. Ord, Eds.). Reidel, Dordrecht.

Hoffmann-Jørgensen, J. (1994): *Probability with a View to Statistics, Vols. I, II.* Chapman and Hall, New York.

Höpfner, R. (1987): A remark on stopping times associated with continuous-time ergodic Markov chains and an application to statistical inference in birth-and-death processes. *Scand. J. Statist.* **14**, 211–219.

Hougaard, P. (1986): Survival models for heterogeneous populations derived from stable distributions. *Biometrika* **73**, 387–396.

Hudson, I.L. (1982): Large sample inference for Markovian exponential families with application to branching processes with immigration. *Australian J. Stat.* **24**, 98–112.

Hwang, S.Y. and Basawa, I.V. (1994): Large sample inference for conditional exponential families with applications to nonlinear time series. *J. Statist. Plann. Inference* **38**, 141–158.

Ikeda, N. and Watanabe, S. (1981): *Stochastic Differential Equations and Diffusion Processes.* North-Holland, Amsterdam.

Irle, A. and Schmitz, N. (1984): On the optimality of the SPRT for processes with continuous time parameter. *Math. Operationsforsch. und Statist., Ser. Statistics* **15**, 91–104.

Ito, K. (1969): *Stochastic Processes*, Lecture Notes Series No. 16, Matematisk Institut, Aarhus University.

Ito, K. and McKean, H.P. (1965): *Diffusion Processes and Their Sample Paths.* Springer, Berlin.

Ito, K. and Nisio, M. (1964): On stationary solutions of a stochastic differential equation. *J. Math. Kyoto Univ.* **4**, 1–75.

Jacobsen, M. (1982): *Statistical Analysis of Counting Processes.* Lecture Notes in Statistics 12. Springer-Verlag, New York.

Jacod, J. (1976): Un théorème de représentation pour les martingales discontinues. *Z. Wahrscheinlichkeitstheorie verw. Gebiete* **34**, 225–244.

Jacod, J. (1979): *Calcul stochastique et problèmes de martingales.* Lecture Notes in Mathematics, Vol. 714. Springer, Berlin.

Jacod, J. (1987): Partial likelihood process and asymptotic normality. *Stoch. Proc. Appl.* **26**, 47–71.

Jacod, J. (1989): Fixed statistical models and Hellinger processes. *Stoch. Proc. Appl.* **32**, 3–45.

Jacod, J. and Mémin, J. (1976): Caractéristiques locales et conditions de continuité absolue pour les semi-martingales. *Z. Wahr. verw. Geb.* **35**, 1–37.

Jacod, J. and Shiryaev, A.N. (1987): *Limit theorems for stochastic processes.* Springer, Berlin.

Jeganathan, P. (1981): On a decomposition of the limit distribution of a sequence of estimators. *Sankhyā Ser.* **A 43**, 26–36.

Jeganathan, P. (1982): On the asymptotic theory of estimation when limit of the log-likelihood ratios is mixed normal. *Sankhyā Ser.* **A 44**, 173–212.

Jeganathan, P. (1995): Some aspects of asymptotic theory with applications to time series models. *Econometric Theory* **11**, 818–887.

Jensen, J.L. (1987): On asymptotic expansions in non-ergodic models. *Scand. J. Statist.* **14**, 305-318.

Jensen, J.L. (1997): A simple derivation of r^* for curved exponential families. *Scand. J. Statist.* **24**, 33–46.

Jensen, J.L. and Johansson, B. (1990): The extremal family generated by the Yule process. *Stoch. Proc. Appl.* **36**, 59–76.

Johansen, S. (1979): *Introduction to the Theory of Regular Exponential Families.* Lecture Notes 3. Institute of Mathematical Statistics, University of Copenhagen, Copenhagen.

Johnson, N.L. and Kotz, S. (1970): *Continuous Univariate Distributions, Vol. 1.* Wiley, New York.

Jørgensen, B. (1982): *Statistical Properties of the Generalized Gaussian Distribution.* Lecture Notes in Statistics 9. Springer, New York.

Jørgensen, B. (1992): Exponential dispersion models and extensions: A review. *Internat. Statist. Rev.* **60**, 5–20.

Jørgensen, B. (1997): *The Theory of Dispersion Models.* Chapman and Hall, London.

Jørgensen, B. and Souza, M.P. (1994): Fitting Tweedie's compound Poisson model to insurance claims data. *Scand. Actuarial J.*, 69–93.

Kabanov, Ju.M., Liptser, R.Sh., and Shiryaev, A.N. (1980): On absolute continuity of probability measures for Markov-Ito processes. In *Lecture Notes in Control and Information Science* **25**, pp. 114–128. Springer, Berlin.

Karlin, S. and McGregor, J. (1957): The differential equations of the birth-and-death processes and the Stieltjes moment problem. *Trans. Amer. Math. Soc.* **85**, 489–546.

Karlin, S. and Taylor, H.M. (1975): *A First Course in Stochastic Processes.* Academic Press, Orlando.

Karlin, S. and Taylor, H.M. (1981): *A Second Course in Stochastic Processes.* Academic Press, Orlando.

Karr, A.F. (1991): *Point Processes and Their Statistical Inference.* Second Edition. Marcel Dekker, New York.

Kaufmann, H. (1987): On the strong law of large numbers for multivariate martingales. *Stoch. Proc. Appl.* **26**, 73–85.

Kazamaki, N. (1972): Changes of time, stochastic integrals and weak martingales. *Z. Wahrscheinlichkeitstheorie verw. Gebiete* **22**, 25–32.

Kazamaki, N. (1978): The equivalence of two conditions on weighted norm inequalities for martingales. In K. Ito (ed.): *Proc. Intern. Symp. SDE Kyoto 1976*, 141–152. Kinokuniya, Tokyo.

Keiding, N. (1974): Estimation in the birth process. *Biometrika* **61**, 71–80.

Keiding, N. (1975): Maximum likelihood estimation in the birth-and-death process. *Ann. Statist.* **3**, 363–372.

Keiding, N. (1978): Stopping rules and incomplete information in estimation theory for stochastic processes. Invited lecture at the 11th European Meeting of Statisticians.

Kendall, D.G. (1952): Les processus stochastiques de croissance en biologie. *Ann. Inst. H. Poincaré* **13**, 43–108.

Küchler I. (1979): On the exponential class of processes with independent increments - analytical treatment and sequential probability ratio tests. *Wiss. Sitzungen zur Stochastic* **WSS-04/79**, Akademie der Wissenschaften der DDR, Berlin.

Küchler I. (1985): A generalized fundamental identity. *Banach Center Publications* **16**, 321–326. Polish Scientific Publishers, Warsaw.

Küchler, I. and Küchler, U. (1981): An analytical treatment of exponential families of processes with independent stationary increments. *Math. Nach.* **103**, 21–30.

Küchler, U. (1982a): Exponential families of Markov processes, Part I: General results. *Math. Oper. Statist., Ser. Statist.* **13**, 57–69.

Küchler, U. (1982b): Exponential families of Markov processes, Part II: Birth-and-death processes. *Math. Oper. Statist., Ser. Statist.* **13**, 219–230.

Küchler, U. (1993): On life-time distributions of some one-dimensional diffusions and related exponential families. Preprint 9/1993, Humboldt-Universität zu Berlin, Institut für Mathematik.

Küchler, U. and Kutoyants, Yu.A. (1997): Delay estimation for stationary diffusion-type processes. Preprint.

Küchler, U. and Lauritzen, S.L. (1989): Exponential families, extreme point models and minimal space-time invariant functions for stochastic processes with stationary and independent increments. *Scand. J. Statist.* **16**, 237–261.

Küchler, U. and Mensch, B. (1992): On Langevin's stochastic differential equation extended by a time-delayed term. *Stochastics and Stochastics Reports* **40**, 23–42.

Küchler, U. and Sørensen, M. (1989): Exponential families of stochastic processes: A unifying semimartingale approach. *Internat. Statist. Rev.* **57**, 123–144.

Küchler, U. and Sørensen, M. (1994a): Exponential families of stochastic processes and Lévy processes. *J. Statist. Plann. Inference* **39**, 211–237.

Küchler, U. and Sørensen, M. (1994b): Exponential families of stochastic processes with time-continuous likelihood functions. *Scand. J. Statist.* **21**, 421–431.

Küchler, U. and Sørensen, M. (1996a): Exponential families of stochastic processes and their envelope families. *Ann. Inst. Stat. Math.* **48**, 61–74.

Küchler, U. and Sørensen, M. (1996b): A note on limit theorems for multivariate martingales. Research Report No. 350, Dept. Theor. Statist., Aarhus University. To appear in *Bernoulli.*

Küchler, U. and Sørensen, M. (1997): On exponential families of Markov processes. To appear in *J. Statist. Plann. Inference*

Kunita, H. (1976): Absolute continuity of Markov processes. *Séminaire de Probabilité X. Lecture Notes in Mathematics* **511**, 44–77. Springer, Berlin.

Lai, T.L. and Siegmund, D. (1983): Fixed accuracy estimation of an autoregressive parameter. *Ann. Statist.* **11**, 478–485.

Lauritzen, S.L. (1984): Extreme point models in statistics. *Scand. J. Statist.* **11**, 65–91.

Lauritzen, S.L. (1988): *Extremal Families and Systems of Sufficient Statistics.* Springer, New York.

Le Breton, A. (1977): Parameter estimation in a linear stochastic differential equation. *Transactions of the 7th Prague Conference on Information Theory, Statistical Decisions and Random Processes*, Vol. **A**, 353–366. Academia, Prague.

Le Breton, A. (1984): Propriétés asymptotiques et estimation des paramètres pour les diffusions gaussiennes homogènes hypoelliptiques dans le cas purement explosif. *C. R. Acad. Sc. Paris, Ser. I* **299**, 185–188.

Le Breton, A. and Musiela, M. (1982): Estimation des paramètres pour les diffusions gaussiennes homogènes hypoelliptiques. *C. R. Acad. Sc. Paris, Ser. I* **294**, 341–344.

Le Breton, A. and Musiela, M. (1985): Some parameter estimation problems for hypoelliptic homogeneous Gaussian diffusions. *Banach Center Publications* **16**, 337–356.

Le Breton, A. and Musiela, M. (1987): A strong law of large numbers for vector Gaussian martingales and a statistical application in linear regression. *Statist. Probab. Lett.* **5**, 71–73.

Le Breton, A. and Musiela, M. (1989): Laws of large numbers for semimartingales with application to stochastic regression. *Probab. Th. Rel. Fields* **81**, 275–290.

Lee, M.-L.T. and Whitmore, G.A. (1993): Stochastic processes directed by randomized time. *J. Appl. Prob.* **30**, 302–314.

Lehmann, E.L. (1983): *The Theory of Point Estimation.* Wiley, New York.

Lepingle, D. (1978): Sur le comportement asymptotique des martingales locales. *Séminaire de Probabilités XII. Lecture Notes in Mathematics* **649**, 148–161. Springer, Berlin.

Letac, G. (1986): La réciprocité des familles exponentielles naturelles sur *R. C. R. Acad. Sc. Paris* **703**, 61–64.

Letac, G. and Mora, M. (1990): Natural exponential families with cubic variance functions. *Ann. Statist.* **18**, 1–37.

Liptser, R.Sh. (1980): A strong law of large numbers for local martingales. *Stochastics* **3**, 217–228.

Liptser, R.Sh. and Shirayev, A.N. (1977): *Statistics of Random Processes.* Springer, New York.

Liptser, R.Sh. and Shiryaev, A.N. (1989): *Theory of Martingales.* Kluwer Academic Publishers, Dordrecht.

Magiera, R. (1974): On the inequality of Cramér-Rao type in sequential estimation theory. *Applicationes Mathematicae* **14**, 227–235.

Magiera, R. (1977): On sequential minimax estimation for the exponential class of processes. *Zast. Mat.* **15**, 445–454.

Magiera, R. (1984): Sequential estimation of the transition intensities in Markov processes with migration. *Zastosowania Matematyki* **18**, 241–250.

Magiera, R. (1991): Admissible sequential estimators of ratios between two linear combinations of parameters of exponential-type processes. *Statistics and Decisions* **9**, 107–118.

Magiera, R. and Wilczyński, M. (1991): Conjugate priors for exponential-type processes. *Stat. Probab. Lett.* **12**, 379–384.

Manjunath, S.M. (1984): Optimal sequential estimation for ergodic birth-death processes. *J. R. Statist. Soc.* **B 46**, 412–418.

Melnikov, A.V. (1986): The law of large numbers for multidimensional martingales. *Soviet Math. Dokl.* **33**, 131–135.

Melnikov, A.V. and Novikov, A.A. (1990): Statistical inferences for semimartingale regression models. In Grigelionis et al. (eds.): *Prob. Theory and Math. Stat., Proc. of the Fifth Vilnius Conference, Vol.* **2**, 150–167. VSP, Utrecht.

Michalevič, V.S. (1961): On a simple sufficient statistic for infinitely divisible processes. *Zbirnik Prac z Občisljuvalnoi Matematiki i Techniki,* **1**. Akademija Nauk Ukrainskoi RSR, Kiev (in Ukrainian).

Mohammed, S.-E.A. (1984): *Stochastic Functional Differential Equations.* Pitman, London.

Mohammed, S.-E.A. and Scheutzow, M.K.R. (1990): Lyapunov exponents and stationary solutions for affine stochastic delay equations. *Stochastics and Stochastics Reports* **29**, 259–283.

Moran, P.A.P. (1951): Estimation methods for evolutive processes. *J. R. Statist. Soc.* **B 13**, 141–146.

Moran, P.A.P. (1953): Estimation of the parameters of a birth-and-death processes. *J. R. Statist. Soc.* **B 15**, 241–245.

Morris, C.N. (1981): Models for positive data with good convolution properties. Memo No. 8949, Rand Corporation, Santa Monica, California.

Neven, J. (1972) *Martingales à temps discret.* Masson & Cie.

Norros, I., Valkeila, E. and Virtamo, J. (1996): An elementary approach to a Girsanov formula and other analytical results on fractional Brownian motions. Preprint 133, Department of Mathematics, University of Helsinki.

Novikov, A.A. (1972): Sequential estimation of the parameters of diffusion-type processes. *Math. Notes* **12**, 812–818.

Novikov, A.A. (1973): On moment inequalities and identities for stochastic integrals. *Proc. Second Japan-USSR Symp. Prob. Theor., Lecture Notes in Math.* **330**, 333–339. Springer, Berlin.

Okamoto, I., Amari, S.-I. and Takeuchi, K. (1991): Asymptotic theory of sequential estimation: Differential geometrical approach. *Ann. Statist.* **19**, 961–981.

Pruscha, H. (1987): Sequential estimation functions in stochastic population processes. In F. Bauer et al. (eds.): *Mathematical Statistics and Probability Theory*, Vol. B, 189–203. D. Reindel.

Puri, P.S. (1966): On the homogeneous birth-and-death process and its integral. *Biometrika* **53**, 61–71.

Rao, R.M. (1969): On decomposition theorems by Meyer. *Math. Scand.* **24**, 66–78.

Revuz, D. and Yor, M. (1991) *Continuous Martingales and Brownian Motion.* Springer, Berlin, Heidelberg.

Roberts, G.E. and Kaufman, H. (1966): *Table of Laplace Transforms.* Saunders, Philadelphia.

Rogers, L.C.G. (1995) Which model for the term structure of interest rates should one use? In Davis, M., Duffie, D., Flemming, W., and Shreve, S. (eds.): *Mathematical Finance*, 93–116. Springer.

Rogers, L.C.G. and Williams, D. (1987): *Diffusions, Markov Processes, and Martingales. Volume 2: Ito Calculus.* Wiley, Chichester.

Rogers, L.C.G. and Williams, D. (1994): *Diffusions, Markov Processes, and Martingales. Volume 1: Foundations.* Wiley, Chichester.

Rothkirch, U. (1993): Schätzung von Parametern in linearen stochastischen Funktionaldifferentialgleichungen. Dissertation, Universität Jena, Fakultät für Mathematik und Informatik.

Różański, R. (1980): A modification of Sudakov's lemma and efficient sequential plans for the Ornstein-Uhlenbeck process. *Zastosowania Matematyki* **17**, 73–85.

Różański, R. (1989): Markov stopping sets and stochastic integrals. Application in sequential estimation for a random field. *Stoch. Proc. Appl.* **32**, 237–251.

Salminen, P. (1983): Mixing Markovian laws. With an application to path decompositions. *Stochastics* **9**, 223–231.

Scheutzow, M. (1984): Qualitative behavior of stochastic delay equations with a bounded memory. *Stochastics* **12**, 41–80.

Sellke, T. and Siegmund, D. (1983): Sequential analysis of the proportional hazards model. *Biometrika* **70**, 315–326.

Sharpe, M. (1988) *General Theory of Markov Processes.* Academic Press.

Shiryaev, A.N. (1978): *Optimal Stopping Rules.* Springer, New York.

Siegmund, D. (1985): *Sequential Analysis. Tests and Confidence Intervals.* Springer, New York.

Soler, J.L. (1977): Infinite dimensional-type statistical spaces (Generalized exponential families). In Barra, J.R. et al. (eds.): *Recent Developments in Statistics.* North Holland, Amsterdam.

Sørensen, M. (1983): On maximum likelihood estimation in randomly stopped diffusion-type processes. *Internat. Statist. Rev.* **51**, 93–110.

Sørensen, M. (1986a): On sequential maximum likelihood estimation for exponential families of stochastic processes. *Internat. Statist. Rev.* **54**, 191–210.

Sørensen, M. (1986b): Classes of diffusion-type processes with a sufficient reduction. *Statistics* **17**, 585–596.

Sørensen, M. (1990): On quasi likelihood for semimartingales. *Stoch. Proc. Appl.* **35**, 331–346.

Sørensen, M. (1991): Likelihood methods for diffusions with jumps. In Prabhu, N.U. and Basawa, I.V. (eds.): *Statistical Inference in Stochastic Processes,* 67–105. Marcel Dekker, New York.

Sørensen, (1994a): On the moments of some first passage times for exponential families of processes. Research Report 302, Dept. Theor. Statist., Univ. of Aarhus.

Sørensen, M. (1994b): On comparison of stopping times in sequential procedures for exponential families of stochastic processes. To appear in *Scand. J. Statist.*

Sørensen, M. (1996a): A semimartingale approach to some problems in risk theory. *Astin Bulletin* **26**, 15–23.

Sørensen, M. (1996b): On exponential families of discrete time stochastic processes. Research Report 351, Dept. Theor. Statist., Univ. of Aarhus.

Sørensen, M. (1996c): The natural exponential family generated by a semimartingale. In Shiryaev, A.N., Melnikov, A.V., Niemi, H., and Valkeila, E. (eds.): *Proceedings of the Fourth Russian-Finnish Symposium on Probability Theory and Mathematical Statistics,* 177–186. TVP Science Publishers, Moscow.

Sørensen, M. (1998): On conditional inference for the Ornstein-Uhlenbeck process. In preparation.

Stefanov, V.T. (1982): Sequential estimation for compound Poisson process, I and II. *Serdica* **8**, 183–189 and 255–261.

Stefanov, V.T. (1984): Efficient sequential estimation in finite-state Markov processes. *Stochastics* **11**, 291–300.

Stefanov, V.T. (1985): On efficient stopping times. *Stoch. Processes Appl.* **19**, 305–314.

Stefanov, V.T. (1986a): Efficient sequential estimation in exponential-type processes. *Ann. Statist.* **14**, 1606–1611.

Stefanov, V.T. (1986b): On the moments of some first passage times and the associated processes. *Stochastics* **19**, 207–220.

Stefanov, V.T. (1988): On some stopping times for dependent Bernoulli trials. *Scand. J. Statist.* **15**, 39–50.

Stefanov, V.T. (1991): Non-curved exponential families associated with observations over finite-state Markov chains. *Scand. J. Statist.* **18**, 353–356.

Stefanov, V.T. (1995a): Explicit limit results for minimal sufficient statistics and maximum likelihood estimators in some Markov processes: Exponential families approach. *Ann. Statist.* **23**, 1073–1101.

Stefanov, V.T. (1995b): On efficient sequential estimation for the simple single server queue. *Brazilian Journal of Probability and Statistics* **9**, 55–65.

Stockmarr, A. (1996): Limits of autoregressive processes with a special emphasis on relations to cointegration theory. Ph.D. thesis, University of Copenhagen.

Sudakov, V.N. (1969): On measures defined by Markovian moments. In *Investigations on the Theory of Random Processes*, 157–164, Memoirs of the Scientific Seminars of the Leningrad Section of the Steklov Mathematical Institute, Vol. 12. (In Russian).

Sweeting, T.J. (1980): Uniform asymptotic normality of the maximum likelihood estimator. *Ann. Statist.* **8**, 1375–1381. [Correction: *Ann. Statist.* (1982) **10**, 320–321.]

Sweeting, T.J. (1986): Asymptotic conditional inference for the offspring mean of a supercritical Galton-Watson process. *Ann. Statist.* **14**, 925–933.

Sweeting, T.J. (1992): Asymptotic ancillarity and conditional inference for stochastic processes. *Ann. Statist.* **20**, 580–589.

Taraskin, A.F. (1974): On the asymptotic normality of vector-valued stochastic integrals and estimates of multi-dimensional diffusion processes. *Theory Prob. Math. Statist.* **2**, 209–224.

Trybuła, S. (1968): Sequential estimation in processes with independent increments. *Dissertationes Math.* **60**, 1–46.

Trybuła, S. (1982): Sequential estimation in finite-state Markov processes. *Zastosowania Matematyki* **17**, 227–247.

Tweedie, M.C.K. (1984): An index which distinguishes between some important exponential families. In *Statistics: Applications and New Directions* (eds. J.K. Ghosh and J. Roy), pp. 579–604. Indian Statistical Institute, Calcutta.

Wald, A. (1945): Sequential tests of statistical hypotheses. *Ann. Math. Statist.* **16**, 117–186.

Wald, A. (1947): *Sequential Analysis.* Wiley, New York.

Væth, M. (1980): A test for the exponential distribution with censored data. Research Report No. 60, Dept. Theor. Statist., Aarhus Univ.

Væth, M. (1982): A sampling experiment leading to the full exponential family generated by the censored exponential distribution. Research Report No. 78, Dept. Theor. Statist., Aarhus Univ.

Watanabe, S. (1975): On time inversion of one-dimensional diffusion processes. *Z. Wahrscheinlichkeitstheorie u. verw. Gebiete* **31**, 115–124.

White, J.S. (1958): The limit distribution of the serial correlation in the explosive case. *Ann. Math. Statist.* **29**, 1188–1197.

Winkler, W. (1980): Sequential estimation in processes with independent increments. *Banach Center Publications*, **6**, 325–331.

Winkler, W. (1985): A survey of sequential estimation in processes with independent increments. *Banach Center Publications*, **16**, 546–554.

Winkler, W. and Franz, J. (1979): Sequential estimation problems for the exponential class of processes with independent increments. *Scand. J. Statist.* **6**, 129–139.

Winkler, W., Franz, J., and Küchler, I. (1982): Sequential statistical procedures for processes of the exponential class with independent increments. *Math. Oper. Statist., Ser. Statist.* **13**, 105–119.

Ycart, B. (1988): A characteristic property of linear growth birth and death processes. *Sankhyā Ser.* **A 50**, 184–189.

Ycart, B. (1989): Markov processes and exponential families on a finite set. *Stat. Probab. Lett.* **8**, 371–376.

Ycart, B. (1992a): Integer valued Markov processes and exponential families. *Stat. Probab. Lett.* **14**, 71–78.

Ycart, B. (1992b): Markov processes and exponential families. *Stoch. Proc. Appl.* **41**, 203–214.

Zabczyk, J. (1983): Stationary distributions for linear equations driven by general noise. *Bull. Acad. Pol. Sci.* **31**, 197–209.

Appendix D

Basic Notation

$\mathbb{R} = (-\infty, \infty)$ the set of real numbers

$\mathbb{R}_+ = [0, \infty)$

$\mathbb{Z}$ the set of integers

$\mathbb{N} = \{0, 1, 2, \cdots\}$ the set of natural numbers

$\mathbb{C}$ the set of complex numbers

$\mathbb{R}^n$ the n-dimensional Euclidian space

$|x|$ the Euclidian norm of $x \in \mathbb{R}^n$, $n = 1, 2, \cdots$

$a \vee b = \max\{a, b\}$

$a \wedge b = \min\{a, b\}$

$x^+ = x \vee 0$ for $x \in \mathbb{R}$

$x^- = -(x \wedge 0)$ for $x \in \mathbb{R}$

1_A the indicator function for the set A

$\operatorname{int} A$ the interior of the set A

$\operatorname{bd} A$ the boundary of the set A

$\operatorname{cl} A$ the closure of the set A

ϵ_a the Dirac measure at the point a

a. s. = almost surely

$\ll$ absolute continuity between measures

$\sim$ equivalence between measures

$N(\mu, \Sigma)$ normal distribution with mean vector μ and covariance matrix Σ.

$\chi^2(k)$ χ^2-distribution with k degrees of freedom

$\mathcal{P}_x$ see Section 6.1

$\mathbb{P}^\theta$ see Section 6.1

$\tilde{\Gamma}_t$ see (7.1.1)

$\dot{l}_t(\theta)$ the score function (score vector), see Section 8.1

$j_t(\theta)$ observed information (matrix), see (8.3.1)

$i_t(\theta)$ expected (Fisher) information (matrix), see (8.3.12)

$J_t(\theta)$ incremental observed information (matrix), see (8.4.3)

$I_t(\theta)$ incremental expected information (matrix), see (8.4.9)

$\langle M \rangle$ the quadratic characteristic of M, see Theorem A.3.6

$\langle X, Y \rangle$ the quadratic cocharacteristic of X and Y, see (A.3.3)

$[M]$ the quadratic variation of M, see Section A.4

$[X, Y]$ the quadratic covariation of X and Y, see Section A.4

$H \cdot X = \int H dX$ the stochastic integral of H with respect to the semimartingale X, see Theorem A.4.2

$Y * \rho$ the stochastic integral of Y with respect to the random measure ρ, see Section A.8

$Y * (\mu - \nu)$ the stochastic integral of Y with respect to the compensated random measure $\mu - \nu$, see Section A.8

Index

Springer Series in Statistics

(continued from p. ii)

Pollard: Convergence of Stochastic Processes.
Pratt/Gibbons: Concepts of Nonparametric Theory.
Ramsay/Silverman: Functional Data Analysis.
Read/Cressie: Goodness-of-Fit Statistics for Discrete Multivariate Data.
Reinsel: Elements of Multivariate Time Series Analysis, 2nd edition.
Reiss: A Course on Point Processes.
Reiss: Approximate Distributions of Order Statistics: With Applications to Non-parametric Statistics.
Rieder: Robust Asymptotic Statistics.
Rosenbaum: Observational Studies.
Ross: Nonlinear Estimation.
Sachs: Applied Statistics: A Handbook of Techniques, 2nd edition.
Särndal/Swensson/Wretman: Model Assisted Survey Sampling.
Schervish: Theory of Statistics.
Seneta: Non-Negative Matrices and Markov Chains, 2nd edition.
Shao/Tu: The Jackknife and Bootstrap.
Siegmund: Sequential Analysis: Tests and Confidence Intervals.
Simonoff: Smoothing Methods in Statistics.
Small: The Statistical Theory of Shape.
Tanner: Tools for Statistical Inference: Methods for the Exploration of Posterior Distributions and Likelihood Functions, 3rd edition.
Tong: The Multivariate Normal Distribution.
van der Vaart/Wellner: Weak Convergence and Empirical Processes: With Applications to Statistics.
Vapnik: Estimation of Dependences Based on Empirical Data.
Weerahandi: Exact Statistical Methods for Data Analysis.
West/Harrison: Bayesian Forecasting and Dynamic Models, 2nd edition.
Wolter: Introduction to Variance Estimation.
Yaglom: Correlation Theory of Stationary and Related Random Functions I: Basic Results.
Yaglom: Correlation Theory of Stationary and Related Random Functions II: Supplementary Notes and References.

MIX
Papier aus verantwortungsvollen Quellen
Paper from responsible sources
FSC® C105338

If you have any concerns about our products,
you can contact us on
ProductSafety@springernature.com

In case Publisher is established outside the EU,
the EU authorized representative is:
Springer Nature Customer Service Center GmbH
Europaplatz 3, 69115 Heidelberg, Germany

Printed by Libri Plureos GmbH
in Hamburg, Germany